# PLASTIC PACKAGING OF MICROELECTRONIC DEVICES

# PLASTIC PACKAGING OF MICROELECTRONIC DEVICES

Louis T. Manzione

*AT&T Bell Laboratories Division*

Library of Congress Catalog Card Number 90-42629
ISBN 0-442-23494-5



Manufactured in the United States of America

Published by Van Nostrand Reinhold
115 Fifth Avenue
New York, New York 10003

Chapman and Hall
2-6 Boundary Row
London, SE1 8HN

Thomas Nelson Australia
102 Dodds Street
South Melbourne 3205
Victoria, Australia

Nelson Canada
1120 Birchmount Road
Scarborough, Ontario M1k 5GA, Canada

16 15 14 13 12 11 10 9 8 7 6 5 4 3 2 1

**Library of Congress Cataloging-in-Publication Data**

Manzione, Louis T.
Plastic packaging of microelectronic devices / by Louis T. Manzione.
p. cm.
Includes bibliographical references and index.
ISBN (invalid) 0-442-23494-5
1. Electronic packaging. 2. Plastics in packaging. I. Title.
TK7870.15.M36 1990
621.381'046—dc20 90-42629
CIP

*To*
*Linda*

# CONTENTS

# Preface

Those of us who work in packaging know that it is a challenging and fascinating field that has a major effect on product quality and competitiveness. Unfortunately, it has not attracted the level of research and development effort commensurate with its value. There are a dearth of published studies on plastic package manufacturing, yet the area is ripe with opportunities for engineers, rheologists, polymer chemists, polymer physicists, and applied mechanicians. The joke among insiders is that for want of a better name, this is a field that research institutes are formed to address. Packaging is an exceptionally problem-rich area whose proper treatment requires a broad multi-disciplined approach.

One purpose of this book is to stimulate interest in packaging in the research communities of both academic institutions and corporate research centers. The time is particularly appropriate because as the technology begins to approach its limits, even small advances translate to major accomplishments in extending plastic packaging to applications that would have gone to more expensive options. Another purpose of the book is to fill a perceived void in applied research on the materials and processes used in packaging. Although there have been a small number of published reports on package molding, my aim was to consolidate the work currently available, add my own work and interpretations, and present it in a way that is comprehensible to a wide audience that either works in the field at present or could contribute if they were more aware of the issues. I have attempted to include sufficient depth for readers already knowledgeable in the field, sufficient breadth to attract readers from other fields who can contribute to packaging, and sufficient explanations for those who have little formal training in engineering. An entire chapter is devoted to providing the fundamentals of polymer science needed to derive full benefit from the rest of the book. In all cases where mathematical relations are used, I have made the effort to describe the terms and dependencies in words. I have spared the reader the more tedious details of the math where I felt its inclusion detracted from the flow of the discussion. With regard to units, I have used the units that are most common in the microelectronics industry for the specific parameter. This has resulted in a mixture of English, metric, and IUPAC units throughout the text. I have provided a second unit in parentheses when I thought that more than one unit was common for a parameter. I apologize for not using the standard IUPAC units throughout, but I believe the text will be more accessible to professionals who work in the field if the units are the ones that they recognize.

No one attempts or realizes an effort of this scale without help and support. I was exceptionally fortunate to have generous amounts of both. Members of my supervisory group at AT&T Bell Laboratories, both past and present, have made important contributions to the principles described here. This book is largely the result of being surrounded by good people who truly advanced the field. I am happy to acknowledge Gary Poelzing, Jules Osinski, Bob Lien, Norris Cole, Tim Sullivan, Ephraim Suhir, and Steve Kulba. Every one of these individuals made a significant contribution to what is contained herein. Similarly, I had exceptional support from my management at AT&T Bell Laboratories. Phil Hubbauer, Dave McCall, Jim Arnold, Dave Lando, and Kumar Patel all took a genuine and supportive interest in this book. Kumar Patel took particular interest in the project, providing me with much needed encouragement. Dave McCall and I had an informal study group going for several months when we were both completing major projects in this area. In addition, I benefited from the help of the large number of experts on packaging that we have in AT&T. These include Shiro Matsuoka, Harvey Bair, John Segelken, Harry Scholz, Dave Crouthamel, Anil Rane, Kurt Steiner, Luis Garcia, Arnold Lustiger, Frank Houlihan, and Mike Zimmerman. The most time consuming work was probably done by those who agreed to review sections of the manuscript for me. I received careful and diligent reviews from Lee Blyler, Ed Chandross, Tim Sullivan, Dave Crouthamel, Mike Zimmerman, Kurt Steiner, Bonnie Bachman, Shiro Matsuoka, and John Segelken. Kristin Engelmann helped me with organization and paper handling. Danuta Sowinska-Khan did an excellent job in managing the artwork. To all of the above, I offer my sincerest gratitude for your time, support, and encouragement.

Personal support was as important as professional support in bringing this project to fruition. I have a family that helps and supports me in any endeavor I undertake. My wife, Linda, makes all this effort worthwhile and to her I have dedicated my efforts on this book. My two boys, Matthew and Michael, contributed to it more than they realize. Their simple belief that I work on things that are important was strong motivation for me. The same sentiments extend to the rest of my family. My mother has played a particularly important role. Her caring, commitment and wisdom have always been an inspiration to me. My grandfather clips out all the newspaper articles on computer chips that he finds and saves them for me. My mother and father in marriage keep a copy of my first book out in their living room. My brothers, aunts, uncles, cousins, and in-laws are always interested in my latest projects and travels. The value of this extended family is extraordinary, and I am truly fortunate to have such a resource.

Murray Hill, NJ

# NOTATIONS

$a_i$ unknowns in the element equations of a finite element solution
$A$ coefficient in Arrhenius relation for temperature dependence of viscosity
$A_g$ wetted flow area of the molding tool gate
$A_W, A$ surface area exposed to flow (wetted area), also heat transfer area
$A_{xs}$ cross section area
$b$ lead width
$B$ thickness of flow channel, also coefficient containing geometric parameters in pressure drop-volumetric flow rate relation derived in Appendix 4-A
$Br$ Brinkman number
$C$ coefficient in the special viscosity-shear rate relation derived in Appendix 4-A, also concentration of reactant
$C_D$ drag coefficient
$C_0$ initial concentration of reactant
$C_p$ heat capacity at constant pressure
$d$ gap thickness variable in a cone and plate viscometer
$D$ diameter, also diffusivity
$Da$ Damkohler number
$De$ Deborah Number
$D_{\text{wire}}$ deformation of wire bond
$E$ activation energy in Arrhenius relations for reaction kinetics or temperature dependence of viscosity; also tensile modulus
$E_r$ relaxation tensile modulus derived from time dependent stress
$E_p$ tensile modulus of plastic molding compound
$E_\eta$ flow activation energy in Arrhenius relation for viscosity
$\dot{E}$ rate of energy dissipation
$f$ functionality of a multifunctional monomer
$f(t)$ failure rate
$F_i$ the forcing terms in a finite element solution
$F(t)$ cumulative failure distribution function
FIT failure unit defined as one failure per $10^9$ device hours
$g$ ratio of radius of gyration of a branched molecule to the linear molecule of the same molecular weight
$G$ shear modulus
$h$ convective heat transfer coefficient, also lead thickness

$h_g$ convective heat transfer coefficient through molding tool gate
$h_g^*$ parameter representing the convective heat transfer coefficient through gate which contains the area, mass flow rate, and heat capacity
$h_0$ half thickness of thick end of wedge in wedge flow geometry
$h_1$ half thickness of thin end of wedge in wedge flow geometry
$h(t)$ instantaneous failure rate which is the failure rate normalized to the fraction surviving
$H$ height of molding compound in transfer pot
$\Delta H$ heat of reaction liberated at some time
$\Delta H_{\text{rxn}}$ heat of reaction for complete conversion
$I$ moment of inertia
$k$ stress optical coefficient
$k_i$ thermal conductivity of a single component in a multicomponent system
$k_c$ thermal conductivity of the continuous phase in a multicomponent material
$k_d$ thermal conductivity of the discontinuous phase in a multicomponent material
$k_T$ thermal conductivity
$\bar{k}_T$ resultant thermal conductivity of a multicomponent material
$k, k_{1,2}$ reaction rate constants for a multiparameter kinetic model, also the fractional length of the lead the flow front has transversed
$K$ viscosity power law coefficient, also coefficient in the pressure drop–volumetric flow rate relation derived in Appendix 4A, also shift parameter on Duane plot of device failure rate, also the geometric constant for stress computation
$K_e$ Einstein coefficient for filled system viscosity
$K_0$ coefficient in temperature dependence relation for gate and runner pressure drop–volumetric flow rate relation
$K_{ij}$ terms of the element equations in a finite element solution
$K_{G,R,x}$ coefficient in pressure drop–volumetric flow relation for the gate, runner, or undetermined
$K_{\text{wire}}$ coefficient for wire deformation
$L$ length; also the bobbin length in a Couette viscometer
$m$ mass
$\dot{m}$ mass flow rate
$m_{A,B}$ average equivalent weight per reactive group of component $A$ or $B$
$m'_{A,B}$ average equivalent weight per reactive group of component $A$ or $B$ normalized with a weighting factor that is the molecular weight multiplied by the mole fraction of the reactive species
$M_{A,B}$ average molecular weight of component $A$ or $B$

$M_w$ weight average molecular weight
$M_{wc}$ critical weight average molecular weight for chain entanglements
$MX$, $MN$ maximum point, minimum point
$m$, $n$ reaction order constants in a multiparameter kinetic model
$n$ viscosity power law index
$n_G$ power law index for pressure drop–volumetric flow rate relation for the gate
$n_R$ power law index for pressure drop–volumetric flow rate relation in the runner
$N$ fringe order in the stress optic law for birefringence
$N_l$ number of leads
$N_c$ number of memory cells in a memory device
$Nu$ Nusselt number
$P$ pressure; also the degree of planarization
$P_i$ pressure at each length increment
$P_0$ inlet pressure
$P_{0,s,e}$ pressure drops due to entrance, shear and elongational losses
$P_L$ pressure at specified length $L$
$P_w$ wetted perimeter
$\Delta P$ pressure drop
$\Delta P_i$ pressure drop increment
$\Delta P_{i,i+1}$ pressure drop between consecutive gates numbered $i$ and $i + 1$
$q$ volumetric flow rate in the runner segments between gates of a mold, also heat flux
$Q$ volumetric flow rate
$Q_i$ volumetric flow rate through the $i$th gate
$r$ radius variable, also reaction rate, also radius of gyration
$r_h$ hydraulic radius
$r_l$ radius of gyration of a linear chain molecule
$r_{gb}$ radius of gyration of a branched chain molecule
$R$ gas constant; also radius of flow channel and other circular objects
$R(t)$ reliability function which is the fraction surviving at time $t$
$R_i$ hydraulic flow resistance parameter for the $i$th gate
$Re$ Reynolds Number
$S$ slope on Duane plot of device failure rate
$S_p$ shape factor for flow channel
$t$ time variable
$t_f$ height of feature on device with regard to planarization
$t_h$ height of feature through the fluid layer with regard to planarization
$t_{16}$ the time to 15.866% cumulative failure in log-normal distribution analysis
$t_{50}$ the median time to failure in a log-normal distribution

$T$ temperature; also the gap thickness of a couette viscometer
$T_b$ ambient temperature
$T_i$ temperature at a length increment
$T_g$ glass transition temperature
$T_m$ mold temperature
$T_r$ temperature in the runner
$T_s$ surface temperature of device
$T_w$, $T_{w0}$ wall temperature, initial wall temperature
$T_0$ initial temperature
$\langle \tilde{T} \rangle$ mixing cup or bulk averaged temperature
$U$ average velocity
$v$ linear velocity as in the cone and plate viscometer
$V$ viscosity in Figure 6-11
$W$ width of flow channel
$x$ dimensional variable
$X$ degree of conversion, also known as extent of reaction
$X_g$ degree of conversion to reach gelation
$y$ dimensional variable
$z$ dimensional variable
$\Delta z$ axial length increment

## GREEK

$\alpha$ coefficient for Rent's rule; also angle of cone in cone and plate viscometry, also coefficient of thermal expansion
$\alpha_{1,2}$ coefficients of thermal expansion below and above the glass transition temperature
$\beta$ exponent for Rent's rule, also exponent for Weibull distribution function, also stress parameter in metal deformation computation
$\gamma$ shear
$\dot{\gamma}$ shear rate
$\dot{\gamma}_r$ shear rate at the wall in tube flow
$\delta$ deformation as used in metal deformation computation
$\epsilon_{1,2}$ principal strains
$\epsilon'$ dielectric constant
$\epsilon''$ dielectric dissipation factor
$\zeta$ dimensionless axial length variable
$\theta$ angle, angular variable
$\eta$ viscosity
$\eta_\infty$ viscosity at infinite shear rate or infinite temperature, also the solvent viscosity in a polymer solution

$\eta_0$ zero shear viscosity; also viscosity of continuous phase fluid in a filled fluid system
$\eta_{b0,b1}$ bulk averaged viscosity at inlet and outlet
$\lambda$ coefficient in the Carreau model of viscosity, also characteristic lifetime of a device population, also wavelength of light, also elongational viscosity
$\lambda_c$ characteristic relaxation time
$\lambda_{c,g}$ time scales of heat conduction and generation
$\mu$ exponent in the mean time to failure function used in the log-normal distribution
$\mu$, $\mu m$ micron
$\pi$ ratio of circumference to diameter of a circle (3.1415)
$\pi_T$ thermal acceleration factor
$\rho$ density
$\sigma$ tensile stress, also a scaling parameter in the log-normal distribution representing the natural log of the ratio of $t_{50}/t_{16}$
$\sigma^*$ tensile stress parameter that includes the geometric constant
$\sigma_p$ tensile stress required to damage the passivation layer
$\tau$ shear stress
$\tau_w$ shear stress at the wall
$\phi$ volume fraction
$\phi_i$ volume fraction of one component in a multicomponent system
$\phi_d$ volume fraction of the discontinuous phase in a multicomponent system
$\omega$ angular velocity

# PLASTIC PACKAGING OF MICROELECTRONIC DEVICES

# 1 Introduction to Plastic Packaging of Microelectronic Devices

The microelectronics industry continues to grow rapidly in size and importance. Estimates indicate that annual world production of integrated circuit (IC) devices reached 28 billion in 1990 [1]. The industry will soon approach the size of other major industries with sales of product, manufacturing, and test equipment totaling tens of billions of dollars a year. Yet, the size and growth of the industry still underrepresent its importance because it does not reflect the tremendous influence integrated circuits have on most other segments of an industrialized society. Leadership in the area of IC devices translates to superior products in the growing list of products that use them including consumer electronics, household appliances, computers, automobiles, telecommunications, robotics, and military equipment. Plastic packaging will continue to play a major role in the growth of the microelectronics industry and have a profound effect on all of the markets where chips are used. Plastic packages are less expensive than other packaging options, yet they provide performance and reliability that is acceptable for a large and growing fraction of applications. Plastic packaging is not hermetic, hence it does not provide the reliability performance of ceramic or metallic packaging, but it is adequate for most products except those with the strictest reliability criteria, notably military hardware. Continued improvements in the passivation and protection layers on the device and further improvements in the packages themselves will continue to promote the use of plastic packages in these reliability weighted applications. Plastic packages account for approximately 80% of the worldwide package share and this percentage has been increasing [1].

Packaging has often been a secondary consideration to silicon device fabrication. For this reason, packaging has received far less attention than its contribution to overall device performance, reliability, and cost should warrant. This situation is changing quickly, however, as packaging limitations begin to impose restrictions on device and system designs. This is most evident in the need to develop and fabricate packages with very large numbers of leads. More sophisticated devices need more input/output leads, and plastic packages with more than 300 leads are anticipated for the early 1990s. The continuing drive toward miniaturization means that more interconnections will have to be made in less space. The design, material, and fabrication technology to achieve this level of interconnection density is not available at the present time. It is at this point that packaging and interconnection of the device, technologies that were often taken for granted, may for the first time become the bottleneck to future system improvements. This book concentrates on the manufacturing considerations of molded plastic packages, the package option that is used for the majority of devices. To treat this topic properly, many other aspects of device and interconnection technology must be discussed. The book is not intended as a general reference work on packaging, however. For this, the reader is directed to References 2 and 3, two books on microelectronics packaging that go beyond the scope of the present work and complement the discussions presented herein. Broader treatments of polymer materials in microelectronics technology are also available in References 4 and 5.

## 1.1 PACKAGING AND INTERCONNECTION HIERARCHY

There is an overall packaging and interconnection hierarchy that couples active elements on a chip to the electronic system in which they function. A description of the author's version of the interconnection levels and their numbering is provided in Table 1-1. Different sources are likely to use slightly different levels and numbering. The plastic packages which are the topic of this book are but one level of this hierarchy, which may include as many as six levels. Some system designs may combine interconnection levels, such as the thermal conduction module mentioned in Section 8.3.1, whereas some other new technologies may add levels between existing levels, such as the multichip modules described in Section 9.2.2.

## 1.2 THE FUNCTION OF THE PACKAGE

Packaging serves several distinct functions in microelectronics, some obvious and others not so obvious. A list of the major package functions is provided below.

1. Protection of the device from mechanical and chemical hazards
2. Signal and power distribution
3. Signal timing (if signal propagation delays affect system performance)
4. Heat dissipation

The ability of a package to adequately perform these functions depends on the properties of the device as well as the properties of the package. It is therefore worthwhile to review the features of integrated circuits devices. As an example, the protection against mechanical and chemical hazards can be less if the protection afforded on the device through passivating and overcoat layers is greater. Similarly with regard to heat dissipation and signal management, specific device

**Table 1-1. The Hierarchy of Interconnection.**

| *Level Number* | *Description (with example in parentheses)* |
|---|---|
| Level 0 | Interconnections on the chip (aluminum conductor lines) |
| Level 1 | Chip to package (wire bonding, Tape Automated Bonding (TAB), solder bumps) |
| Level 2 | Package to circuit board (plastic or ceramic packages on printed circuit boards) |
| Level 3 | Circuit board to system board (the backplane of a system) |
| Level 4 | Cabinet wiring and cabling (intracabinet wiring of a complex system such as a telephone private branch exchange (PBX)) |
| Level 5 | Intercabinet connections (local area network) |
| Level 6 | Coupling of local area networks (communications network of a large corporation) |

types and configurations have greater needs and mandate more active roles for the packages in meeting these needs.

Integrated circuit devices are composed of a large number of interconnected metal oxide semiconductor (MOS) field effect transistors (FET). These devices are formed with layers of conducting and semiconducting material separated by insulating layers. The lowest semiconducting layer is the diffusion layer, where dopants create an electron mobility channel. The gate of the transistor is formed by a second semiconducting polysilicon layer crossing over the diffusion layer forming the source and drain zones of a transistor, as shown in Figure 1-1 [6]. A *p*-channel FET is based on hole mobility, whereas *n*-channel FET utilizes electron mobility. Metallic conducting layers are used to interconnect the discrete transistors, integrating them into more complex electronic functions such as inverters and logic elements. A three-dimensional view with cross section of a transistor is shown in Figure 1-2 [7]. References 6, 7, and 8 are good sources of information on the structure and manufacture of integrated circuits. The protection of these structures is one important function of the package.

Several different types of semiconductor technology are used to make integrated circuits. Bipolar technology has the longest history of development, as it was pivotal in the growth of the early computer industry. For a long time, bipolar transistors were produced as discrete devices. They have always been known and valued for their speed. The development of very thin base regions with correspondingly short transit times is the basis of this fast gate switching. The invention of the Schottky clamp and emitter-coupled logic (ECL) in the mid-1950s improved bipolar performance and minimized saturation problems. In many instances, bipolar technology is also known as ECL. Bipolar devices are relatively high heat generators compared to the other IC technologies.

The metal oxide semiconductor field effect transistor (MOSFET) is the dominant device used for VLSI technology and it is the basis of the MOS technol-

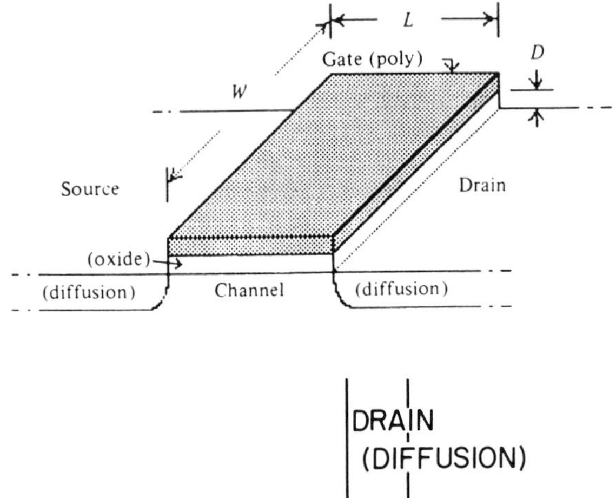


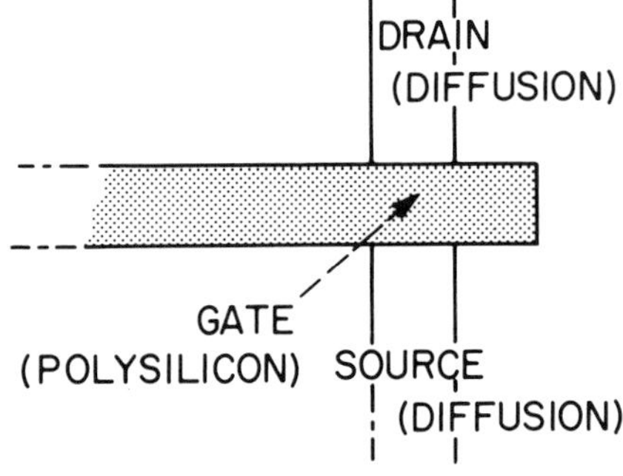


**FIGURE 1-1.** Source and drain of an MOS transistor, the basic element of an integrated circuit device. [6]. (C. Mead and L. Conway, Introduction to VLSI Systems, Copyright © 1980 by Addison-Wesley Publishing Co., Inc., Reprinted by Permission of Addison-Wesley Publishing Co., Inc. Reading, MA.)

ogy. The early ICs using MOSFET's were based on *p*-channel (PMOS) devices, but *n*-channel (NMOS) devices soon displaced them because of the higher mobility of electrons compared to holes. They have dominated the IC market since the early 1970s [8]. Complementary MOS (CMOS) technology provides both PMOS and NMOS transistors on the same chip. They consume less power than NMOS circuits and far less than bipolar devices. Early CMOS processes were more complex, however, and required larger chip areas than NMOS for the same device function. Improvements in CMOS processing have now simplified the manufacture so that NMOS and CMOS manufacturing processes are equivalent, with CMOS still enjoying the advantage of lower power usage. CMOS has emerged as a major VLSI technology and it will most likely become the dominant technology by the mid-1990s. BiCMOS technology is a hybrid of combining bipolar and CMOS elements on the same chip. Its importance will also continue to grow as process and design improvements are made.

Passivation of the structures themselves can be used to remove the active

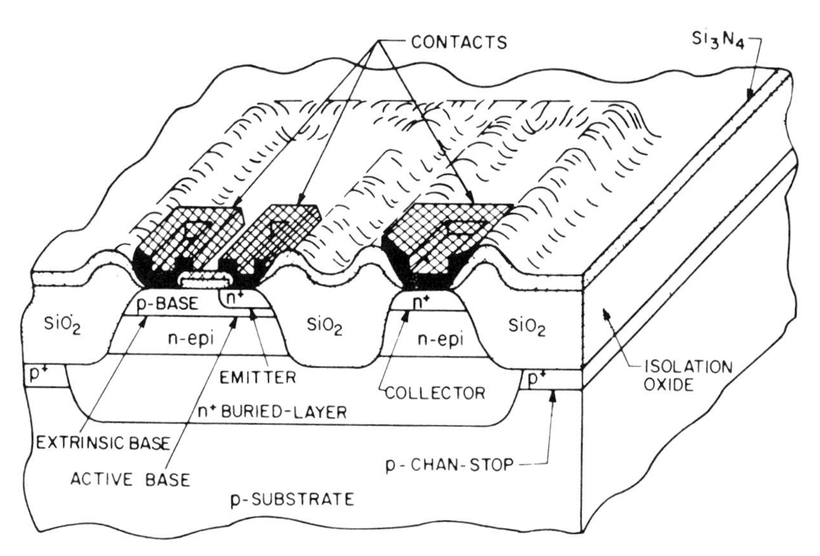


**FIGURE 1-2.** Three-dimensional view with cross section of an MOS device showing the multi-layer structure of doped semiconductors, signal and passivating layers. [7]. (E. F. Labuda and J. T. Clemens, "Integrated Circuit Technology" in R. E. Kirk and D. F. Othmer, Editors, Encyclopedia of Chemical Technology, Copyright © 1980 by John Wiley and Sons, Inc. Reprinted with permission of John Wiley and Sons.)

transistor junctions from the surface of the device. There have been many significant improvements in this area. A silicon oxide layer affords both electrical insulation and passivation protection, yet ionic impurities in the oxide could induce spurious transistor action, cause junction leakage current, or cause gain degradation [3]. One solution was to use a gettering layer in the oxide to trap, or getter, ionic impurities such as the alkali metals. Phosphosilicate glass is an effective gettering agent when the phosphorous concentration is held within certain bounds [9]. A more recent solution has been the use of silicon nitride, typically $Si_3N_4$, as a passivation layer which has several important advantages over silicon oxide. These include a higher breakdown voltage, a higher dielectric constant, and a higher density providing a better moisture barrier. The benefits of silicon nitride are not without complications, in that mechanical stresses in the glazed layers can lead to passivation layer cracking and attendant loss of the barrier layer function. Also, incorporation of free protons during the deposition of the SiN layer can cause malfunctions as they drift during operation of the device.

The development of a mechanically, chemically, and electrically stable barrier layer will still leave a corrosion problem where the metal conductor paths have to emerge from the passivation layer at the bonding pads. Aluminum is the principal metal for conductors on IC devices and it is exceptionally prone to corrosion. Important developments have been made to improve the passivity of the bonding pads, and most of this technology involves the use of multiple layers of deposited metals to achieve conductivity, passivity, and barrier layer properties. The first layer may have an affinity for oxygen, an intermediate barrier layer prevents reaction with gold, and a top layer of gold is used for interconnection and wire bonding.

Ionic impurities combined with water provide the essential components of an electrochemical cell that can rapidly accelerate the corrosion of the metallic features of the device. Hence the principal packaging requirement for early discrete and integrated circuit devices was hermeticity [9] and the first packages were metal cans constructed of nickel lids and gold-plated bases. The introduction of the silicon planar technology, which as described above afforded some passivation to the active surface of the device, and the drive toward higher pin counts made the costly metal packaging inefficient. Ceramic packaging was developed to provide hermeticity at lower costs and with more flexibility than was achievable with metal housings. Improved passivation of the surface of the chip and improved moisture barrier layers then allowed the use of plastic packages in all but the most demanding and high-reliability applications of IC chips.

## 1.3 THE TYPES OF SECOND-LEVEL INTERCONNECTS (PACKAGES)

This section describes the interconnect designs that have been developed for IC devices. In most instances these designs are known as *packages.* There are, however, some designs where interconnection is achieved without a package, for example, in chip-on-board interconnection and some applications of TAB. The following sections describe the various interconnected designs that are of commercial importance.

### 1.3.1 Direct Chip Attach

There are several alternatives to formed body packages. The lowest-cost option is *chip-on-board.* In this packaging scheme, the silicon device is attached directly to the circuit board with an adhesive, and then wire bonded to the metallized circuitry. The bare surface of the chip and the bonding pads are protected with an encapsulant, usually known as a *glob top* coating. An illustration of chip-on-board interconnection is shown in Figure 1-3. The encapsulant is applied in the liquid state at low delivery pressures, so it cannot contain the high filler loadings that provide the low coefficient of thermal expansion achieved in transfer-molded packaging compounds. Adhesive properties of the glob top may exceed those of a molded body package because the molding compound must include mold release agent, yet reliability of the glob top coat-

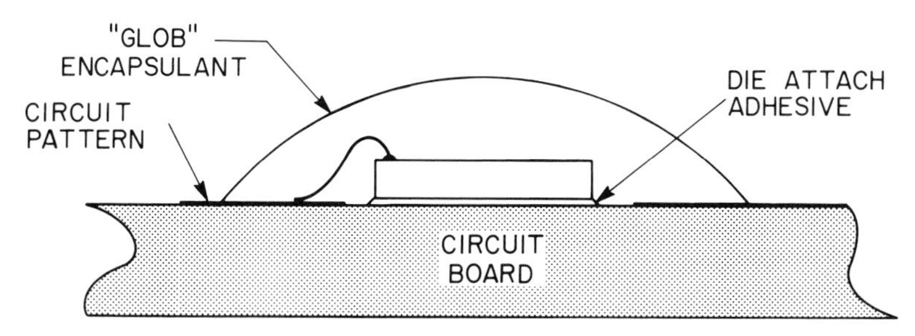


**FIGURE 1-3.** Chip-on-board attachment.

ing is typically lower. This type of packaging is most often used in low-cost consumer products such as toys and digital watches where cost considerations are dominant and repairability is not an issue.

A variation of chip-on-board is used in the manufacture of more sophisticated hybrid integrated circuits, commonly known a HICs. These are essentially miniature circuit boards using a film integrated circuit (FIC) on a ceramic or epoxy-glass substrate. These circuits contain silicon devices and a mixture of thin- and thick-film components, such as thin-film capacitors, chip capacitors, and thick-film resistors. The components are usually tunable, with laser trimming operations, for example, to provide a high performance device that has a specific function such as a digital filter or signal generator. The small size of most HICs has prohibited the use of prepackaged silicon, thereby creating need for direct chip attach interconnection. One type of chip attach used on HICs is *beam leads* [10], where short metal stubs, or beams, are attached to the bond pads of the device. The device is mounted to the substrate with the active surface down, commonly known as a *flip-chip* configuration. The beam leads are then thermal compression bonded to the circuit pattern on the substrate, acquiring a crimp from the bonding tool which props up the chip. A drawing of this chip attach technology is shown in Figure 1-4. It appears that beam leads, though never a major interconnection medium, are losing ground to the solder bump and TAB technologies discussed in the following sections.

The introduction of small outline packages (SOP) offers the possibility of attaching prepackaged silicon devices to the surface of hybrid circuits. Although there is some sacrifice in area, there are other benefits such as full testing of the

**FIGURE 1-4.** Beam-leaded chip attachment.

components prior to attachment, greater protection for the silicon device during circuit processing, and greater robustness of the hybrid itself.

Another direct chip attach option is *solder bump* technology [11, 12] developed in the early 1960s for three-point attachment of discrete transistors. Continued improvements and refinements now enable the bonding of hundreds of pads on 4-mil centers, providing very high interconnection densities. Either edge pads or grid array pads are feasible with this technology. An illustration of grid array solder bump attachment is shown in Figure 1-5. Solder bumps also provide excellent thermal dissipation and a good thermal expansion match with silicon and ceramic substrates, but attachment to an epoxy printed circuit board will cause significant expansion mismatch.

### 1.3.2 Package Types

Packages are a type of second level interconnection where a formed body is created around the device to provide the functions of signal distribution, heat dissipation, protection and signal timing that were mentioned earlier. The different types of packages are described in the following subsections.

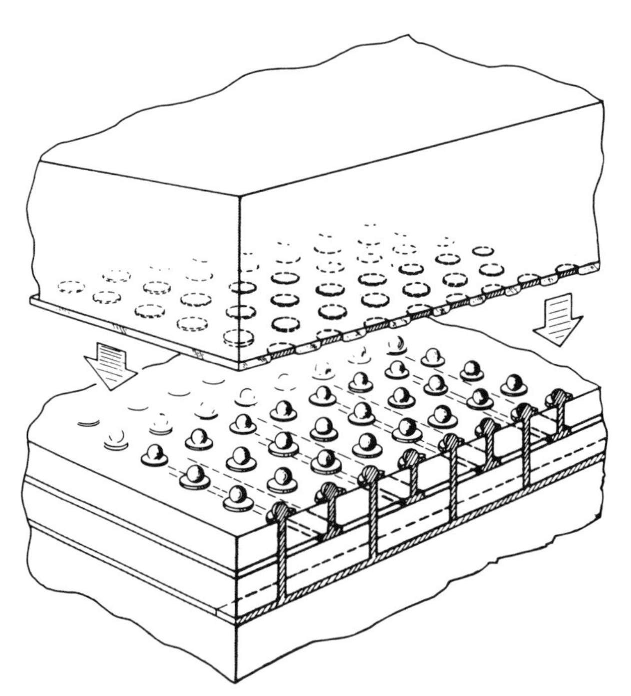

**FIGURE 1-5.** Solder bump attachment of a chip to the next level of interconnection which could be either a circuit board or a multichip module substrate, either ceramic or silicon.

**1.3.2.1 Lead Attachment Type.** A distinguishing feature of package design is the way in which the leads attach to the circuit board. Through-hole packages are restricted to 0.100 inch lead spacings because this is the practical minimum for drilling holes in a circuit board. Surface mount packages are not restricted to the through-hole spacing; they can accommodate much finer lead pitch and a higher interconnection density. In addition, components can be placed on both sides of the board, again increasing interconnection density, reducing system size, and lowering costs. An illustration comparing the attachment of through-hole and surface mount devices to a multilayer circuit board is provided in Figure 1-6. Through-hole packages have dominated interconnection, but the growth of surface mount technology (SMT) will ultimately eclipse the volume of through-hole packages. The value of surface mount packages probably exceeded that of through-hole product by the late 1980s, since higher-pin-count, more expensive devices are placed in surface mount packages. In surface mounting, there are three types of lead designs: the J lead, the gull wing lead, and the butt lead, all of which are illustrated along with the through-hole configuration in Figure 1-7. The J lead and the butt lead are the most efficient with respect to area on the circuit board, but the gull wing provides the greatest wetted area for soldering. Both the gull wing and the butt lead allow for visual inspection of the solder joints, whereas the J lead is more difficult to inspect.

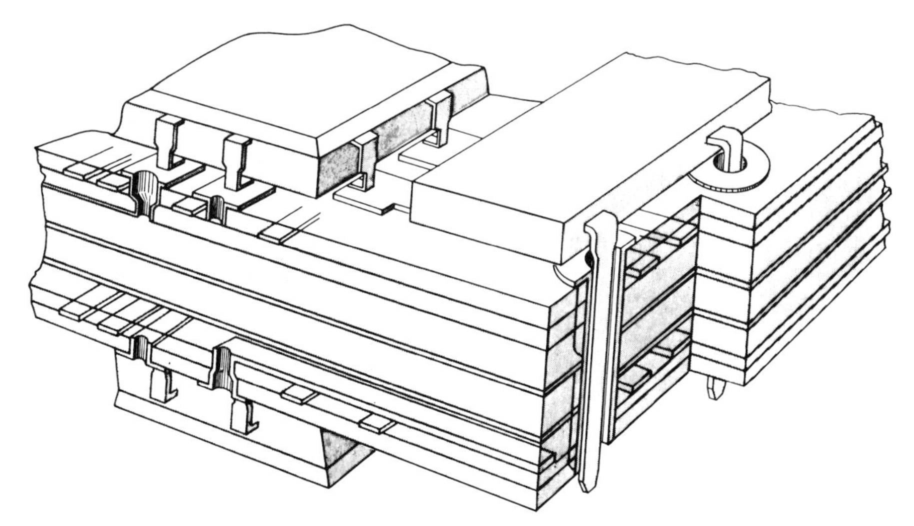

**FIGURE 1-6.** A multilayer circuit board showing the attachment of through-hole and surface mount packages. Surface mounting allows attachment of components to both sides of the board and a higher wiring density in the board since it permits blind vias that do not occupy the grid point throughout the entire board depth.

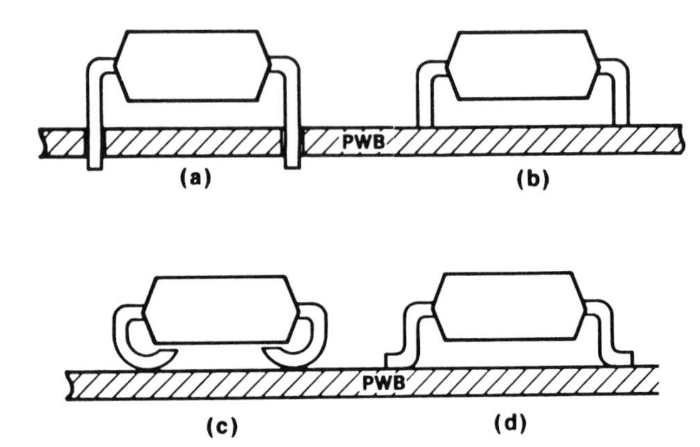


FIGURE 1-7. The different types of package leads and their attachment to the circuit board: (a) through-hole, (b) surface mount butt lead, (c) surface mount J lead, (d) surface mount gull wing lead. (After Ref. 8.)

**1.3.2.2 Single-In-Line Package (SIP).** The single-in-line package, commonly known as SIP, uses a single row of leads extending from one edge of a formed body. These are usually through-hole packages where the leads are inserted into metallized vias in the printed circuit board. A newer variation of this design is called a zigzag-in-line package (ZIP), where the leads emerge from a single edge of the body but are formed into a zigzag pattern, thereby increasing the lead density that can be realized in a given linear distance. SIPs are attractive because they use minimal circuit board area, but the close packing of circuit boards in most systems limits SIP height and their utility.

**1.3.2.3 Dual-In-Line Package (DIP).** The most popular package design is a dual-in-line package (DIP), where leads extend in straight lines from two edges of the formed body. The outline of a DIP is usually rectangular, with the leads on the long sides. Most DIPs are designed for through-hole interconnection, but surface mount DIPs are also possible. Lead counts for DIPs range from 8 to 64, with 28 to 40 being the most common for logic and processor devices and 14 to 20 for memory devices, depending on the size and configuration of the memory.

**1.3.2.4 Chip Carrier.** When the number of leads exceeds 48, a dual-in-line configuration becomes impractical and wasteful of circuit board area. Packages with leads on all four sides, known as chip carriers or quads, are preferable for these higher-pin-count devices. The name *chip carrier* is probably a holdover from the time when most packages with enough pins to warrant four-sided leads were packaged in pre-molded carriers. The advancement of postmolding technology and the improvement in the reliability of plastic packaging have now made molding of high-pin-count quad packages routine. Some other common acronyms distinguish whether the package has leads or bond pads for interconnection, and whether it is a plastic or ceramic body package. The names leaded chip carrier (LLC) and leadless chip carrier (LLCC) make the distinction on lead type. The PLCC, or plastic leaded chip carrier, is probably the most common of the four-sided packages. The lead spacing of PLCCs is 0.050 inch, providing for a significant advantage in compactness compared to DIPs.

**1.3.2.5 Pin Grid Array (PGA).** A pin grid array (PGA) is a type of chip carrier package where the leads are not restricted to the edge of the package, but are instead distributed in an array over the area on one side of the package body. PGAs are not amenable to postmolding operations because the leads cannot be configured on the parting line of a molding tool. Although postmolded PGAs can be made, most are ceramic packages, premolded plastics or packages based on multilayer wiring boards. The device is located in a well in the center of the package body and interconnected to the leads through a metallized fan-out pattern and wire-bonding. Solder bumps, beam leads, or TAB may also be used in place of wire bonding in some designs. The leads are typically on 0.100 inch centers in through-hole packages. Newer surface mount grid arrays can accommodate finer pad spacings and these packages are more appropriately called pad grid arrays, with the same PGA acronym. An illustration of a pin grid array is provided in Figure 1-8.

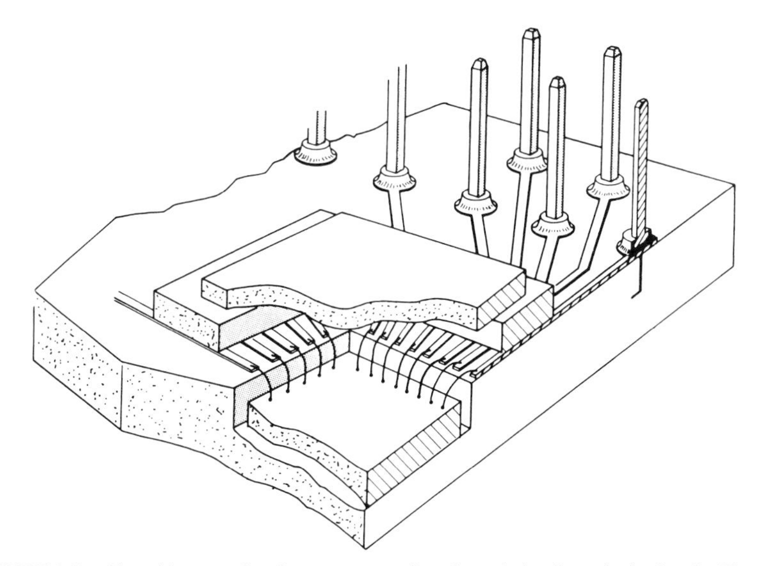

**FIGURE 1-8.** Pin grid array, showing a cutaway view through the die and wire bonds. The body of the package can be either ceramic, premolded plastic, or multilayer wiring board.

**1.3.2.6 Small Outline (SO).** Yet another distinguishing feature of packages is the degree of compactness of the formed body. Small outline packages, commonly referred to as SO, SOP, or SOIC, have an exceptionally small footprint on the circuit board for the size of the device they package and the number of leads they encompass. They achieve this compactness through small lead spacing, special lead frame design, and an extra-thin body. Both two- and four-sided lead designs are made in the SO configuration. These packages are characterized by an exceptionally small amount of molding compound surrounding the device. Therefore, the performance of the molding compound in terms of crack resistance is a key element of SO package development and reliability.

**1.3.2.7 Quad Flat Pack (QFP) and Plastic Quad Flat Pack (PQFP).** The quad flat pack (QFP) is the fine-pitch, thin-body version of the LCC, with leads on all four sides of a square or rectangular package. Lead spacings are finer than the 0.050 inch of the PLCCs. The leads are gull wing design as opposed to the J leads of PLCCs. QFP can be either plastic or ceramic bodies. A plastic QFP carries the popular designation of PQFP. There are two major standards of PQFP. The Joint Electronic Device Committee (JEDEC) of the Electronic Industry Association (EIA) has registered a PQFP package with bumpers in the corners to protect the leads during shipping and handling. The lead spacing is the same 0.025 inch for all pin counts and a different body size is used for each. The Electronic Industry Association of Japan (EIAJ) has registered a PQFP package that does not include bumpers. The lead spacing is metric and three different pitches are available: 1.0 mm, 0.8 mm and 0.65 mm. There are eight different body sizes, from 10 mm × 10 mm to 40 mm × 40 mm, spread irregularly over the three different lead spacings, providing 15 different package types up to a pin count of 232. Other package types will be added as the pin count continues to grow. The option of one body size accommodating several different pin counts is considered to be an important advantage in that the same mold and trim and form tools can be used over a wide range of pin count. The lack of bumpers on the EIAJ PQFP poses a problem, however, because the packaged devices have to be shipped in special injection-molded shipping trays, whereas JEDEC PQFPs can be shipped in conventional tubes because of the robustness imparted by the bumpers.

The rectangular geometry of the EIAJ PQFP can also provide an advantage in interconnection density in future high-pin-count packages. A rectangular geometry can accommodate a greater interconnection density on a circuit board with 0.100 inch via spacing for pin counts over 256. This is because some of the peripheral leads can be routed to plated through holes, or vias, underneath the package, providing the interconnection density of a PGA. In a square geometry not all vias underneath the package can be accessed, forcing some chip connections to be routed to vias outside the package outline, increasing its effective interconnect area. A rectangular geometry allows access to all vias underneath the package provided that the number of leads on the short side does not exceed 64 and the lead spacing is not larger than 0.025 inch (1 mm).

Figure 1-9 [13] is an illustrated guide to the packages of commercial importance, listing details of lead spacing, typical number of lead pins, common name or acronym, and other distinguishing features.

## 1.4 PACKAGE MATERIALS

There are several types of package material for silicon integrated circuits: hermetic packages made from ceramic or glass materials, and nonhermetic plastic packages. Metal packaging was an early form of hermetic packaging that has essentially been eliminated from commercial products. In hermetic packaging, there are fired ceramic packages and CERDIP packages. Plastic packages are nonhermetic packages but they command the largest share of the production volume.

### 1.4.1 Ceramic Packages

Ceramic packages are constructed of laminated sheets prepared from a slurry of ceramic powder and a liquid binder, usually a polymer, plasticizer, or solvent. The sheets are cut to the proper size and via holes and cavities for the devices are punched out. Circuit patterns are imaged onto the sheets using a metal powder slurry and the via holes are also filled with metal. The sheets are then laminated and registered prior to sintering. As many as 30 layers can be used in state of the art packages. The circuit lines are then metallized and the leads are attached. Several ceramic package designs are illustrated in Figure 1-10 with both drawings and photographs.

Other benefits of ceramic packages besides hermeticity include the large number of signal, ground and power planes that can be accommodated and the overall complexity of the package that can be constructed. Drawbacks include the poor dimensional precision of the fired assembly and high dielectric constant [14].

### 1.4.2 Refractory Glass Packages

Lower-cost hermetic packages can be made with refractory glasses. These packages do not have the complexity of multilayer ceramics, but they afford excellent reliability at a lower cost. The more expensive multilayer ceramic packages still have application for very high reliability packaging because refractory glass packages, although they are hermetic, can trap moisture and impurities within

| Type | Name | | Shape | Features | | |
|---|---|---|---|---|---|---|
| | Abbreviation | Full name | | Material | Lead pitch | Number of I/O pins |
| Insertion mounting type | DIP | DUAL INLINE PACKAGE |  | P<br>C | • 2.54mm (100MIL) | 8 ~ 64 |
| | SIP | SINGLE INLINE PACKAGE |  | P | • 2.54mm (100MIL)<br>• One-direction lead | 3 ~ 25 |
| | ZIP | ZIGZAG INLINE PACKAGE |  | P | • 2.54mm (100MIL)<br>• One-direction lead | 16 ~ 24 |
| | S-DIP | SHRINK DIP |  | P | • 1.778mm (70MIL) | 20 ~ 64 |
| | SK-DIP | SKINNY DIP |  | C<br>P | • 2.54mm<br>• Half-size pitch in the width direction | 24 ~ 32 |
| | PGA | PIN GRID ARRAY | 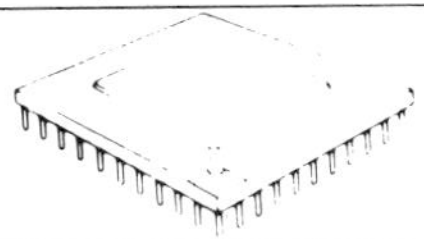 | C<br>P | • 2.54mm (100MIL) | |

| | | | | | | |
|---|---|---|---|---|---|---|
| Surface mounting type | SOP | SMALL OUTLINE PACKAGE |  | P | • 1.27mm (50MIL)<br>• Two-direction lead | 8 ~ 40 |
| | MSP | MINI SQUARE PACKAGE |  | P | • 1.27mm (50MIL)<br>• 1.016mm (40MIL)<br>• Four-direction lead | 18 ~ 84 |
| | QFP | QUAD FLAT PACKAGE |  | P | • 1.0mm<br>• 0.8mm<br>• 0.65mm<br>• Four-direction lead | |
| | FPG | FLAT PACKAGE OF GLASS |  | C | • 1.27mm (50MIL)<br>• 0.762mm (30MIL)<br>• Two-direction lead<br>• Four-direction lead | 20 ~ 80 |
| | LCC | LEADLESS CHIP CARRIER | 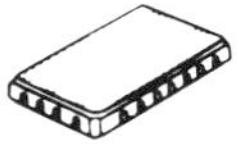 | C | • 1.27mm (50MIL)<br>• 1.016mm (40MIL)<br>• 0.762mm (30MIL) | 20 ~ 40 |
| | PLCC | PLASTIC LEADED CHIP CARRIER | 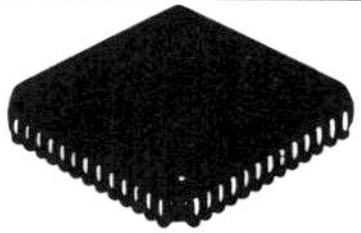 | P | • 1.27mm (50MIL)<br>• J-shaped bend<br>• Four-direction lead | 18 ~ 124 |
| | SOJ | SMALL OUTLINE J-LEAD PACKAGE | | P | • 1.27mm (50MIL)<br>• J-shaped bend<br>• Two-direction lead | 20 ~ 32 |

**FIGURE 1-9.** Commercially important package types, with the common names or acronyms used to identify them. (From Reference 13. Reprinted with permission.)

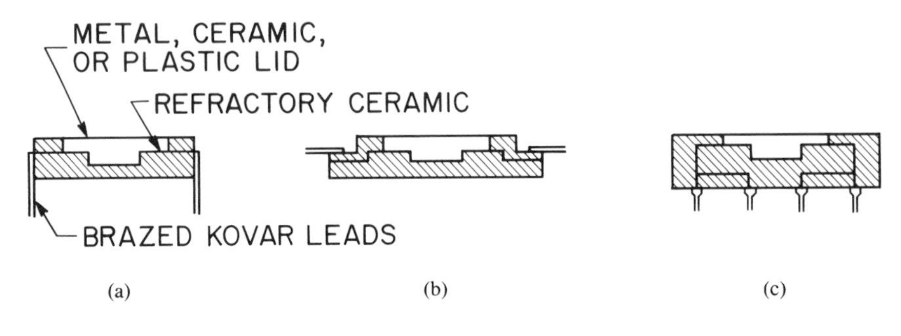


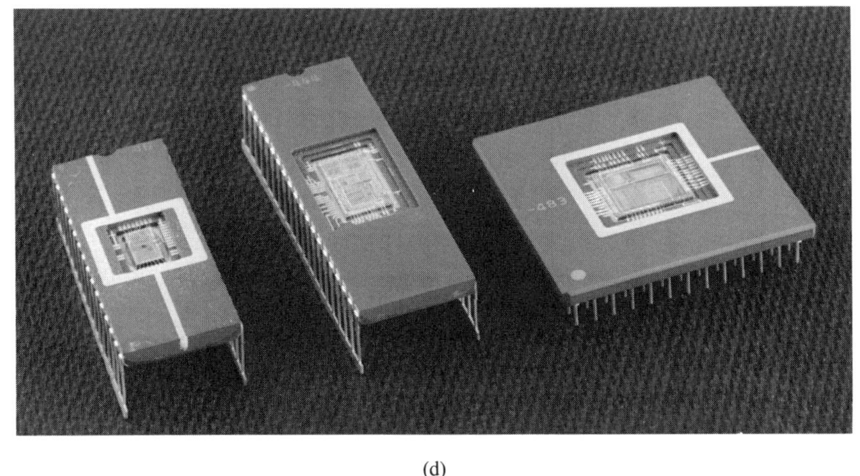

(d)

**FIGURE 1-10.** Cross sections of several different ceramic packages of increasing complexity: (a) one plane of metallization with side brazed leads, (b) metallization on several planes with signals routed to top mount leads, (c) metallization along internal vias to provide a pin grid array format. The photograph in (d) shows three different ceramic packages without their lids: a 32-pin package with one signal layer, a 40-pin DIP with two signal layers, and a pin grid array with two signal layers.

the package during processing, thereby lowering their reliability performance in comparison to fired ceramics.

Refractory glasses can be used to produce a wide variety of package types including DIPs, chip carriers, and pin grid arrays. More complicated multilayer packages are manufactured by sealing ceramic sections together with glass and then brazing metal pins, typically Kovar (an iron-nickel alloy) since it closely matches the thermal expansion behavior of ceramics and glasses, to the sides or bottom of the package. The circuit paths are metallized onto the ceramic sections. A metal or ceramic lid seals the package in most designs.

A common form of refractory glass package uses a metal leadframe sandwiched between ceramic slabs sealed with refractory glass. The glasses used are typically of the $PbO-ZnO-B_2O_3$ type with melting points in the vicinity of 400°C [14]. The leadframe material is Kovar, but aluminum wires replace gold for the wire bonding between the lead frame and the chip. Intermetallic growth between gold and Kovar would be severe at the glass sealing temperatures [14]. The leadframe assembly and processing for this type of packaging is much more amenable to automation than multilayer ceramic or refractory packages with metallized circuitry. They offer significantly lower costs that in some cases approach the costs of plastic packages. This technology has acquired the name CERDIP because it has been used predominantly for DIP packages, but CERQUADs are also available.

### 1.4.3 Plastic Packaging

Plastic packaging utilizes a metal leadframe or metallized circuit pattern to mount the silicon device and to provide a fanout pattern of leads to the pins of the package. Molded plastic is then used to protect the chip and leadframe from physical damage and contamination, although the plastic itself is not hermetic. There are two types of molded plastic packages: premolded packages and postmolded packages.

*Premolded Packaging.* In premolded packaging, the device and the fanout pattern may be inserted into a molded plastic part, typically a base. The chip is connected to the fanout pattern through wire bonds or direct chip attach. The fanout pattern connects to pins extending from the package providing a leaded package, or to bonding pads providing a leadless package. An encapsulating material, such as epoxy or silicone, may be used to fill the open cavity and offer further protection to the device. A lid made from plastic or metal, or a molded plastic package top, is then adhered to the base sealing the package. Premolded packages are most often used for high-pin-count devices or pin grid arrays that are not amenable to flat leadframes and simple fanout patterns. Premolded packaging is more expensive than postmolding because of the greater number of parts and assembly steps. Important advances in design and molding capability are enabling molded plastic packages to reach the very high pin counts that have been the preserve of premolded packages. Figure 1-11 is an illustration of a premolded plastic package in the gull wing configuration.

*Postmolded Packaging.* Post-molded packages are the most common form of packaging accounting for more than 80% of all integrated circuit devices [1]. Postmolded packaging uses a flat metal leadframe upon which the die is attached. Wire bonding is then used to interconnect the bond pads of the device to the fanout pattern of the leads. Tape automated bonding is an emerging op-

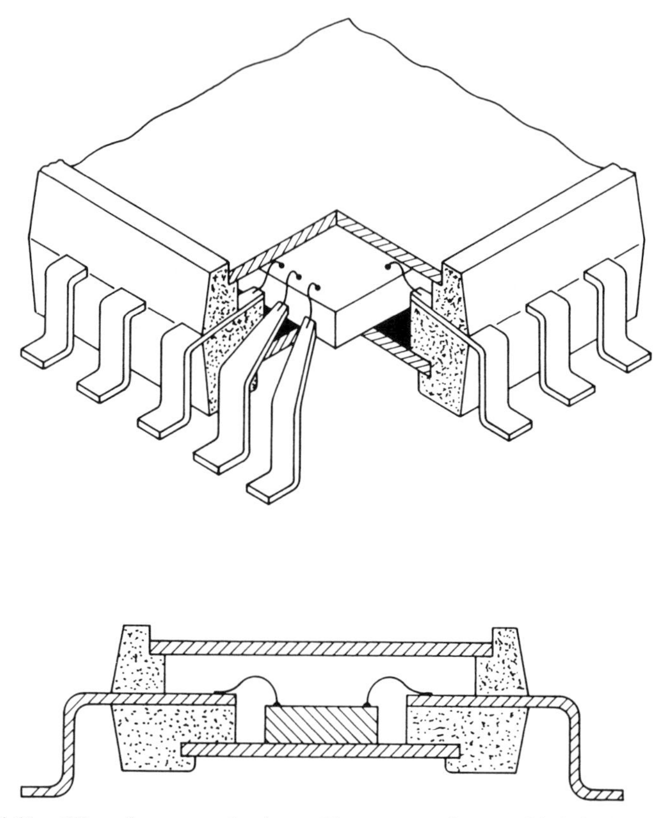

**FIGURE 1-11.** Side and axonometric views with cutaways of a premolded plastic package with four-sided, gull wing leads.

tion for interconnection to the leadframe that is discussed in Section 9.2.4. An illustrative drawing of a postmolded DIP package is shown in Figure 1-12. A drawing with cutaway view of a postmolded plastic package with gull wing leads is provided in Figure 1-13. The leadframes are loaded into a multicavity molding tool and encapsulated in thermoset molding compound using a transfer molding process. A high loading of inert filler such as ground silica is used to lower the coefficient of thermal expansion of the molding compound so that it approaches the low expansion values of the leadframe and the silicon device. In some instances, an intermediate encapsulant is used directly over the IC device to shield it from the molding compound, prevent water clustering at the active surface of the device, and lower stress at the die surface. For memory devices, this coating serves as a barrier layer to alpha particles, which can cause soft errors. An encapsulant such as silicone rubber is occasionally used with microprocessor and logic devices as well, since silicon-silicone interfacial

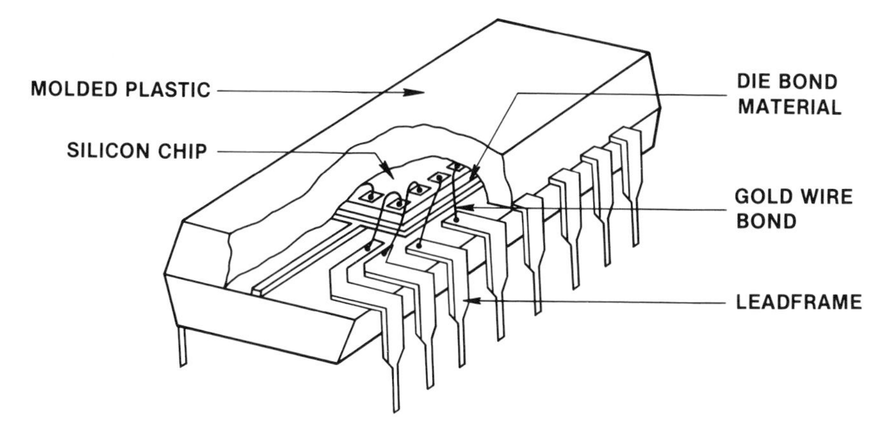


**FIGURE 1-12.** Cutaway view of a postmolded plastic package in the configuration of a dual-line package (DIP).

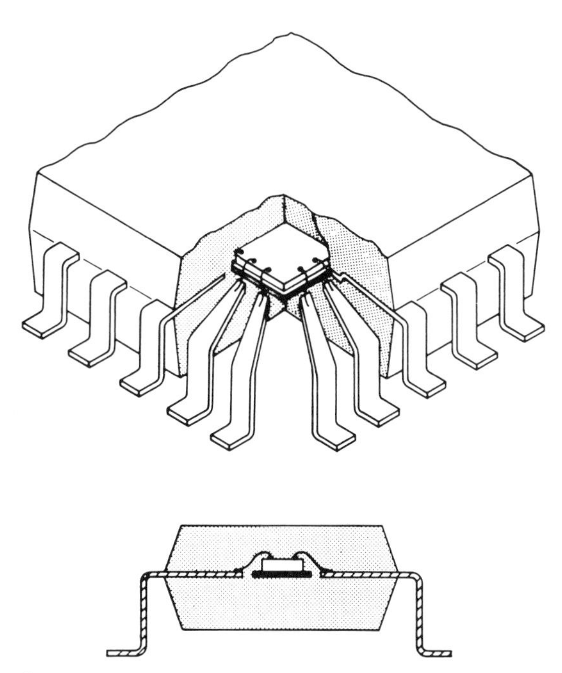

**FIGURE 1-13.** Cutaway view of a postmolded plastic package with quad configuration and gull wing leads.

bonding is believed to improve the moisture barrier properties at the chip surface.

Plastics packaging is efficient since the assembly and process operations are conducted on the leadframes, which typically cluster four to twelve devices together, as shown in Figure 1-14. This configuration facilitates handling and makes the process more amenable to automation, thereby helping to keep costs low and market share high. The penalty paid for this fixturing is that the devices cannot be tested once they have been wire bonded to the frame. The many process steps after wire bonding are then conducted without any yield feedback, or opportunity to remove failed devices from further processing. This puts greater importance on conducting plastic packaging operations at near 100% yield.

A thermoset polymer material is one that polymerizes during the processing operation, forming a crosslinked molecular structure that can no longer flow as a true liquid, although it can sustain deformation. The thermoset character of the materials used for postmolded plastic packaging means that they can have exceptionally low viscosity when filling the mold and flowing over the leadframe and chip. The polymerization then continues in the mold, forming a high-molecular-weight polymer that is vitrified at room temperature providing a hard plastic casing. An assortment of plastic packages is shown in Figure 1-15.

There have been many improvements in the thermoset molding compounds that have significantly improved the reliability of plastic packaging. One of these improvements is the reduction in ionic impurities. State-of-the-art purity in the late 1980s was approximately 15 ppm, compared to levels of 50–70 ppm in the early 1980s. Other major improvements concern the process characteristics. A major contributor to loss of yield during packaging is flow-induced stress damage to the leadframes and wire bonds. These stresses are proportional to the product of the molding compound viscosity and velocity during flow in the mold as described in Chapters 6 and 7. Viscosity of these materials depends on both the molecular weight of the base resin and the filler loading as discussed in Chapter 2. Molding compound viscosities have been significantly reduced, while preserving the desirable mechanical properties attendant with high molecular weight and high filler loading. Moisture resistance has also been improved substantially.

Another important improvement in molding compounds has been the introduction of low-stress materials. These materials generate exceptionally low mechanical stresses caused by the mismatch in the coefficients of thermal expansion among plastic, leadframe, and silicon die. Novel polymer chemistry, discussed in Chapter 2, has been utilized to achieve these improvements. The result of these efforts has been that plastic packaging has emerged as the dominant type of second level interconnection, and its capabilities and market share continue to increase.

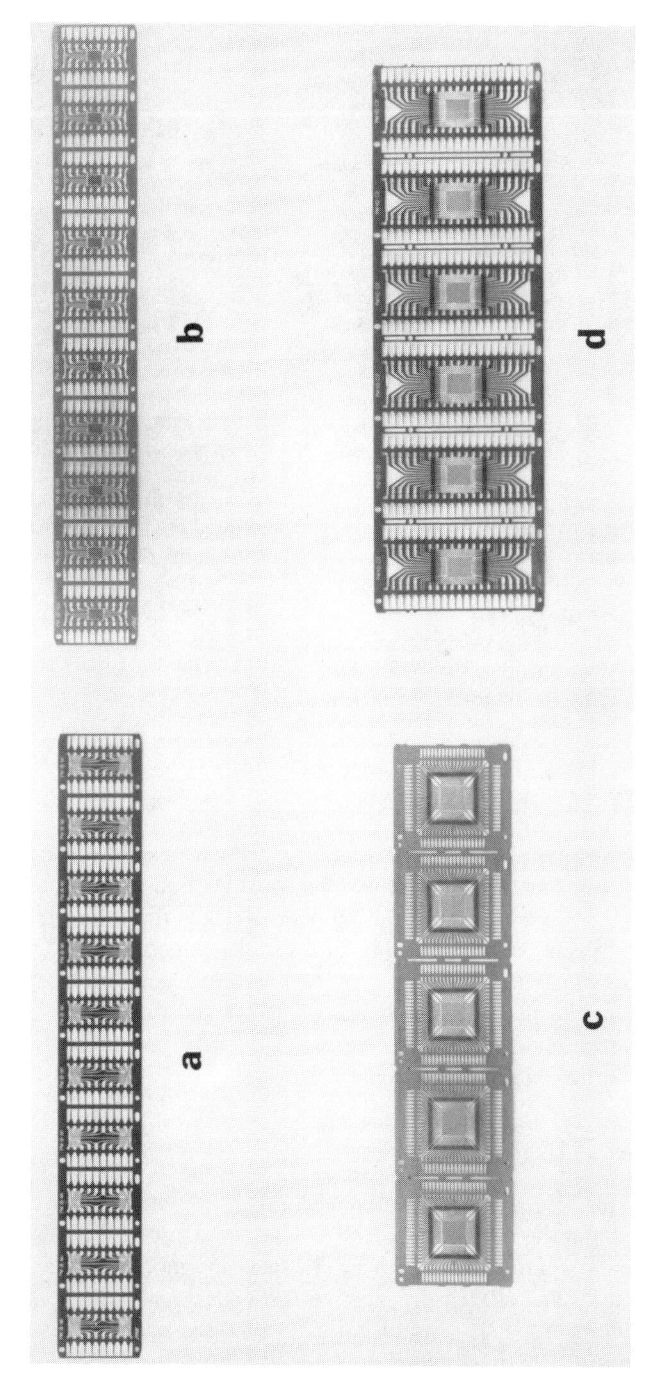


**FIGURE 1-14.** Some common leadframes used in post-molded plastic packaging: (a) 18-pin DIP with leads routed underneath the chip, (b) 18-pin DIP with conventional die support paddle, (c) 68 PLCC, (d) 40-pin DIP.

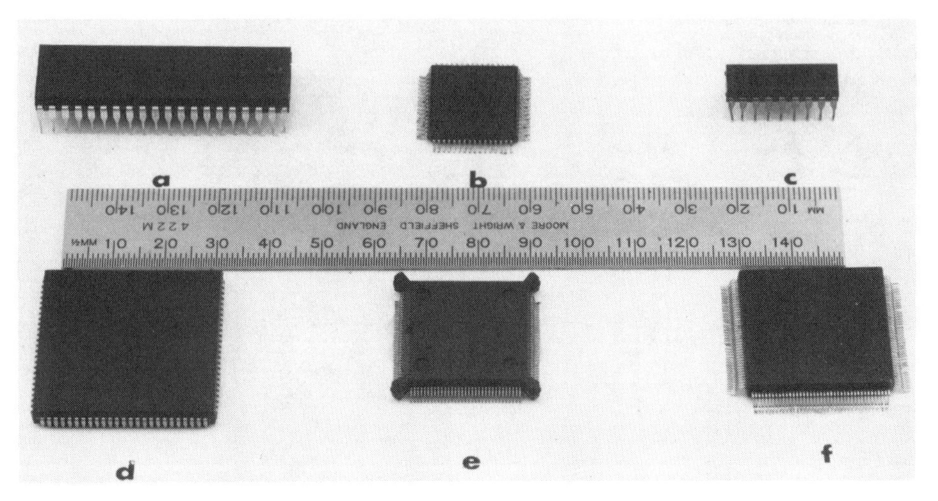


FIGURE 1-15. An assortment of molded plastic packages: (a) 40-pin DIP (0.100 inch pitch), (b) 64 quad (2 mm pitch), (c) 18-pin DIP (0.100 inch pitch), (d) 100 PLCC (0.050 inch pitch), (e) 132 PQFP with bumpers, JEDEC standard (0.025 inch pitch), (f) 164 PQFP without bumpers, EIAJ standard (1 mm pitch).

A bar graph showing the production volume and percent of market share of the major package types is provided in Figure 1-16 [1].

## 1.5 CRITERIA FOR SELECTING AMONG DIFFERENT TYPES OF PACKAGES

There are numerous types of packages that have been designed to meet different I/O, reliability, cost, and system needs. One consideration that determines the type of package chosen is simply the number of leads the integrated circuit needs to communicate with the next higher level of interconnection. Other major considerations include reliability, cost, and system criteria such as how the packaged device will be placed on the circuit board and the amount of circuit board area that can be devoted to the device. These criteria are examined more thoroughly in the following subsections.

### 1.5.1 Number of Leads Required

The number of leads required for interconnection varies with the type of device being packaged and the device complexity. The pin count of the package can exceed the number required by the device. In many instances there are package pins that are excess and electrically inactive. To understand the criteria for selecting a specific package, it is useful to understand the interconnection needs of different types of integrated circuits.

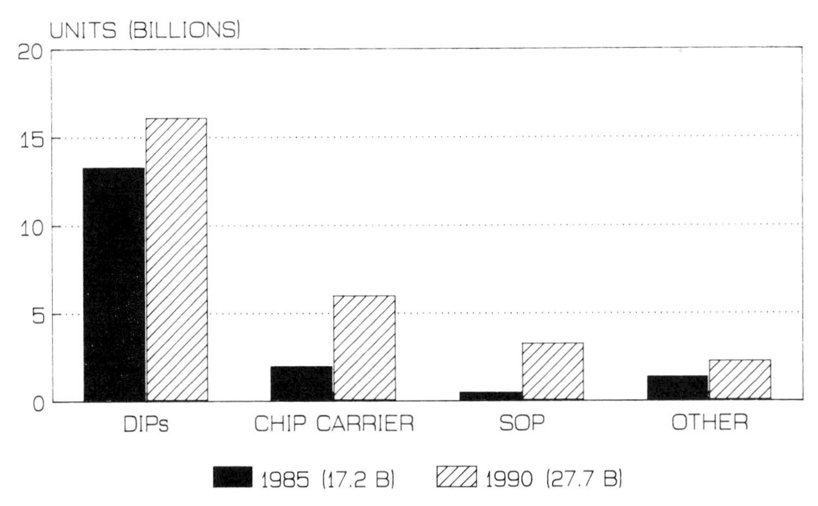


**FIGURE 1-16.** The market share of the major types of packages. (Data from Reference 1.)

**1.5.1.1 Logic Devices.** The number of pins required for a logic or microprocessor device depends on the type, function and complexity of the specific device. There is an empirical rule that has been formulated to estimate the number of interconnects required by a device of a given level of sophistication, measured as function of the number of logic elements it encompasses. This relation, known as *Rent's Rule*, states that the number of leads needed for a device to communicate with the next level of interconnection is proportional to the number of logic gates raised to some power which varies between 0.4 and 0.7 based on the type of device [15].

Rent's Rule: $$N_{\text{leads}} = \alpha(N_{\text{gates}})^{\beta} \tag{1-1}$$

The $\alpha$ and $\beta$ coefficients vary with the device type. For microprocessors, the most accepted coefficients are $\alpha = 4.5$ and $\beta = 0.5$. With these parameters, Rent's Rule predicts that gate arrays with thousands of gates will require hundreds of leads; 5000 gates require approximately 300 leads. Comparisons with actual gate arrays of this size indicate that the number of leads predicted is slightly high in most cases. Although this rule was formulated empirically in 1960, it continues to be relevant despite the improvements in technology and design that have occurred over the intervening three decades. In the absence of some radical new interconnection technology, the number of lead pins will continue to scale to the number of logic elements on the device raised to the 0.5 power (although some citations supports exponents as low as 0.2 for chips with less partitioning). The number of gates per unit area on a device scales with the square of the feature size reduction ratio, which means that the number of leads required scales directly with the feature size reduction ratio. Actually, logic devices have been growing larger as well as more densely packed, so the lead count has been increasing even faster than the feature size reduction ratio.

Rent's Rule can be broken at any level of interconnection. Ultimately, it is broken in any system because for the system to be useful, the number of interconnects must be reduced to some manageable number. As an example, a personal computer interconnects dozens of chips that have nearly 100 leads each, to circuit boards that have tens of edge connections each, to wire harnesses that have 10–20 wires, to local area network lines that have either a single optical fiber or several copper wires. The multichip module is an example of breaking Rent's Rule at a fairly high level of integration. It is generally advantageous to break Rent's Rule at the highest possible level of integration because it preserves parallel data paths into the higher levels of interconnection; thereby providing greater system performance and system architecture flexibility.

**1.5.1.2 Memory Devices.** Memory devices require much fewer lead pins than processors because the memory cells can be accessed more efficiently owing to the column and row structure of the device. As an example, a 4K static ram has 4096 memory cells configured in a 64 × 64 cell grid. The maximum number of pins to address these cells would then be 64 column pins and 64 row pins, for a total 128. The number of package pins is greatly reduced, however, by using address decoding [16], in which a decoder routine is incorporated in the device. Address decoding allows $N$ pins to address $2^N$ memory cells by setting up a binary truth table, an example of which is shown in Figure 1-17 [16], which demonstrates how 3 pins can access $2^3$ memory cells. The relation to compute the number of leads, $N_l$, to access a memory location in a grid of $N_c$ cells is:

$$N_c = (2^{N_l}) \tag{1-2}$$

| PIN 0 | 0 | 1 | 0 | 1 | 0 | 1 | 0 | 1 |
|---|---|---|---|---|---|---|---|---|
| PIN 1 | 0 | 0 | 1 | 1 | 0 | 0 | 1 | 1 |
| PIN 2 | 0 | 0 | 0 | 0 | 1 | 1 | 1 | 1 |
| MEMORY CELL NUMBER | 1 | 2 | 3 | 4 | 5 | 6 | 7 | 8 |

**FIGURE 1-17.** Binary truth table used for address decoding of memory devices. The table shows how 3 pins can be used to address $2^3$ memory cells providing a significant reduction in the number of I/O required for memory device packages. (After Howes and Morgan, Reference 16.)

The 4K memory locations can be accessed by only 12 pins since $2^{12} = 4096$. An actual device requires additional pins for power, ground, data in, and data out, but the accessing of information on the grid is accomplished with very few leads through the decoding algorithm.

A further reduction in the number of pins is attainable by multiplexing the signals on the pins such that one pin reads both the column and row location on two clock sequences. An additional pin must be used to specify whether column or row signals are being received, but this added pin decreases the number of pins to access the memory by a factor of 2. As an example, a 256K DRAM should require 18 pins to access its $2^{18}$ (262,144) memory cells. With auxiliary pins for power, ground, and data control, the 256K package should have a least 22 leads. Essentially all 256K packages used multiplexing to reduce the number of access pins to 9, allowing the device to be packaged in 16-pin DIPs. There is a small penalty in device performance for using multiplexing because the access time is increased, but it appears that most manufacturers accept this drawback.

### 1.5.2 Reliability Criteria

The wide variety of products that use ICs includes musical greeting cards, space satellites, home appliances, aircraft, automobiles and telephones. These products have vastly different reliability criteria and expectations. They operate under different ambient conditions, ranging from the high temperatures and vibrations under the hood of a car to the relatively benign atmosphere in an office building. The reliability of a packaged device depends on both the intrinsic protection afforded on the silicon itself, and the protection provided by the package. For this reason, the reliability expectation of an IC product has a major role in the selection of a packaging technology.

Ceramic packages are the most reliable means of protecting an IC device. These are hermetic packages that can adequately protect the device from moisture intrusion, and therefore essentially eliminate corrosion failure. United States military specifications (Mil. Std. 38510 Group D) which pertain to integrated circuit devices have allowed only hermetic packages. They are also the most expensive type of commercial package. CERDIPs also have excellent reliability with a lower cost than ceramic packaging, yet they are more expensive than non-hermetic plastic packaging.

The presence of water at the active surface of the device introduces a host of failure mechanisms. Studies show that water vapor quickly permeates plastic packages [17], transporting impurities from the surface of the package and leaching impurities from the molding compound. The second and much slower step is the permeation of water through the passivation layers of the chip. If the passivation layer is cracked, the permeation occurs much more rapidly. It has been reported that the penetration of water through the passivation layer determined the rate of failure of the packaged device [17]. Electrochemical corrosion will occur quickly once water and ionic impurities reach the metallization layers of the device.

Plastic packaging has shown significant improvements in reliability to the point where performance rivals that of the more costly ceramics and CERDIPs. In fact, plastic packages of the mid-1980s were as reliable as hermetic packages of the mid-1970s [18]. These improvements in reliability have resulted from improvements in the passivation layers on the device, particularly the increasing use of silicon nitride, and the improvements in the plastic molding compounds used to make the package. The ionic impurities in the molding compound, which cause dendritic migration of gold contact layers on the device [19], have been significantly reduced. This makes the resolution of reliability criteria simpler, since plastics are suitable for all but the most demanding applications.

Electronic systems are often composed of hundreds or thousands of IC devices. System failures are caused by individual failures of one or more IC devices, depending on the degree of redundancy built into the system, so system reliability depends on the number of devices and the expected reliability of each device. Device failures are usually defined in terms of *F*ailure un*IT*s, or FITs which are defined as 1 failure per $10^9$ device hour [20]. Table 1-2 [20] shows the effect of device failure on system performance for a small system with 200 ICs and a large system with 10,000 ICs. The mean time to failure for the smaller system is 5 years when the FIT is 100. At this degree of reliability, Table 1-2 shows that 1.6% of the systems fail each month. If the failure rate is reduced to 10 FIT, the mean time to failure is extended to 51 years. The system (circuit pack) failures for the large system are much greater at the same FIT rate. The

**Table 1-2. Effect of Device Failure on Modern System Performance.**

| *Data set: 150 to 225 ICs* | | |
|---|---|---|
| *Failure rate (FIT)* | *Mean time to failure (year)* | % of sets failing per month |
| 10 | 51 | 0.16 |
| 100 | 5 | 1.6 |
| 1000 | 0.5 | 16 |
| *Private telephone system: bulk of installations 5000 to 10,000 ICs; large installations 50,000 ICs* | | |
| *Failure rate (FIT)* | *System failure/month* | *% of circuit packs failing in 10 years* |
| 10 | 0.07 | 1 |
| 100 | 0.7 | 10 |
| 1000 | 7 | 65 |

(From W. J. Bertram, "Yield and Reliability" in S. M. Sze, *VLSI Technology* McGraw-Hill Book Company (1983) Copyright © 1983 by Bell Telephone Laboratories)

manufacturer must decide what level of reliability is desirable, what is acceptable, and choose packaging commensurate with those reliability criteria.

### 1.5.3 Cost Criteria

There are significant cost differences associated with the different types of packages. Hermetic ceramic packages are the most expensive and they provide the greatest reliability, as well as improved performance with power and ground distribution. A ceramic package costs approximately five to ten times as much as a plastic package. The lower-cost hermetic CERDIP packaging costs approximately five times as much as plastic packaging, but trapped moisture and impurities in the CERDIP lowers the reliability performance compared to ceramic packaging. There is a significant cost difference for the small improvement in reliability that is provided by hermetic packaging. For state-of-the-art products or systems that have unusually long life expectancies, such as mainframe computers and telecommunications systems, or systems which require exceptional reliability, such as military and astronautical applications, the added cost of ceramic packaging does not deter their use. These applications place a far greater emphasis on reliability than on cost. For the remainder of the IC device applications, including most consumer products, the performance and reliability of plastic packaging is more than adequate.

Cost criteria should also include the productivity and yield for the manufacture of different packages. Due to differences in size and aspect ratio of the packages, the number of cavities in a molding tool can vary from 50 to 400. This is a direct metric of productivity because the time per molding cycle is essentially the same. There are also yield considerations. Although most package molding operations will have to run at near 100% yield to be competitive, the ramp-up time to satisfactory yield can be significantly different. This ramp-up time for package yield is often an important factor in product introduction time and cost projections. Profitability often depends on these initial cost projections, particularly if they are locked into large contracts.

### 1.5.4 Systems Criteria

The systems criteria in selecting integrated circuit packaging concern system-wide issues such as assembly, testing, performance, and repairability. Assembly of the packaged device on the printed circuit board has emerged as a primary consideration during the time that through-hole, surface mount, and fine-pitch packages will coexist. This time of mixed assembly technologies is expected to extend to the turn of the century.

For a microprocessor that requires 60 leads, there are several packaging options that can be exercised: 68- and 84-lead PLCC packages or an 84-lead PQFP fine-pitch package. The decision between the 25-mil spacing and conventional 50-mil spacing is a systems and manufacturing decision. The smaller footprint of the fine-pitch package would always be welcome, but penalties in the cost of more precise placement equipment and possible lower yields in solder attachment could favor the larger packages. The introduction of fine-pitch packages in the late 1980s appears to be one of the first cases where assembly technology was the bottleneck to continued improvement in interconnection density and miniaturization. Using the 84-lead PLCC package instead of the 68 PLCC sacrifices some area on the circuit board, but also accommodates device and systems upgrading without reworking the board.

It is also helpful if the design engineer responsible for selecting the packaging is aware of the manufacturing considerations in producing the packages under consideration. For example, the cost of molding the 84 PLCC package may be greater because it is a larger package and fewer devices are packaged in a molding cycle. Also there could be yield differences among the options that may influence the decision. In general, the more the design engineer knows about the systems and manufacturing considerations of different packages, the more likely he or she is to make the best choice.

## 1.6 PROCESS STEPS IN ASSEMBLY AND PACKAGING OF IC DEVICES

It is important to know the entire assembly and packaging process sequence to be able to design, manufacture, and optimize molded plastic packaging. In many cases, the devices are packaged at a different factory than where the wafers were produced. Although there has been a movement toward integrating wafer fabrication and packaging in the same production line, most operations still require the transport of the wafers to the packaging facilities, in a different section of the plant, at a different location, or on a different continent. Wafers travel thousands of miles for the many companies that do packaging in countries where the cost of labor is low. The devices are usually tested at the wafer line and defective devices are marked, often with a black dot that will cause the chip to be overlooked by a vision system.

The process steps conducted prior to encapsulation in the plastic molding compound are known as the assembly or front end operations; steps after molding are called back end. Clean rooms of the Class 100 to 1000 scale are used for most front-end assembly operations. Some molding operations are conductive in clean rooms as well, although the machinery hydraulics and dust from the preforms makes it very difficult to achieve anything better than Class 10,000. In general, the increasing complexity and miniaturization of the silicon devices will drive more of the assembly and molding operations to dust controlled environments. A more detailed discussion of the process steps of plastic packaging is provided in Chapter 6.

*Wafer Separation.* The first assembly step is to divide the wafer into the individual chips. This process is known as *sawing* or *dicing*. Older dicing equipment was manually operated, whereas present equipment is automatic with vision system registration. Some techniques scribe the silicon surface then break it into separate chips. Partial scribing may be accomplished with a pulsed laser beam, a diamond-tipped scribing tool, or a diamond-impregnated saw blade [14]. Sawing is preferred for partial scribing or full wafer separation because it produces a straight edge with less chipping and cracking [21]. The silicon device is commonly referred to as a *die* at this point of the assembly sequence.

*Die Attach.* The sawed wafers are transferred to a placement machine that, using a vision system or computer tape, sorts the good devices and removes them from the polyethylene support. The chips are then positioned on the paddle of the leadframe. Upstream of the sort-and-place machine, a die attach material has been applied to the paddle. The means of attaching the die to the support include eutectic solders, polymer adhesives, and silver-filled glass resins. All of these materials are conductive, allowing electrical contacts to be made to the back of the chip. In eutectic die bonding, the back of the die and the bonding area of the copper alloy leadframe are both metallized for wettability. Probably the most common die attachment materials are conductive epoxies containing metallic silver filler. The die is sometimes scrubbed onto the paddle to spread the adhesive and improve the bonding. The leadframes then pass through an oven to cure the adhesive. After die attach, the leadframe strips are usually handled in cassettes in batch packaging lines, whereas automated in-line operations process the leadframes on a continuous track.

*Die Interconnection.* Wire bonding and tape automated bonding (TAB) are the two principal ways of connecting the silicon device to the leadframe in molded plastic packaging. Direct chip attachment using solder bumps is not common in molded packaging, but is more feasible with premolded packaging where the fan out pattern is supported on the premolded base.

*Wire Bonding.* Gold or aluminum wire is used to connect the bonding pads of the die with the leads of the frame. Typical wire diameters are 0.0025 to 0.0032 cm (1.00 to 1.25 mil) for gold and up to 0.005 cm (2.0 mils) for aluminum, because of its lower electrical conductivity. Gold wire can be bonded by thermocompression, ultrasonic, or thermosonic methods, whereas aluminum wire can only be ultrasonically bonded. For gold wire, the bond at the die pad is usually a ball bond formed by flame softening the tip of wire and then mashing it against the pad. The wire is looped out and formed into a wedge bond on the leadframe finger. Aluminum wire requires wedge bonds on both the bonding pad and the lead finger. Wire bonding is done with automated machinery capable of forming up to 200 bonds per minute. Both the overall length of the wire bond and the shape of the loop can be carefully controlled. Typical bond lengths range from 0.15 to 0.30 cm (60 to 120 mils) and loop heights can be 0.075 cm (30 mils) above the plane of the device. Longer wire spans, caused by placing a small chip on an oversized paddle, create greater opportunities for wire displacement during plastic molding.

*Tape Automated Bonding.* Most often known by the acronym TAB, tape automated bonding was developed in the mid-1970s, but did not begin to reach commercial importance until the late 1980s [22, 23, 24]. The development of very high I/O packages seems to have provided the strong motivation for a batch interconnection process. The fanout pattern, in the form of an etched metal frame on a film carrier, interconnects the bonding pads of the device through thermocompression bonding or eutectic methods [14]. The pads on the device often have solder bumps to aid the thermocompression bonding. In another variation of the process, the bumps are etched onto the tape rather than deposited on the device. The patterns are attached to a continuous polymer tape to facilitate high productivity assembly. TAB can be used to attach a fine fanout pattern to a larger leadframe in a process known as inner lead bonding, or it can be used as the leadframe itself.

There are several types of tapes. Single-layer copper is the simplest and the most precise, but it is lacking in ability to support longer thinner lead fingers. Copper tape is electrically continuous and cannot be tested after bonding. Two-layer tapes of copper on polyimide film are produced by imaging and etching either the copper or the polymer. Two-layer tapes are more rigid than single-layer metal and therefore allow larger patterns to be bonded. They may also be tested since the film supports the electrically isolated lead fingers. Three-layer tapes use an adhesive layer, such as epoxy, to bond the copper and polymer film. Materials other than polyimide can be used, and stiffer substrates facilitate testing after bonding by allowing the extension of fingers to probe test points [10]. A double-thickness copper tape, typically 66 microns, with etched bumps provides improved rigidity over single-layer copper and eliminates the need to bump the device, but it is not testable. A further discussion of TAB which also considers its implications in affecting future packaging technology is provided in Section 9.2.4.

*Molding.* After die attachment and interconnection, the leadframes are sent to the molding facilities. Molding equipment consists of a transfer molding press with the appropriate molding tool, a dielectric preheater, and assorted handling equipment such as frame and cassette loaders. The devices cannot be tested after interconnection because of the electrical continuity of the frame, but it would be impractical to eliminate defective devices from packaging since all handling is done on the leadframe strips. In a semi-automated molding operation, the cassettes are loaded into a frame loader which places them on the loading frames of the molding tool. An operator then manually carries the frame loader over to the molding tool and places it over the lower mold half. The loading frame positions the leadframes onto registration pins on the mold. The mold halves are then brought together in what is known as the *clamped* position.

The thermoset molding compound is dielectrically preheated in a separate piece of equipment. At a preset time or temperature, the molding compound is removed and placed into the transfer pot of the molding tool. A transfer cylinder, also known as the *plunger*, is then activated and it pushes the molding compound out into the runner system of the mold. A drawing of the runner system and cavities for a production-scale molding tool is shown in Figure 1-18.

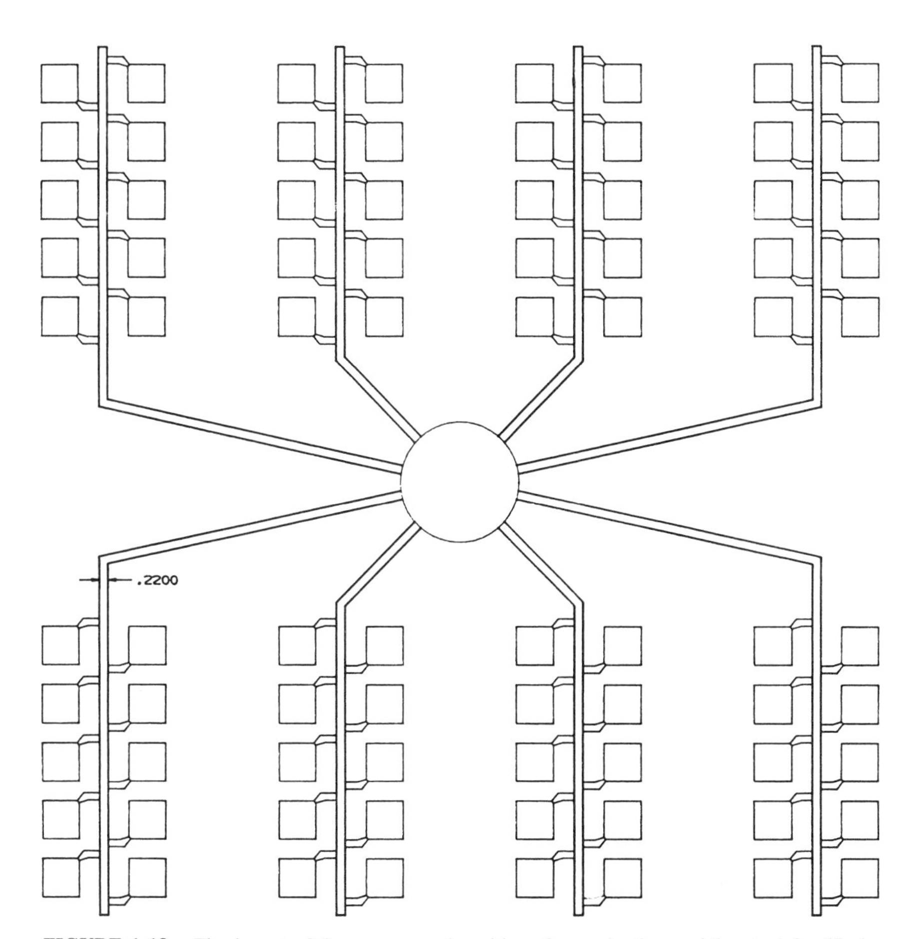


**FIGURE 1-18.** The layout of the runners and cavities of a production molding tool for 68-pin PLCC packages.

The viscous molding compound flows over the chips, wire bonds, and lead frames encapsulating the device with incipient plastic. Yield losses can be high if the flow-induced stresses are excessive. The thermoset molding compound must polymerize, or cure, in the molding tool for a period of 1–3 minutes. The press opens automatically after the prescribed time interval, and the operator removes the molded leadframe strips from the tool, loading them into cassettes for the next process operation.

In an automated molding line, the cassettes of leadframes are loaded into one end of the machine and cassettes of molded strips are removed from the other end. All frames handling, preform heating, and loading operations are performed with automation.

*Deflashing/Dejunking.* The molded leadframe strips may require removal of a thin flash of molding compound known as *resin bleed* or *flash* that flows onto the lead fingers. Also, the molding compound caught behind the dam bars of the frame may have to be removed prior to trim and form to reduce abrasive wear on the trimming tool in an operation known as dejunking. Removal of excess molding compound can be accomplished in several different ways. Media deflashers use hard, tough projectiles such as polymer beads or ground hardwood in a pneumatic stream to blast the molding compound from the leads. Solvent deflashers soften and loosen the thin flash. Water deflashers use a high-pressure water jet to blast the molding compound off the leads.

Quality molding practices can minimize flash to the point that deflashing is not needed. Poor design, construction, or maintenance of molding tools can promote flashing that may increase in area and thickness until conventional equipment will have trouble removing it, leading to downstream problems in trim-and-form and solder dipping.

*Code Marking.* Legible information such as the device code, date of manufacture, the manufacturer, and the country of origin is put on the top surface of the package in the code marking operation. This marking is done either with polymer-based inks or through laser writing.

*Postcure.* A postcure procedure is required for essentially all molding compounds. The molding compound does not reach complete conversion in the several-minute cycle in the molding tool, and it is inefficient to utilize expensive machine and human resources to wait for full conversion. Instead, the devices are ejected after they have achieved an ejectable degree of conversion and then large batches are placed in ovens for 4–8 hours at 170–180°C.

*Trim-and-Form.* The lead fingers are electrically isolated and the leads are formed during the trim-and-form process. This operation is conducted in a punch press in which a trim-and-form tool is mounted. The tool usually comprises several stages which sequentially step through the cutting and forming of the lead fingers. Most tools and presses operate automatically: cassettes of molded strips are loaded at one end and completed units are collected in tubes at the other end.

Any discrepancies or inaccuracies in the dimensions of the leadframe or the molded body can cause trim-and-form problems. Such problems could arise from buckling of the leadframe due to excessive shrinkage of the molding compound, or a shift in the location of the molded body on the leadframe caused by damage or wear of the molding tool. Excessive and irremovable flash of molding compound on the leads can cause jamming and premature wear of the trim tool.

*Solder Dipping/Solder Plating.* After the leads have been formed, they often require solder coating to facilitate high-yield attachment to the circuit board. This thin coat of solder, typically less than 0.0005 inch, is applied using either a solder dipping or solder plating operation. Solder dipping exposes the molded body to a greater thermal shock, whereas solder plating exposes the package to the corrosive plating chemicals. Solder dipping is done after singulation, whereas solder plating is conducted on the entire strip and performed before the trim-and-form step discussed above.

*Burn-in.* Burn-in is the common name for screening for premature failure at high temperature and high electrical loading. The packaged devices are often tested before burn-in and failed devices are removed. The devices are mounted on special circuit boards that apply the electrical loading, and these boards are then plugged into large oven chambers that maintain a constant high temperature. Typical burn-in cycles are 120°C for a time period of 9–24 hours at a voltage loading 1–2 volts above the nominal 5-volt operating range typical for most devices. The plant area required for burn-in is considerable, given the large number of devices and the long residence time. The cost associated with this operation is proportionally high, since every device must be handled, either manually or automatically. Burn-in of a statistical sampling of devices is also used instead of 100% burn-in.

*Testing.* In most cases 100% of the molded product is tested for electrical performance. The procedures and sampling sizes for burn-in and testing differ among manufacturers, however. Typically, there will be a pretest, burn-in, and then a second test cycle to determine the failure rate. Special test boards and custom evaluation software are required for testing. The analysis is performed by sophisticated computer systems that keep records on the number and type of defects, information that is vital to process analysis and improvement. Some manufacturers use a molded-in marking on the package to indicate the cavity location where a package was molded thereby facilitating mold mapping studies that can indicate certain mold design or processing problems. The amount of equipment and personnel required for testing is large and adds a significant fraction to the cost of the device. In general, burn-in and testing can account for as much as 50% of the plant area needed for assembly and packaging.

## REFERENCES


1. *Semiconductor Int'l*, January 1987.
2. R. R. Tummala and E. J. Rymaszewski, Editors, *Microelectronics Packaging Handbook*, Van Nostrand Reinhold Publishers, New York, NY (1989).
3. F. N. Sinnadurai, Editor, *Handbook of Microelectronics Packaging and Interconnection Technologies*, Electrochemical Publications Ltd., Ayr, Scotland (1985).
4. M. T. Goosey, Editor, *Plastics for Electronics*, Elsevier Applied Science Publishers, London and New York (1985).
5. D. S. Soane and Z. Martynenko, *Polymers in Microelectronics: Fundamentals and Applications*, Elsevier Press, New York and Amsterdam (1989).
6. C. Mead and L. Conway, *Introduction to VLSI Systems*, Addison-Wesley Publishing Co., Reading, MA (1980).
7. E. F. Labuda and J. T. Clemens, "Integrated Circuit Technology," in R. E. Kirk and D. F. Othmer, Editors, *Encyclopedia of Chemical Technology*, John Wiley and Son, New York (1980).
8. S. M. Sze, Editor, *VLSI Technology*, Second Edition, McGraw-Hill, New York (1988).
9. J. C. Harrison and E. F. Rickard, "Plastics for Long-Life Microcircuit Encapsulation, Part I: Materials Properties and Possible Failure Mechanisms," *Proceedings of the IEEE Conference on Reliability in Electronics*, 129, (Dec. 1969).
10. J. L. Dais and F. L. Howland, *IEEE Trans. Components, Hybrids and Manufacturing Tech.*, 158–166, (June 1978).
11. L. F. Miller, "Controlled Collapse Reflow Chip Joining," *IBM J. Research and Development*, **13,** 239–250, (1969).
12. P. A. Totta and R. P. Sopher, "SLT Device Metallurgy and Its Monolithic Extension," *IBM J. Research and Development*, **13,** 226 (1969).
13. N. Nagashima, "A View of VLSI Packaging Technology," *Nitto Technical Reports*, Special Report on Semiconductor Encapsulants (September 1987).
14. C. A. Steidel, "Assembly Techniques and Packaging," Chapter 13 in S. M. Sze, Editor, *VLSI Technology*, McGraw-Hill Book Company, New York (1983).
15. Rent never published an account of the rule that bears his name. An early reference and description of the rule is: R. L. Russo, "On a Pin Versus Block Relationship for Partitions of Logic Graphs," *IEEE Trans. on Computers*, **C-20,** 1467 (1971).
16. M. J. Howes and D. V. Morgan, *Large Scale Integration*, John Wiley and Sons, New York (1981).
17. J. E. Gunn and S. K. Malik, "Highly Accelerated Temperature and Humidity Stress Test Technique (HAST)," *Reliability Physics, 19th Annual Proceedings*, 48–51 (1981).
18. "Special Report: Plastic Package Update," *Linear and Interface Circuits Reliability Report*, Texas Instruments Corporation, Dallas TX (1983).
19. F. Gruthaner, T. Griswold, and H. Bright, "Migratory Gold Resistive Shorts (MGRS) Failures—Chemical Aspects of a Failure Mechanism," *Proceedings of the IEEE 13th Annual Reliability Physics Symposium*," 99 (1975).

20. W. J. Bertram, "Yield and Reliability," in S. M. Sze, Editor, *VLSI Technology*, McGraw-Hill, New York (1983).
21. R. C. Cook, S. Madden, and H. Williamson, "New Approaches to Sawing Microelectronic Materials," *Semiconductor Int'l*, (Dec. 1980).
22. G. Dehaine and K. Kurzweil, "Tape Automated Bonding Moving into Production," *Solid State Technology*, 46 (Oct. 1975).
23. A. Keizer and D. Brown, "Bonding Systems for Microinterconnect Tape Technology," *Solid State Technology*, 59 (Mar. 1978).
24. J. Lyman, "NEPCON Highlights: The Dominant Role Tape-Automated Bonding is Taking," *Electronics* (February 18, 1988).

# 2 Polymer Science and Engineering for Microelectronics Packaging

Familiarity with the principles of polymer science and engineering is necessary for the improved understanding of plastic packaging of integrated circuit devices. Phenomena such as material rheology, glass transition temperature, crosslink density, and viscoelasticity play important roles in the production and performance of plastic packages. The concepts needed to understand plastic package production, properties, and performance are described in this chapter.

## 2.1 THERMOSETTING POLYMERS

Many of the polymer materials used in plastic packaging are thermoset polymers. These are materials that are initially monomers or prepolymers, which then polymerize during processing. The polymerization of thermoset polymers is often known as *curing*, a term derived from rubber technology. Materials that have a functionality greater than two are able to form a three-dimensional molecular structure. Such a material is said to be crosslinked because there are physical links or branch points that tie the polymer chains together. The relative number of branch points is known as the *crosslink density*, and materials with very high crosslink densities tend to be stiffer but more brittle. The degree of chemical conversion at which the three-dimensional structure first appears is known as the *gel point* and the phenomenon is known as *gelation* [1]. The conversion to gel depends on the functionality of the material system, as is shown in Equation (2-1) for a single component system:

$$X_g = \frac{1}{(f - 1)} \tag{2-1}$$

In multicomponent systems such as epoxy molding compounds, there is a mixture of reactive species with different functionalities and competing crosslinking reactions making it difficult to assign a single functionality for the system.

The material cannot flow as a true liquid after gelation; therefore forming operations such as molding must be completed before the material gels. Viscosity depends upon molecular weight and both increase during the polymerization reaction, effectively going to infinite molecular weight and infinite viscosity at the gel point. Figure 2-1 is a plot of viscosity versus time of polymerization for a thermoset molding compound, showing the gel point.

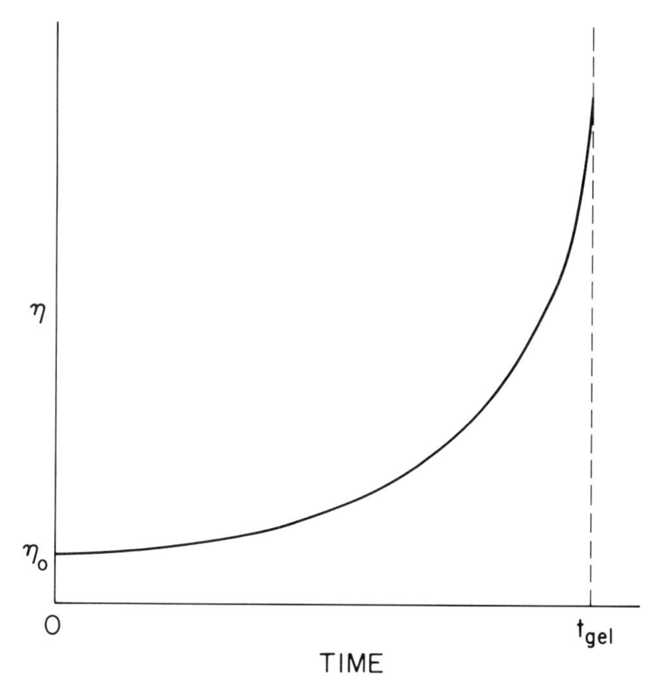


**FIGURE 2-1** Plot of viscosity versus time at isothermal conditions for a thermoset polymer during polymerization. $\eta_0$ denotes the initial viscosity and $t_{gel}$ is the gel time.

The term *thermoset* derives from the fact that these materials do not flow at high temperatures because of the interlocked structure. The use of thermoset materials in plastic packaging takes advantage of their low initial viscosity, and their ability to sustain very high temperature without flow when polymerized. These materials can withstand solder temperatures without gross deformation. Epoxies, silicones, and some polyimides are examples of thermoset polymers used in electronics packaging.

Thermoset polymers are processed in special equipment that is capable of handling the transformation from low-molecular-weight material to crosslinked plastic. This equipment essentially conducts a polymerization reaction, hence temperature and residence time control are important. The reactivity and stoichiometry of the starting components must also be carefully controlled. The time for polymerization is an important productivity consideration when using thermoset polymers. Cure times in the mold of up to four minutes are not unusual for epoxy molding compounds, and a postcure heat treatment out of the mold may be required to bring the reaction to completion. For this reason, large multicavity molds are used in most plastic packaging operations to improve productivity. The injection pressures of thermoset materials are often much lower than for conventional plastics; therefore, the mold clamping equipment is usually smaller and less expensive.

## 2.2 THERMOPLASTIC POLYMERS

The other major organic materials used in plastic packaging are the thermoplastic polymers. These materials consist of long polymer chains that have few if any side branches, and the chains themselves are not physically linked. Most of these materials can flow at a high temperature since they do not possess the three-dimensional network structure that precludes flow of thermoset polymers. The properties of these materials are dependent on the structure of the backbone chain. High-temperature engineering thermoplastics have been introduced that rival the performance of thermosets. They can withstand the temperatures of soldering operations without physical deformation and flow because they possess exceptionally rigid chain structures that resist large-scale molecular motions until very high temperatures. Such materials are considered to be melt-processable. They are amorphous glasses similar in physical state to window glass. Materials such as poly(etherimide) and poly(ethersulfone) are examples of high-temperature amorphous polymers.

Another class of thermoplastic materials that is useful in electronic applications are semi-crystalline polymers. These materials have backbone chains that can crystallize into a lattice structure. Maximum crystallinities in commercial plastics usually do not exceed 65% because a substantial amount of material is isolated outside of the crystalline regions as tie molecules between the crystallites. The high crystalline melting points of these polymers impart high-temperature properties and solvent resistance that make them useful for electronic applications. Poly(phenylene sulfide), poly(butylene terephthalate), and the liquid crystal polymers are examples of semi-crystalline thermoplastics used in electronic applications.

Thermoplastic materials are purchased as granular pellets and then softened and formed in special equipment. Injection molding is a common process method for thermoplastics. The material is mechanically and thermally worked from the solid pellets into a melt prior to being injected into a molding tool, where it solidifies to form the plastic article. Solidification takes only about 20 seconds, compared to the several minutes it can take to harden a thermoset plastic part by polymerization. The high molecular weight and high viscosity of thermoplastic materials provides injection pressures up to 18,000 psi for an injection molded part compared to only 1000 psi for molding of a thermoset material.

## 2.3 GLASS TRANSITION TEMPERATURE

An important parameter of polymeric materials, both thermoset and thermoplastic, is the glass transition temperature, usually denoted $T_g$. This is the tem-

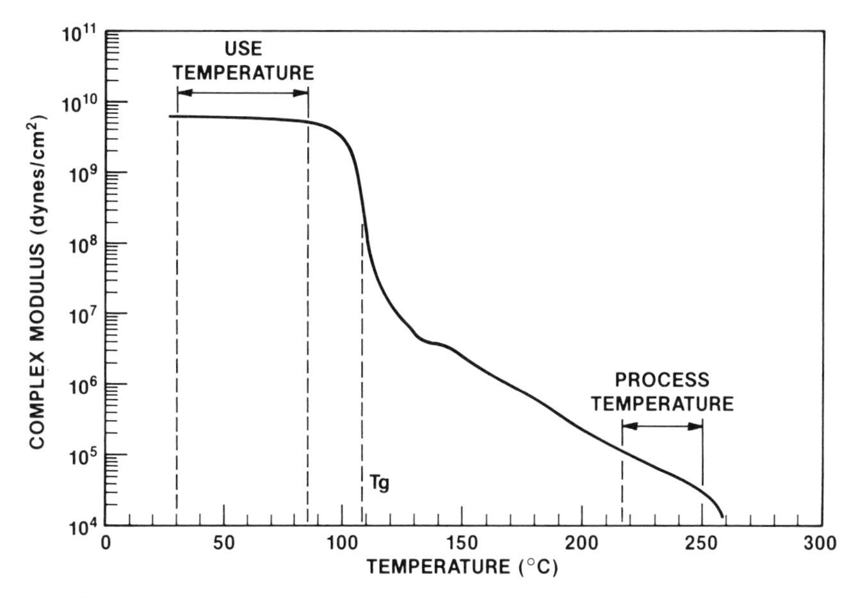


**FIGURE 2-2** Plot of modulus versus temperature for a thermoplastic material, showing the glass transition temperature.

perature at which longer-range cooperative motions of the polymer chains become sufficient to allow the material to deform in response to an external load. On a plot of the mechanical tensile modulus versus temperature, shown in Figure 2-2, the glass transition temperature is characterized by a steep drop in modulus. The region below the glass transition temperature is known as the *glassy plateau*. If the use temperature falls onto this glassy plateau, the material will be a hard solid. Thermoplastic and thermoset materials are different in the region above the glass transition temperature. Thermoplastics drop to a rubbery plateau where temporary molecular entanglements provide a mechanical response that is similar to a lightly crosslinked material. Under constant strain, however, these entanglements disengage, allowing the material to flow. At still higher temperatures, the relaxation time for disentanglement is shorter, the viscosity is lower, and the material flows more readily. Thermoset materials also have a glass transition temperature, but their mechanical behavior above the glass transition temperature is that of a crosslinked rubber: they can sustain deformation, often with high elastic recovery, but cannot flow. (For comparison, a plot of glass transition temperature for thermoset epoxy materials is provided in Figure 4-16.) Higher degrees of crosslink density will increase the modulus level of the rubbery plateau of a crosslinked material and may also increase the glass transition temperature. The number of chain atoms involved in the glass transition is approximately 50 for most materials. $T_g$ increases with increasing crosslink density when the number of chain atoms between crosslinks is in the vicinity of 50. For example, the epoxy materials used for molding compounds have crosslink densities high enough that increases in crosslink density strongly influence $T_g$, whereas some lightly crosslinked silicone materials will exhibit very little effect. The glass transition is obliterated in cases of extremely high crosslink density.

Other properties besides the modulus undergo an abrupt change at the glass transition temperature, and these can also be used to determine $T_g$. These include the heat capacity, the dielectric loss factor, and the coefficient of thermal expansion. $T_g$ is often measured according to ASTM D 3418-82 (American Society of Testing Materials), but any instrument that can detect the large mechanical, thermal, and dielectric changes that accompany $T_g$ can be used for noncrystalline polymers. A common measured parameter associated with the glass transition temperature is the heat deflection temperature. This is determined by a standard ASTM test designated D 648-72, which measures the deformation of a plastic rectangular bar under a specified flexural loading. The temperature at which this deformation exceeds some arbitrary value is the heat deflection temperature, often denoted $T_{hdt}$. It is usually several degrees lower than the glass transition temperature because of the loading imparted to the specimen.

## 2.4 VISCOSITY AND FLOW BEHAVIOR

Polymeric materials are characterized by a complex rheology that strongly influences their processing. Their rheology is intrinsically dependent on their molecular weight and molecular architecture, but it is also strongly influenced by process parameters such as the shear rate, temperature, and residence time for curing materials. High-molecular-weight polymers can show viscoelasticity, where their mechanical behavior shows both viscous and elastic response. Time dependencies and memory effects may also be present in high-molecular-weight materials.

### 2.4.1 Shear Rate Dependence

Newton's law of viscosity states that for simple fluids such as water and low molecular weight solvents the shear stress is linearly proportional to the shear rate in laminar flow with the constant of proportionality being defined as the viscosity:

$$\tau = \eta \dot{\gamma} \tag{2-2}$$

In the cgs system (centimeters, grams, seconds) the unit of viscosity is the dyne-sec/cm$^2$: the IUPAC unit of viscosity is the pascal-sec (Pa-s), and probably the most common measure of viscosity is the poise (one poise equals 0.1 Pa-s). A

Newtonian fluid has a constant viscosity over all shear rates; it provides a straight line on a plot of the shear stress versus shear rate. Viscosity behavior of more complex polymeric materials can be characterized according to the shear rate dependency. Materials whose viscosity increases with shear rate are known as *dilatant*, whereas the shear thinning behavior characteristic of polymer melts and solutions is called *pseudoplastic*. Both dilatancy and pseudoplasticity are non-Newtonian behaviors. Another type of non-Newtonian behavior is yield stress; the material will flow only after a minimum threshold stress level is applied, after which the response may be Newtonian, pseudoplastic, or dilatant. These materials are known as Bingham fluids, the most common examples of which are ketchup, dripless paint, and toothpaste. Flow curves for the different types of shear rate dependencies are shown in Figure 2-3.

Rheologically complex fluids usually display different dependencies over different ranges of shear rate. Most polymeric materials show Newtonian behavior at very low shear rates, providing what is known as a *Newtonian plateau* on a plot of viscosity versus shear rate, as shown in Figure 2-4. High-molecular-weight polymers often show shear thinning behavior in that their viscosity decreases with increasing shear rate after the Newtonian plateau, providing the pseudoplastic portion of the curve. At very high shear rates, there is usually a second upper Newtonian plateau. The exact origins of these behaviors are not completely understood, but are believed to be related to chain orientation and chain entanglements. High degrees of orientation at high shear rates and a high degree of disorientation at low shear rates are accepted explanations for the shear rate independent viscosities at the shear rate extremes.

It is useful to relate shear stress and shear rate of non-Newtonian materials in a manner similar to Newton's Law. These expressions are known as *constitutive relations*. The most popular relation for purely viscous non-Newtonian

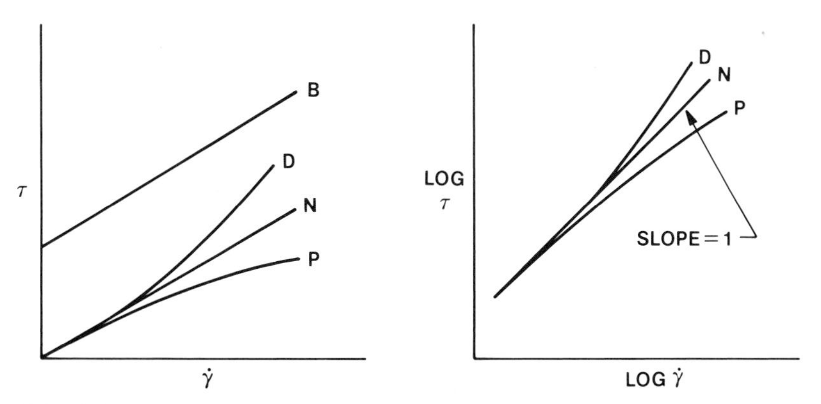


**FIGURE 2-3** Rheological behavior of polymer materials showing the different types of shear rate dependencies; N—Newtonian, P—pseudoplastic, D—dilatant, B—Bingham.

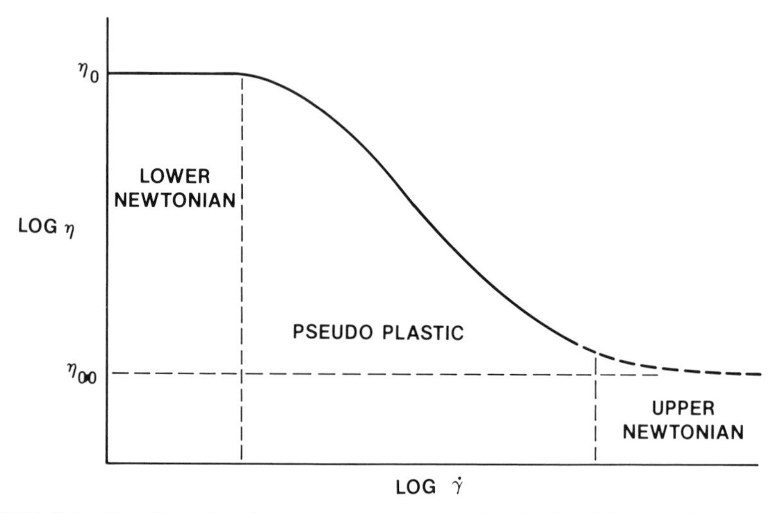


**FIGURE 2-4** Plot of log viscosity versus log share rate, showing the different regions of shear rate dependency.

flow is the two-parameter Ostwald–deWaele model, also known as the power-law model [2]:

$$\tau = K(\dot{\gamma})^n \tag{2-3}$$

The parameters $K$ and $n$ are empirical. The flow index $n$ is dimensionless, whereas the units of $K$ depend on the value of $n$ to provide the correct units of the shear stress. An expression for viscosity is easily derived from the power-law model:

$$\eta = \frac{\tau}{\dot{\gamma}} = K\frac{\dot{\gamma}^n}{\dot{\gamma}} = K(\dot{\gamma})^{n-1} \tag{2-4}$$

A plot of log viscosity versus log shear rate for a power-law fluid will provide a straight line with a slope of $(n - 1)$. The limitation of the power-law model is that it provides an infinite viscosity at zero shear rate. Therefore the power law does not represent all of the regions of the general flow curve shown in Figure 2-4. Better representations are obtained by employing more complicated three- and four-parameter models such as the Carreau model [3]:

$$\frac{\eta(\dot{\gamma}) - \eta_\infty}{\eta_0 - \eta_\infty} = \left[1 + (\lambda\dot{\gamma})^2\right]^{(n-1)/2} \tag{2-5}$$

where $\eta_0$ is the zero shear viscosity, and $\eta_\infty$ is the infinite shear rate viscosity that can be taken to be zero for polymer melts or the solvent viscosity for polymer solutions. Constitutive relations like the Carreau model require extensive experimental data to fit the adjustable parameters and are cumbersome in numerical and analytical solutions of flow behavior, thereby limiting their use for routine engineering analysis. Materials that can be described by algebraic relations between shear stress and shear rate, independent of strain history, are termed *generalized Newtonian fluids* (GNF), even though their shear rate behavior may be non-Newtonian.

### 2.4.2 Time-Dependent Viscosity

Some materials display a reversible, time-dependent viscosity. Materials that increase in viscosity as a result of physical phenomena are known as *thixotropic*, whereas materials exhibiting time-dependent viscosity decrease are termed *rheopectic*. Rheopectic behavior is most often found in highly filled systems where certain flow properties such as dripless application are specifically designed into the material. Another type of time-dependent viscosity is that of thermosetting polymers. As previously described, these are polymerizing systems whose viscosity increases irreversibly with time due to chemical conversion, differentiating them from the reversible increase of thixotropy. This behavior has acquired the name *chemorheology*. Figure 2-5 illustrates the types of time-dependent viscosity behavior.

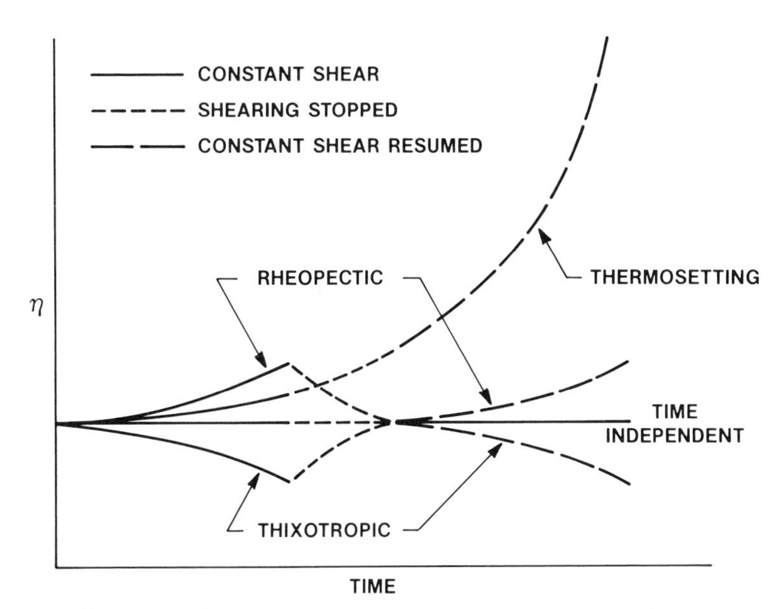


**FIGURE 2-5** The flow curves for materials with time-dependent rheology.

### 2.4.3 Molecular Weight Dependence

The dependence of viscosity on molecular weight of polymers has been found experimentally to show a linear dependence at low molecular weight and a power-law dependence at high molecular weight, with the break point between the linear and nonlinear dependencies thought to be related to the onset of polymer chain entanglement. Compilation of the zero shear viscosity–molecular weight behavior of different polymer melts [4] shows that the exponent on molecular weight in the nonlinear region is 3.4 [4, 5]:

$$\begin{aligned} \eta &= KM_W^1 \quad \text{for} \quad M_W < M_{Wc} \\ \eta &= KM_W^{3.4} \quad \text{for} \quad M_W > M_{Wc} \end{aligned} \tag{2-6}$$

Equation (2-6) is for the zero shear viscosity. $M_W$ is the weight average molecular weight, and $M_{Wc}$ is the critical molecular weight for chain entanglements. Most commercial thermoplastic materials are above the critical molecular weight for chain entanglements. Exceptionally high-molecular-weight polymer will exhibit very high viscosity and will require high pressures and high temperatures for processing which can damage electronic components. Materials with $M_W > M_{Wc}$ will show shear thinning behavior so that the break in the viscosity will be less distinct as the shear rate increases, as shown in Figure 2-6.

Equation (2-6) is most applicable to linear chain polymers. Thermoset polymers can show different dependencies on molecular weight due to the nature of the branching and chain entanglements. In general, the exponent for high molecular weight branched polymer falls somewhere between 1.0 and 3.4. This is because a unit increase in molecular weight of a liner polymer adds directly to chain length and more directly to the molecular size, often characterized through the radius of gyration of the molecule, than does a unit increase in a branched molecule. For this reason, Equation (2-6) should only be used for thermosets if the parameters are judiciously modified through experimental data. Graessley [6] reviewed the literature for the molecular weight dependence of star and branched polymers and concluded that Equation (2-6) can be used if $M_W$ is replaced by $gM_W$ where $g$ is defined as the ratio of the radius of gyration of the branched molecule to that of the linear molecule of the same molecular weight:

$$g = \frac{r_{gb}}{r_{gl}} \tag{2-7}$$

Reference 1 has a thorough treatment of radius of gyration for linear and branched molecules.

Typically, thermoset polymers would initially be below the critical molecular weight for chain entanglements. They often show a Newtonian viscosity that is linearly dependent on the molecular weight of the starting material. The mo-

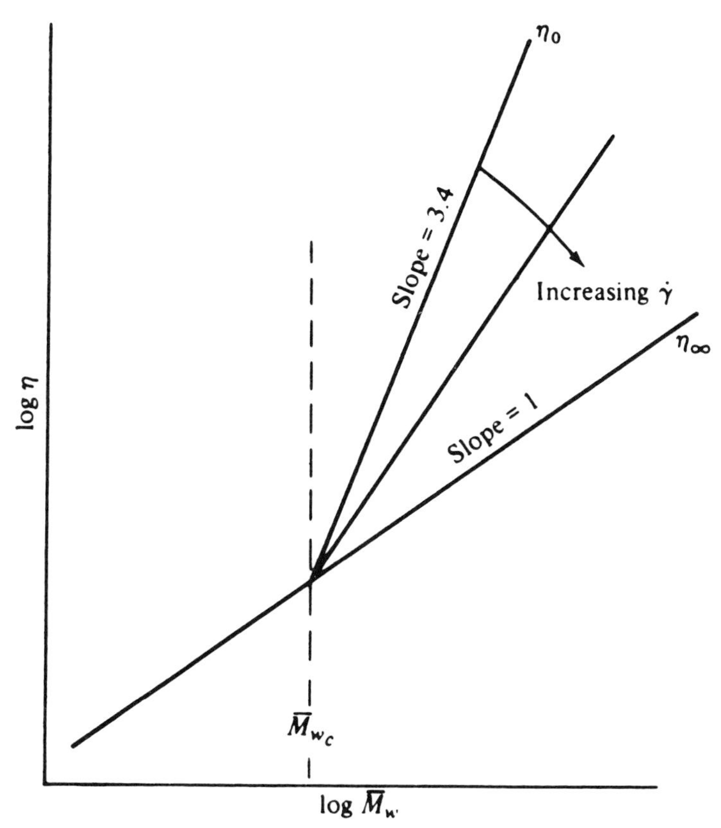


**FIGURE 2-6** Generic behavior of log zero shear viscosity versus log $M_w$ for linear chain polymers showing the break from linear dependence to power-law dependence at $M_{Wc}$. The slope of the viscosity decreases as the shear rate increases. (From *Fundamental Principles of Polymeric Materials* by S. L. Rosen. Copyright © 1982 by John Wiley and Sons, Inc. Reprinted by Permission of John Wiley and Sons, Inc.)

lecular weight increases during polymerization according to the functionality of the starting materials and the stoichiometry of the starting mixture. The increase in molecular weight with chemical conversion in branched and crosslinked systems has been derived [1, 6, 7]. The precise form of the expression depends on the functionalities of the components of the system, their initial molecular weights, and the stoichiometry of the mixture. Some systems with a single multifunctional reactive species are relatively easy to treat in this way, whereas more complicated systems such as thermoset molding compounds that include two or three multifunctional species and competing reactions with different reactivities are far more difficult. One form of this expression for a two component system with functionalities $f_A$ and $f_B$ is provided below [6, 7]. The reader is cautioned that it is difficult in most cases to determine the functionalities and molecular weights of commercial molding compounds.

$$M_W = \frac{X_B m'_A + X_A m'_B}{X_B m_A + X_A m_B} + \frac{X_A X_B \{X_A(f_A - 1)M_B^2 + X_B(f_B - 1)M_A^2 + 2M_A M_B\}}{(X_B m_A + X_A m_B)\{1 - X_A X_B(f_A - 1)(f_B - 1)\}} \quad (2\text{-}8)$$

In Equation (2-8), $M_A$ is the average molecular weight of component $A$ and $m_A$ is the average equivalent weight per reactive group for component $A$; the mole fraction of reactive group $A$ is the weighting factor for these molecular weights [7]; $m'_A$ is the average equivalent weight per reactive group of component $A$ where the weighting factor is the molecular weight times the mole fraction of reactive species $A$ [7]. $M_B$, $m_B$, and $m'_B$ are defined similarly for component $B$. The molecular weight expression for actual molding compounds is difficult to determine, since the exact chemistry, functionality, and stoichiometry of these materials is not completely known in most cases. The use of well characterized model systems is one approach to the study of these materials. A relation for the molecular weight as a function of degree of conversion X for an epoxy molding compound model system has been reported [8]:

$$M_W = 903 + X\left\{\frac{6.15 \times 10^6 X + 1.296 \times 10^6}{343\,(1 - 15.6X^2)}\right\} \quad (2\text{-}9)$$

Expressions such as the ones shown in Equations (2-8) and (2-9) can be used to relate the degree of chemical conversion to the molecular weight. It may also be possible to relate the molecular weight to viscosity through a relation similar to Equation (2-6), noting the cautions that have been discussed on applying it to thermoset polymers and the correction suggested in Equation (2-7).

### 2.4.4 Temperature Dependence

Of great importance to electronics packaging is the dependence of viscosity on temperature. Most processing operations are nonisothermal, with large changes in temperature during the process. For this reason, it is important to quantify the temperature dependence of the viscosity so as to properly control flow induced forces. Arrhenius relations that have a flow activation energy in an exponential expression are most common for polymers that are more than 100°C above their glass transition temperatures, or for low-molecular-weight liquids such as the thermoset materials at low degrees of conversion (4).

$$\eta_0 = Ae^{E/RT} \quad (2\text{-}10)$$

where $E$ is the flow activation energy, $A$ is the viscosity coefficient, and $R$ is the gas constant. This activation energy can be obtained from experimental data by plotting the log of viscosity versus the reciprocal of temperature. The slope is $E/R$, as is shown in Figure 2-7.

High-molecular-weight polymers that are processed at temperatures within 100°C above their glass transition temperature are more properly described by the corresponding state law which states that polymers equally removed from their glass transition temperatures have similar mechanical behavior. The relation that describes this dependence on temperature is the Williams-Landel-Ferry or WLF equation [9]:

$$\log \frac{\eta(T)}{\eta(T_g)} = -\frac{17.44(T - T_g)}{51.6 + T - T_g} \tag{2-11}$$

Replacing $T_g$ in Equation (2-11) with a reference temperature, usually taken to be $T_g + \Delta T$ where $\Delta T$ is often near 50°C, is sometimes found to improve the fit to the viscosity data. The two numerical constants are related to the free volume fraction (fractional volume not occupied by polymer chains) at the glass

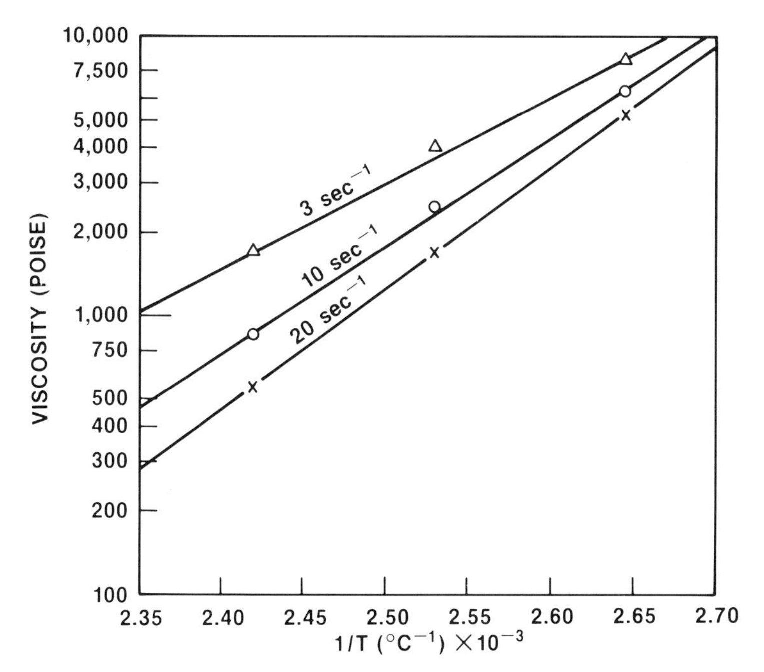


**FIGURE 2-7** Plot of log viscosity versus inverse temperature. The slope of this line is the Arrhenius activation energy for flow. Three lines for three different shear rates are shown, indicating that the flow activation energy may also depend on shear rate.

transition temperature. WLF is most appropriate for high-molecular-weight polymers at temperatures between their $T_g$ and $T_g + 100°C$.

### 2.4.5 Filled-System Rheology

Fillers are used often in electronics packaging to lower the coefficient of thermal expansion, increase the thermal conductivity, or to modify the material rheology. For low loadings of filler Einstein's Law can be applied to predict the viscosity of the filled system:

$$\eta = \eta_0(1 + 2.5\phi) \tag{2-12}$$

where $\eta_0$ is the viscosity of the unfilled fluid and $\phi$ is the volume fraction of filler. This relation is most accurate for loadings below 15% by volume and spherical filler particles. For higher filler loadings with particle interactions, and for non-spherical filler particles, the two-parameter Mooney Equation [10] can be used to achieve a better correlation:

$$\ln \eta = \frac{K_e\phi}{1 - \phi/\phi_m} \tag{2-13}$$

where $\phi_m$ is the maximum filler loading that can be achieved with the particular filler shape used, and $K_e$ is the Einstein coefficient for the filler particle and particle size distribution. Spherical filler particles have a maximum loading of approximately 60%. High filler loadings are achieved in thermoset molding compounds by using spherical and flake fillers. The highest loadings are achieved with a combination of filler shapes and sizes so that all the interstitial spaces are filled. Ultimate filler loadings of over 90% can be achieved in this way, but the physical properties of these low-resin systems are degraded to the point where they are no longer suitable for packaging applications. The Mooney equation can describe the nonlinear behavior of the viscosity increase as the filler loading approaches the maximum loading. The steep viscosity increase as the filler loading approaches the maximum limits the feasibility of using very high loadings to minimize the difference in the coefficients of thermal expansion because the high viscosity promotes flow-induced stresses on the encapsulated devices. The coatings and coupling agents used on the fillers have a significant effect on the viscosity and shear rate dependence. In most cases, a coupling agent that improves the wetting of the filler by the liquid resin reduces the viscosity compared to a system without coupling agent.

### 2.4.6 Viscometry

There are several important methods for measuring the viscosity behavior. Techniques that measure only the ratio of the shear stress to shear strain, the definition of viscosity, are known as viscometers. More complex instruments that can measure elastic normal forces and conduct oscillatory strain studies are more properly termed rheometers. Normal forces arise in high-molecular-weight polymers and are responsible for such behavior as extrudate swell and rod climbing upon stirring. Normal forces are usually not a factor in most thermoset polymers used in the electronics industry. In epoxy molding compounds, the high filler loadings, the asymmetric filler shape, and the high deformation rate through the gates of the molding tool could generate normal stresses, but the consideration of these effects is beyond the objectives of this book. Comprehensive references on polymer rheometry are available [11].

*Couette Viscometer.* In many industrial applications, a Couette viscometer is used for material characterization and quality control. The fluid is held within the gap of two concentric cylinders. In this configuration the technique is a viscometric measurement, meaning that there is a defined relationship between the shear rate and shear stress:

$$\tau(r) = \frac{T}{2\pi r^2 L} \tag{2-14}$$

In most commercial designs of this instrument, the inner cylinder is shorter, resulting in a cup and bobbin geometry, as shown in Figure 2-8. This design is no longer viscometric due to the intractable shear stress dependence in the reservoir region, although most of the torque is generated by the thin gap between the cylinders. Most commercial instruments assign viscosity through a calibration curve derived from known viscosity standards. The technique is most suited for low viscosity, Newtonian materials.

*Cone-and-Plate Viscometer.* An important characterization technique is the cone-and-plate viscometer. A small-angle cone is brought into contact with an oversized plate, creating the geometry shown in Figure 2-9. For small cone angles, less than 3°, the increasing gap thickness with increasing angular velocity moving away from the apex of the cone cancels the radial dependence of the shear rate, providing a constant shear rate over the radius and a viscometric measurement:

$$\dot{\gamma} = \frac{v}{d} = \frac{\omega r \cos \alpha}{\alpha r} \sim \frac{\omega}{\alpha} \quad \text{for small } \alpha \tag{2-15}$$

For this reason, the cone and plate viscometer is useful for characterizing the shear-rate-dependent viscosity of non-Newtonian fluids. It is less amenable to highly filled systems because the filler particles will bind at the apex of the cone. A truncated cone can avoid this problem, but it introduces imprecision that can become important in a very small angle cone. There is also a tendency

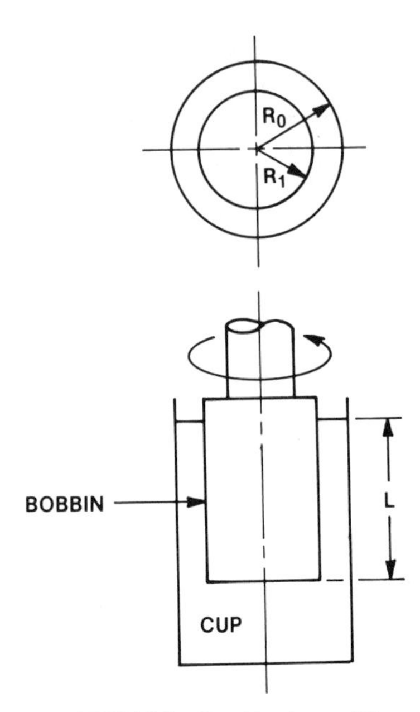


**FIGURE 2-8** Couette viscometer.

for highly filled materials to slip against the metal surfaces at higher shear rates. Oscillatory shearing over a small amplitude can be used to minimize slippage if the instrument is so equipped.

*Plate-Plate Viscometer.* This type of viscometer, also known as parallel plate, is similar in design and function to a cone-and-plate viscometer with the cone replaced by a second parallel disk that is also rotated. The gap between

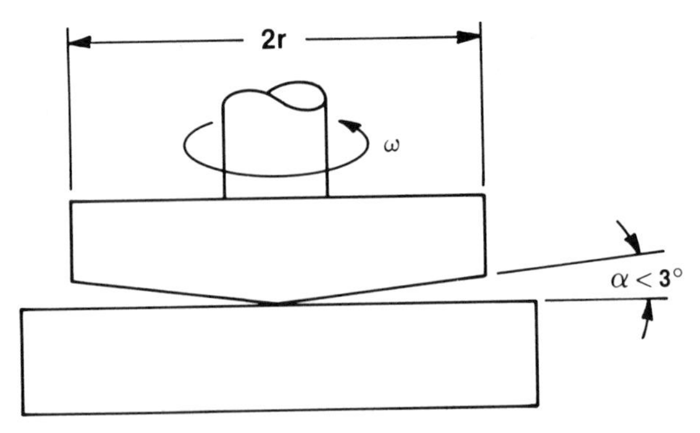


**FIGURE 2-9** Cone-and-plate viscometer.

the disks is much less than the diameter. This geometry is used in place of the cone-and-plate in some instances because it is easier to instrument without having to maintain the contact between cone and plate, and because solid samples can be convenient molded disks. It is also a resolvable flow field providing a viscometric technique, with the shear rate as shown below [5]:

$$\dot{\gamma} = \frac{r\omega}{d} \tag{2-16}$$

The shear rate depends linearly on radius and is not uniform throughout the specimen as in the cone and plate. This makes data reduction more difficult, especially when the flow curve is not known as is the case with a shear thinning or viscoelastic material. In these cases, a plot of torque versus shear rate is used to relate the shear stress to the shear rate to obtain the viscosity [5]. Parallel plates can also be used in oscillatory mode, as with the cone-and-plate viscometer, to study the behavior at higher frequencies (shear rates) and lower strains. The low strains are important in probing the rheology of highly filled systems where structuring of the filler particles can dominate the response.

*Capillary Viscometer.* Capillary viscometers are based on the relation between pressure drop as a function of volumetric flow rate through a precision bore tube. The design of a typical capillary rheometer is shown in Figure 2-10. This type of instrument can provide useful data up to shear rates where melt fracture or viscous heating are encountered. Shear rates and shear strains con-

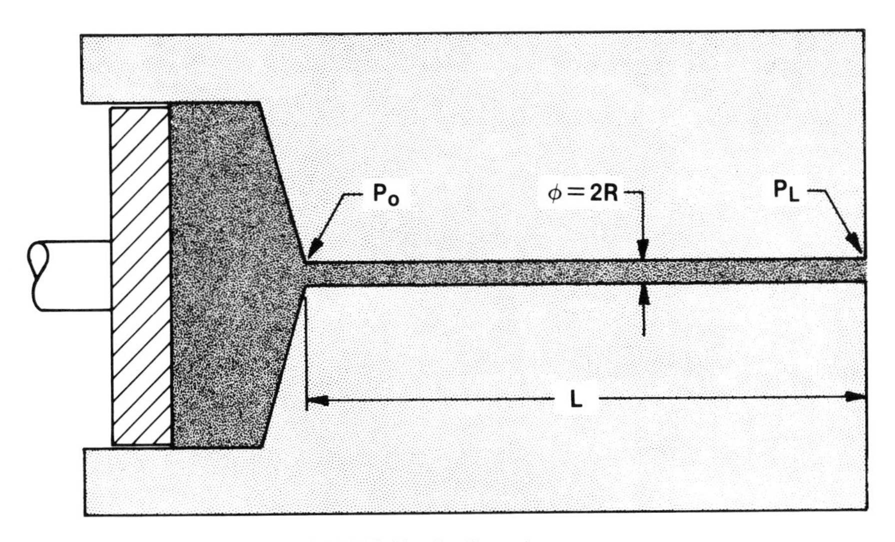


**FIGURE 2-10** Capillary viscometer.

siderably higher than those attainable with cone and plate and parallel plate rheometers can be achieved on highly filled systems.

Tube flow is not a constant shear rate geometry so a relation must be developed to relate the shear stress and shear rate to extract the viscosity from the experiment. The shear stress at the wall is:

$$\tau_w = \left(\frac{P_0 - P_L}{2L}\right)R \qquad (2\text{-}17)$$

where $P_0$ and $P_L$ are the pressures at the inlet and outlet of the tube, $L$ is the tube length, and $R$ is its radius. The Newtonian shear rate at the wall can be expressed in terms of experimentally measured quantities:

$$\dot{\gamma} = \frac{1}{\pi R^3}\left(3Q + \Delta P \frac{dQ}{d(\Delta P)}\right) \qquad (2\text{-}18)$$

Equation (2-18) is known as the Rabinowitsch Equation; it can be used to determine the shear rate at the wall by measuring the volumetric flow rate $Q$ and the pressure drop; thus the viscosity can be obtained as the ratio of the wall shear stress and the wall shear rate.

Capillary viscometers should have a sufficient $L/D$ ratio to minimize the contributions of end effects and surface tension forces, and there should be no slip at the wall. This technique has been developed to study the viscosity of the epoxy molding compounds used in plastic packaging [12]. For these highly filled materials, an $L/D$ ratio of 20 was found to provide good results [12]. For polymer melts, the $L/D$ needed to minimize end effects to the point where they can be neglected is 100 [5]. If shorter capillaries must be used, the end effects can be estimated and accounted for to isolate the viscosity [5].

### 2.4.7 Flow Behavior

The velocity profile has important consequences on the temperature increase during flow, the pressure drop, and ultimately the flow-induced stresses generated on encapsulated devices. Different flow geometries and materials with different rheologies give rise to different velocity profiles. The common configurations encountered in microelectronics packaging are the flow of shear thinning fluids in channels of rectangular or circular cross sections. There are also many free surface flows where polymers are applied over surfaces.

One of the simpler geometries to analyze is the flow in a circular-cross-section tube. Although this geometry itself is not often encountered in microelectronics packaging, many other flow geometries can be approximated as tube flow to obtain a solution that is useful for engineering purposes. The runners in a transfer mold are usually trapezoidal in cross section, but can be approximated as a circular channel with a representative hydraulic radius. The hydraulic radius is a useful concept to translate noncircular-cross-section channels to circular in order to obtain a mathematical solution or to use plotted or tabulated solutions of circular cross sections. The hydraulic radius is defined as the cross sectional area of flow divided by the wetted perimeter.

$$R_h = \frac{A_{xs}}{P_w} \tag{2-19}$$

Note that for a circular pipe this provides $R_h = D/4$ instead of $D/2$. Some definitions of hydraulic radius, such as the one shown in Equation (6-3), will use a modified relation that does provide $D/2$ for a circular tube, therefore always check the definition that applies to the particular results.

For flow of a power-law fluid in a circular tube, the velocity profile can be derived [11]:

$$u_x = \frac{nR}{1+n}\left(\frac{R\Delta P}{2KL}\right)^{1/n}\left[1 - \left(\frac{r}{R}\right)^{1+1/n}\right] \tag{2-20}$$

There is no shear along the centerline of the flow. The maximum shear rate, often called the nominal shear rate, is given by

$$\dot{\gamma}_R = \frac{2(1+3n)}{n}\frac{U}{D} \tag{2-21}$$

There is a significant difference in the velocity profile with increasing shear thinning behavior. The profile becomes more plug flow as the power law index approaches 0, as can be seen in Figure 2-11 [5]. There is, however, very little difference for power law indices greater than $\frac{2}{3}$, which provide velocity profiles that are nearly indistinguishable from Newtonian flow behavior.

For flow of a power-law fluid between parallel plates separated by a distance $B$, the velocity profile in the flow direction can be related to the driving pressure through the following relation:

$$u_x = \frac{nB}{2(1+n)}\left(\frac{B\Delta P}{2KL}\right)^{1/n}\left[1 - \left(\frac{2y}{B}\right)^{1+1/n}\right] \tag{2-22}$$

Flow in rectangular-cross-section ducts can be treated a similar way, but the velocity profile and the volumetric flow rate depend on the shape factor for the duct. For a power-law fluid, an analytical solution is not possible. The numerical solution is most conveniently expressed as an analytical solution with a numerically determined coefficient, known as the shape factor which depends on the duct geometry [13]. Section 6.4.2.1 describes treatment of flow in rectangular ducts.

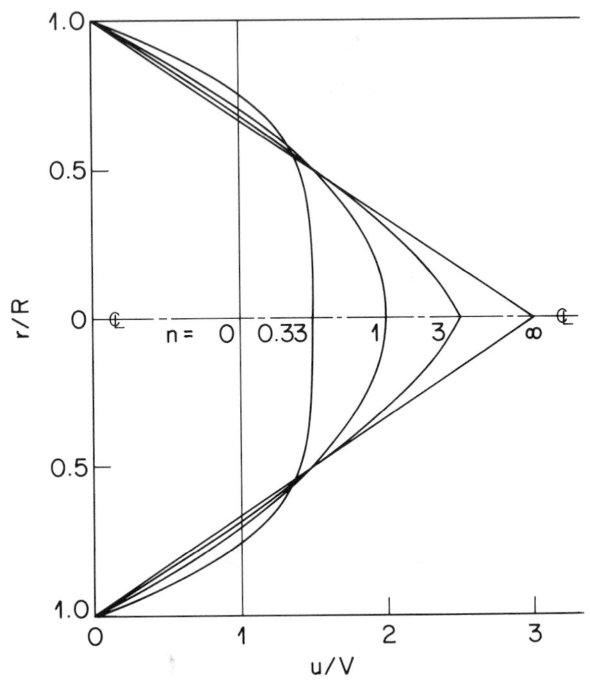


**FIGURE 2-11** The velocity profiles obtained from the flow of power-law fluids with various power law indexes in a circular tube (From *Fundamental Principals of Polymeric Materials* by S. L. Rosen. Copyright © 1982 by John Wiley and Sons, Inc. Reprinted by Permission of John Wiley and Sons, Inc.)

## 2.4.8 Shear-Induced Heating

Shear heating can be a significant effect in the flow of polymeric materials in viscometric studies or flow into a molding tool. In viscometric work they are unwanted because they create nonisothermal conditions that make data analysis difficult. In other situations such as the flow of thermoplastic materials through the gates of the molding tool, they can be beneficial in increasing the material temperature through the gate, thereby lowering its viscosity and the pressure needed to fill the mold. Excessive shear heating, however, can cause thermal degradation of thermoplastics and accelerated reaction rates in thermosets. The rate of energy dissipation due to shear heating, $\dot{E}$, can be computed through the following relation [4]:

$$\dot{E} = \tau\dot{\gamma} \tag{2-23}$$

If the units of shear stress are dynes/cm$^2$, and the shear rate is in reciprocal seconds, the heating rate has the units of ergs/cm$^3$-sec. Estimates of the shear rate and shear stress of the molding compound through the gates are needed to predict the temperature increase that can occur. The shear rate through typical gates of a transfer mold are in the range of 3,000–10,000 $sec^{-1}$. The shear stresses through the gate can be estimated from the pressure drop, typically 100–300 psi. The temperature rise is then computed by factoring in the volumetric flow rate through the gate and the heat capacity of the molding compound. The shear heating through the gate in transfer molding of plastic packages is usually minimal, typically less than a few degrees Celsius. Shear heating through the gate is discussed further in Section 6.4.3.2.

## 2.5 VISCOELASTICITY

The mechanical properties of many polymer materials include both viscous and elastic response over both their liquid and solid states. Polymer liquids display the expected viscous response of flow (strain) under applied stress, but high-molecular-weight materials whose polymer chains can coil and entangle also show a recoverable elastic response. Similarly, solid polymers can show viscous response and flow under applied stress, typically over longer time periods. Many other common solids, particularly nonequilibrium glasses and metals near the melting points, also display this viscous response such as the flow of window glass over hundreds of years. With polymers this behavior, known as *creep*, is much more pronounced and far more important in design considerations. Another significant difference in the mechanical response of polymers is their rate dependence; the stress depends not only on the strain but also on the rate of strain. One further distinction is that the response of polymers is often nonlinear, the ratio of overall stress to overall strain, the modulus, $G$, is not a function of time only, but also depends on the degree of strain:

$$G = \frac{\tau}{\gamma} \tag{2-24}$$

$G(t)$: linear response $\quad$ $G(t, \gamma)$: nonlinear response

$G$ usually denotes the shear modulus, the ratio of shearing stress to shear strain. The tensile modulus is also of importance in microelectronics packaging, and it is probably more common since it is easier to measure. The flexural modulus is obtained through a flex test of the specimen and is considered to be equivalent to the tensile modulus if the tensile and compressive moduli are similar. The shear modulus and tensile modulus are related through the Poisson's ratio of the material, $\nu$, which is the ratio of the lateral strain and the tensile strain:

$$G = E(1 + \nu) \tag{2-25}$$

Undilatable materials such as elastomers have a Poisson's ratio of 0.50. Epoxy molding compounds have a $\nu$ of approximately 0.35, copper is 0.35 and silica is 0.28.

Two mechanical tests that are used to assess viscoelastic behavior are the creep test and the stress relaxation test. In the creep experiment illustrated in Figure 2-12, a constant stress is instantaneously applied to a material and the strain is recorded as a function of time. There will be some immediate strain, as well as some continuing increase in strain due to viscous response. After some time, the stress is removed and the degree of elastic recovery is measured; the unrecoverable strain is known as the *permanent set*. The effects of molecular weight and crosslinking on creep behavior are explored in Figure 2-13.

In the stress relaxation experiment shown in Figure 2-14, an instantaneous strain is applied to the material, and the stress is measured as a function of time with the strain held constant. If the decrease of the stress can be fit to a first-order exponential decay, the response is said to be first-order. Actual polymers can rarely be fit in this way, however. The time dependent stress divided by the constant strain can be expressed as a time dependent relaxation modulus, $E_r(t)$. The effects of $M_W$ and crosslink density on the relaxation modulus are explored in Figure 2-15.

There are many mathematical models that attempt to describe the mechanical response of viscoelastic materials [14]. One of the simpler and more popular models is the *Maxwell element*, a series combination of springs and dashpots. (A dashpot is a viscous dissipation element such as an automobile shock ab-

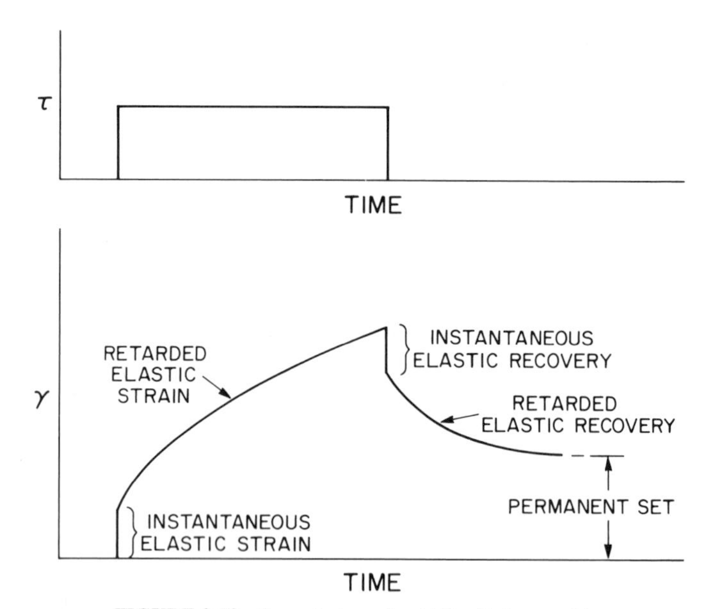


**FIGURE 2-12** Creep test results. (After Reference 5.)

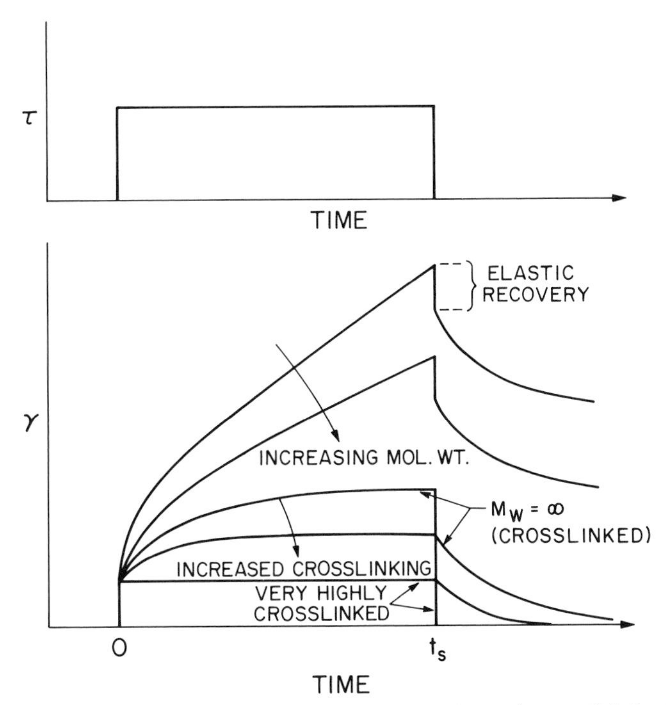


**FIGURE 2-13** The effects of increasing molecular weight and increasing crosslink density on the creep behavior of polymeric materials. (From *Fundamental Principals of Polymeric Materials* by S. L. Rosen. Copyright © 1982 by John Wiley and Sons, Inc. Reprinted by Permission of John Wiley and Sons, Inc.)

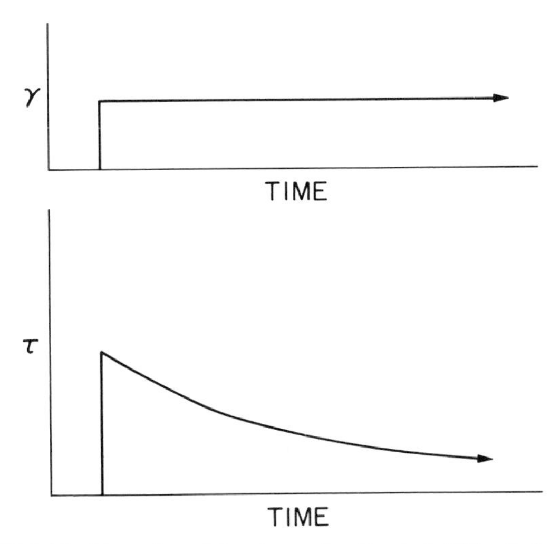


**FIGURE 2-14** Illustration of the stress relaxation test. (After Reference 5.)

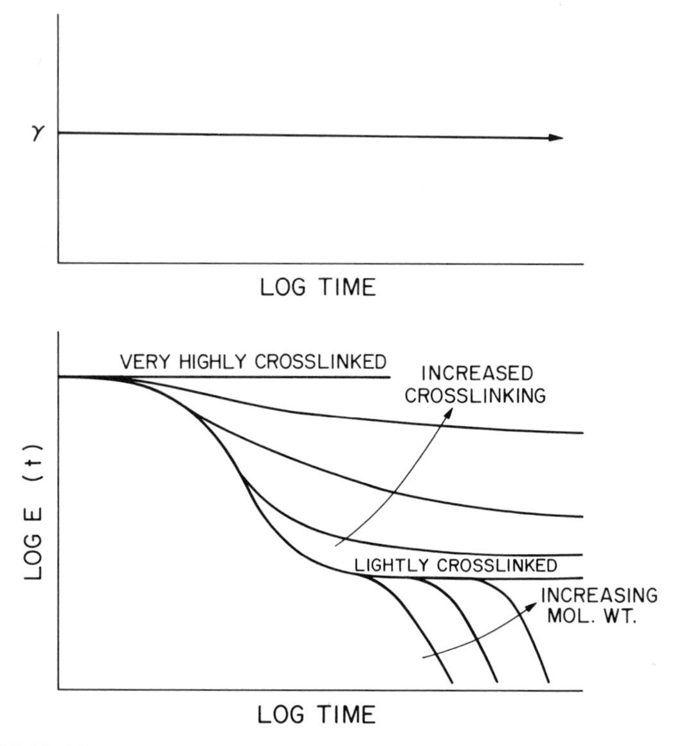


**FIGURE 2-15** The effects of increasing molecular weight and increasing crosslink density on the stress relaxation experiment plotted in terms of the relaxation modulus. (From *Fundamental Principles of Polymeric Materials* by S. L. Rosen. Copyright © 1982 by John Wiley and Sons, Inc. Reprinted by Permission of John Wiley and Sons, Inc.)

sorber.) Full derivations of the response of the Maxwell element are available in textbooks [5, 14]; only the result is provided here:

$$\tau = \eta\dot{\gamma} - \left(\frac{\eta}{G}\right)\dot{\tau} = \eta\dot{\gamma} - \lambda\dot{\tau} \tag{2-26}$$

The ratio of the viscosity over the modulus ($\eta / G$) is known as the relaxation time, a convenient parameter to characterize the viscoelastic response of polymers. The mechanical response of real polymers is often too complicated to be characterized by a single relaxation time because there a number of dissipative and elastic processes occurring. Some of these include the elastic deformation of bond angles, uncoiling of polymer molecules, chain slippage (entanglement release), and the recovery due to thermal motions of the chains. A characteristic time for real materials is still a useful parameter, however, and it is often taken to be time required for the material to reach $1 - 1/e$ (63.2%) of its ultimate elastic response to a step change; a parameter obtained from a stress relaxation experiment. This characteristic relaxation time is denoted $\lambda_c$; small $\lambda_c$ values denote rapid, elastic response, whereas a larger value of $\lambda_c$ indicates a more viscous response.

Another important parameter is the ratio of the characteristic relaxation time to the time scale of the process, denoted $t_s$. This ratio is known as the Deborah number, $De$, a name of biblical origin relating to the prophecy that the mountains will flow to the sea:

$$De = \frac{\lambda_c}{t_s} \tag{2-27}$$

A Deborah number approaching zero indicates greater viscous response, whereas Deborah numbers greater than one indicate increasing elasticity of the material in the particular process. The time scale is derived from a characteristic time interval of the deformation. For example, in a temperature cycle test of an electronic component, the time scale would be the time interval between the temperature extremes.

For plastic packaging of microelectronics, it is important to be aware of the mechanical response of polymer materials, particularly their creep and stress relaxation behavior, and the parameters such as the characteristic relaxation time and the Deborah number that are used to quantify these responses in a simple way. Adding fillers to polymer materials produces a composite response of both the filler and the polymer. Since most fillers are inorganic glasses or crystalline materials, which have very high moduli and very little viscous response, filled materials such as the transfer molding compounds show significantly reduced creep and stress relaxation behavior.

## 2.6 OTHER PHYSICAL PARAMETERS RELEVANT TO PLASTIC PACKAGING

There are several physical parameters of polymer materials that are important to microelectronics packaging. It is useful to have an understanding of the molecular origins, time and temperature dependencies of these parameters.

### 2.6.1 Thermal Conductivity

Thermal conductivity is a measure of the material's ability to conduct heat energy. Polymers have significantly lower thermal conductivities than most other materials used in electronics, such as metals, ceramics, and glasses. The low thermal conductivity of polymers derives from their low atomic density and specific density compared to metals and glasses. This can cause problems in maintaining a moderate temperature in heat generating devices. Devices with

**Table 2-1. Thermal Conductivity of Various Materials**

| | | | |
|---|---|---|---|
| Diamond | 4.70 | Alumina* | 0.070 |
| Silver | 1.00 | Silica (cryst.) | 0.003-.030 |
| Copper | 0.90 | Silica (amorphous) | 0.002-.003 |
| Aluminum | 0.50 | Soda glass | 0.0017 |
| Beryllium oxide | 0.50 | Concrete | 0.0020 |
| Magnesium | 0.40 | Epoxy molding compound | 0.0010-0.0030 |
| Brass | 0.30 | | |
| Mercury | 0.20 | Epoxy resin | 0.0003-0.0005 |
| Steel | 0.10 | Water | 0.0014 |
| Lead | 0.08 | Cork | 0.00012 |
| | | Air | 0.000057 |

Units are cal/cm-°C-s $\times 10^{-4}$
*$\alpha$ crystals
All values at 20°C. (After Reference 13.)

heat outputs exceeding 2 Watts/cm$^2$ may require some auxiliary heat dissipation mechanism such as external cooling fins. A listing of the thermal conductivities of some common electronic materials is provided in Table 2-1. For precise calculations, the change in the thermal conductivity with temperature should also be taken into account. Also, the purity of a material has a significant effect on the thermal conductivity, particularly for the high-conductivity metals such as copper and aluminum whose conductivity can be decreased by more than 40% by alloying with small amounts of other metals.

The thermal conductivity of multiphase systems depends on whether the individual components are configured in series, parallel, or dispersed. The conductivities of these composite materials can be estimated from the thermal conductivities of the components through the following relations [15]. Bear in mind that no relation can actually predict the thermal conductivity of a complicated composite material because it depends on many more factors than the volume fraction alone. Relations like those shown below are nonetheless useful in estimating these parameters. Experimental measurements should be considered if more accurate thermal diffusivities are needed for design decisions.

Parallel: $$\bar{k}_T = \sum k_i \phi_i \tag{2-28}$$

Series: $$1/\bar{k}_T = \sum \phi_i / k_i \tag{2-29}$$

Dispersed: $$\bar{k}_T = k_c \left[ \frac{1 + 2\phi_d \left\{ \dfrac{1 - k_c/k_d}{2k_c/k_d + 1} \right\}}{1 - \phi_d \left\{ \dfrac{1 - k_c/k_d}{2k_c/k_d + 1} \right\}} \right] \tag{2-30}$$

In Equations (2-28)–(2-30), $\bar{k}_T$ is the thermal conductivity of the composite, $k_i$ is the thermal conductivity of the $i$th species, $k_c$ and $k_d$ are the thermal conductivities of the continuous and dispersed phases, respectively, and $\phi$ is the volume fraction, with the subscripts $c$ and $d$ denoting continuous and dispersed phases. The parallel configuration where the direction of heat transfer aligns with the layers of the composite provides the highest thermal conductivity when there is a substantial difference between the thermal conductivities of the components. The series configuration provides the lowest, and the dispersed configuration falls somewhere in between depending on the thermal conductivities of the components.

Manipulation of the thermal conductivity of packaging materials is often used to control the heat transfer rate from a packaged device. In general, ceramic devices have better heat transfer characteristics than plastic packages, as can be seen in Table 2-1 by comparing epoxy and silica to alumina. Epoxy molding compounds with crystalline silica fillers offer some improvement in thermal conductivity, but they still do not approach the thermal performance of ceramics. The thermal conductivity of leadframes also plays an important role in thermal management of plastic packaged devices. There are large and significant differences among the thermal conductivities of the various leadframe materials in use, some of which are listed in Table 2-2. The compositions of the various alloys are provided in Table 3-4 in Section 3.7 on leadframe materials.

### 2.6.2 Heat Capacity

Heat capacity, also known as specific heat, is a measure of the quantity of heat a material must absorb to raise its temperature a specified increment. If the volume of the material is held constant, the ratio of change of the internal energy with changing temperature is known as the heat capacity at constant volume, denoted by the notation $C_v$. Similarly, processes occurring at constant pressure define the heat capacity at constant pressure, or $C_p$. For solids and liquids where the volume change with temperature is often negligible, $C_v$ and $C_p$ become indistinguishable. Most texts and handbooks will list the more common $C_p$ values for electronic materials. Plastics have higher heat capacities than most other packaging and device materials; they require more heat energy to raise their temperature. A listing of the heat capacities of common materials encountered in electronics is provided in Table 2-3.

**Table 2-2. Thermal Conductivity of Leadframe Material**

| Material | Value |
|---|---|
| Copper | 0.90 |
| CDA 151 | 0.64 |
| CDA 155 | 0.62 |
| CDA 194 | 0.46 |
| Alloy 42 | 0.03 |

Units are cal/cm-°C-s. All values at 25°C.

**Table 2-3. Heat Capacities of Some Materials Used in Plastic Packaging**

| | |
|---|---|
| Aluminum | 0.215 |
| Copper | 0.092 |
| Epoxy molding compound | 0.288 |
| Gold | 0.038 |
| Iron | 0.106 |
| Lead | 0.038 |
| Nickel | 0.106 |
| Silicon | 0.168 |
| Silver | 0.057 |
| Tin | 0.052 |

Units are cal/gram-°C. All values at 25°C.
Data, with the exception of epoxy molding compound, from *Handbook of Chemistry and Physics*, 50th Edition, The Chemical Rubber Company, Cleveland, OH (1969).

Heat capacity is often a strong function of temperature, so the values provided in Table 2-3 should be used for comparison purposes only, or for computations at isothermal conditions near the specified temperature. More detailed work on heat transfer analysis and thermomechanical stress will require $C_p(T)$ data that may have to be acquired experimentally.

### 2.6.3 Coefficient of Thermal Expansion

The coefficient of thermal expansion (CTE) is one of the more important physical parameters of packaging materials because disparities in thermal expansion among the different materials within the package can contribute to significant restrained shrinkage stresses. These stresses can reach levels that cause fracture failures of the molding compound, the passivation layers of the device, or the device itself. The CTE changes on thermodynamic transitions, such as the glass transition temperature of polymeric materials, or phase transitions such as crystallization. The different regions of the plot of CTE versus temperature are denoted by $\alpha$ with the appropriate subscript. The region at lowest temperature is $\alpha_1$, and the next higher region is $\alpha_2$, and so on. For plastics, the CTEs above and below the glass transition temperature are the ones of importance. These are shown in the plot of Figure 2-16. The coefficients of thermal expansion of some common materials used in electronics are provided in Table 2-4.

Matching the coefficients of thermal expansion as closely as possible minimizes the thermal shrinkage forces that are encountered. This strongly influences the selection of leadframe materials, epoxy molding compounds, filler types, and filler levels. It is also apparent that selecting polymeric materials that will remain below their glass transition temperature provides a material with the lower $\alpha_1$ values. The modulus is higher below $T_g$, however, so in some

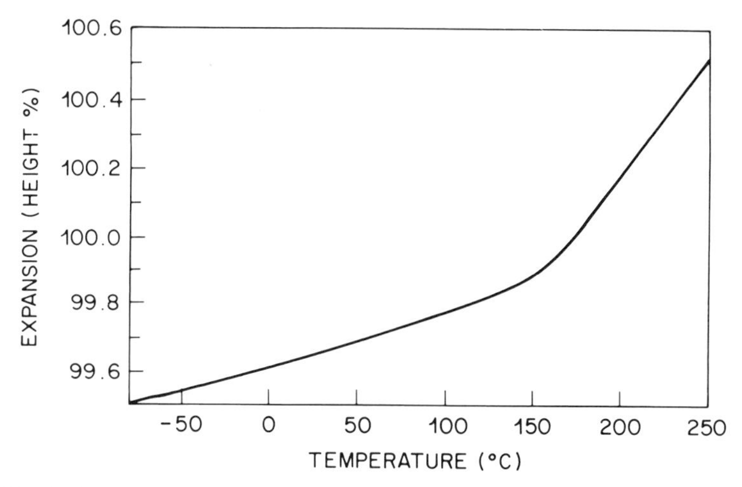


**FIGURE 2-16** Plot of expansion versus temperature for an epoxy molding compound, showing the two regions of different coefficient of thermal expansion above and below the glass transition temperature. (Data of H. E. Bair, AT&T Bell Laboratories.)

cases the thermal induced stresses can actually be higher, since the shrinkage stresses depend on the product of the difference in the coefficients of thermal expansion, the modulus of the plastic, and the temperature excursion.

### 2.6.4 Dielectric Properties

The dielectric properties of the materials used in plastic packaging play an important role in device performance. Both the dielectric constant $\epsilon'$ and the dissipation factor $\epsilon''$ influence the signal carrying capacity and propagation speed

**Table 2-4. Coefficients of Thermal Expansion**

| Material | |
|---|---|
| Alumina | 7 |
| Aluminum | 23 |
| Copper | 17 |
| Diamond | 0.9 |
| Epoxy molding compound | 14–20 |
| FR-4 ($x - y$ plane) | 15 |
| Gold | 14 |
| Lead | 29 |
| Polyimide | 35–50 |
| Silica | 0.6 |
| Silicon | 2.6 |
| Silicon nitride | 0.8 |

Units are cm/cm-°C × $10^{-6}$

**Table 2-5. Dielectric Properties of Packaging Materials**

| *Material* | $\epsilon'$ | $\epsilon''$ |
|---|---|---|
| Alumina | 9.6 | 0.003 |
| Aluminum nitride | 8.9 | 0.004 |
| Silica (fused) | 3.8 | $<0.0001$ |
| Silicon | 12.0 | |
| Diamond | 16.5 | |
| Epoxy molding compound | 4.2 | 0.03 |
| FR-4 | 4.5 | 0.05 |
| Polyimide | 4.2 | 0.01 |
| Triazine | 2.8 | 0.001 |

of the device. In general, materials with low dielectric constant and low dissipation factor provide better device performance, since they allow faster signal propagation with less attenuation. The signal speed is the velocity of electromagnetic radiation in the transmitting medium, which is inversely proportional to the square root of the dielectric constant of the medium, i.e., $\epsilon^{-0.5}$. In a plastic packaged integrated circuit, the substrate of the conductor path is usually silicon and the conductor is embedded in the passivation layer, which is in turn surrounded by the molded body. The dielectric properties of the molding compound play a minimal role in the propagation of the signal on the chip in this geometry. The leads of the device and the wire bonds are completely embedded in molding compound, hence the propagation characteristics of signals moving off the chip are directly dependent on the dielectric properties of the molding compound. In ceramic packages, the leads and interconnection to the device are usually embedded in ceramic, hence the propagation characteristics of the signal are directly dependent on the ceramic dielectric properties, which are slightly inferior to those of epoxy molding compound. (See compilation of dielectric properties of packaging materials in Table 2-5.) This advantage over ceramic will become increasingly important as devices become more sophisticated and operate at higher clock rates. Chapter 9 includes a more thorough discussion of the material implications of packaging advanced high-frequency devices. The effects of moisture and temperature on the dielectric properties of the packaging materials are also important considerations that should not be overlooked.

## REFERENCES


1. P. J. Flory, *Principles of Polymer Chemistry*, Cornell University Press, Ithaca, NY (1953).
2. W. Ostwald, *Kolloid-Z*, **36,** 99 (1925).
3. P. J. Carreau, Ph. D. Thesis, Department of Chemical Engineering, University of Wisconsin (1968).

4. G. C. Berry and T. G. Fox, *Adv. Polym. Sci.*, **5,** 261-357 (1968).
5. S. L. Rosen, *Fundamental Principles of Polymeric Materials*, John Wiley & Sons, New York (1982).
6. W. W. Graessley, "Entangled Linear, Branched and Network Polymer Systems—Molecular Theories," *Adv. Poly. Sci.*, **47,** 67-118 (1982).
7. C. W. Macosko and D. R. Miller, *Macromolecules*, **9,** 199 (1976); also D. R. Miller and C. W. Macosko, *Macromolecules*, **9,** 206 (1976).
8. A. Hale, M. Garcia, C. W. Macosko, and L. T. Manzione, "Spiral Flow Modelling of a Filled Epoxy Novolac Molding Compound," Soc. Plastics Engineers, ANTEC Papers, p. 796 (1989).
9. M. L. Williams, R. F. Landel, and J. D. Ferry, *J. Am. Chem. Soc.*, **77,** 3701 (1955).
10. M. Mooney, *J. Colloid Sci.*, **6,** 162 (1951).
11. J. M. Dealy, Rheometers for Molten Plastics, Van Nostrand Reinhold, New York (1982).
12. L. L. Blyler, Jr., H. E. Bair, P. Hubbauer, S. Matsuoka, D. S. Pearson, G. W. Poelzing, R. C. Progelhof, and W. G. Thierfelder, *Polym. Eng. Sci.*, **26** (20), 1399 (1986).
13. S. Middleman, *Fundamentals of Polymer Processing*, McGraw Hill Book Company, New York (1977).
14. J. D. Ferry, *Viscoelastic Properties of Polymers*, 3rd Edition, John Wiley and Sons, New York (1980).
15. Y. Uhara and K. Miki, "High Thermal Conductivity Encapsulant," *Nitto Technical Reports*, 32 (Sept. 1987).

# 3 Materials for Molded Plastic Packaging

The number and importance of polymer materials used in plastic packaging has been growing steadily. Polymers are now used in several process steps during assembly and packaging, including interlayer dielectric layers on the device, die attach adhesives, tape reinforcement of the lead frames, dielectric layers within the molded body, alpha-particle barriers, and finally as the molding compound to form the body of the package. The following sections describe the materials used in microelectronics packaging, their structures and properties, and relevant processing considerations. The sections are divided according to function rather than materials, providing a means to explore material alternatives to satisfy a specific function.

## 3.1 DIELECTRIC LAYERS ON THE DEVICE

The increasing demands of large scale integration require extensive use of multilayer technology with attendant intermetal insulator layers. The position and purpose of the polymeric interlayer is illustrated in Figure 3-1 [1]. Besides providing electrical isolation between conductors, the requirements of an interlevel insulator are to smooth underlying topography at crossovers so that the second-level metal is not thinned by shadowing effects in the case of evaporated metal or by irregular growth in the case of sputtered metal [1]. An illustration of this problem is provided in Figure 3-2. Planarization of a polymeric interlayer can require an extraordinarily long time depending on the rheology of the applied fluid; in the case of Bingham behavior (see Section 2.4.1, Figure 2-3), the forces of planarization my be below the yield stress of the material, so that complete planarization will never occur. Typically, some degree of planarization less than complete is acceptable. The degree of planarization can be defined as shown in the following relation with the parameters defined in Figure 3-3 (other definitions can also be used):

$$P = 1 - \frac{t_f}{t_h} \tag{3-1}$$

High temperature polymers are used as dielectric and interconnection layers on integrated circuit devices with excellent results. Polyimides are the predominant material for this function. They have excellent planarizing properties, making them well suited for multilevel insulators.

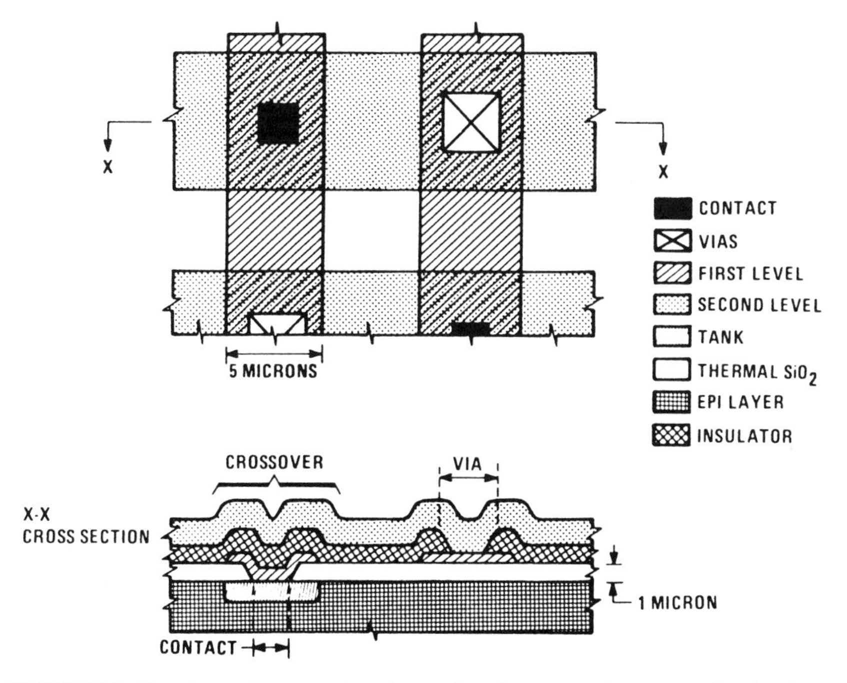


**FIGURE 3-1** Top view and cross section of a two-layer interconnection system, showing the use of dielectric interlayers. (From Reference 1. Copyright Plenum Press. Reprinted with Permission.)

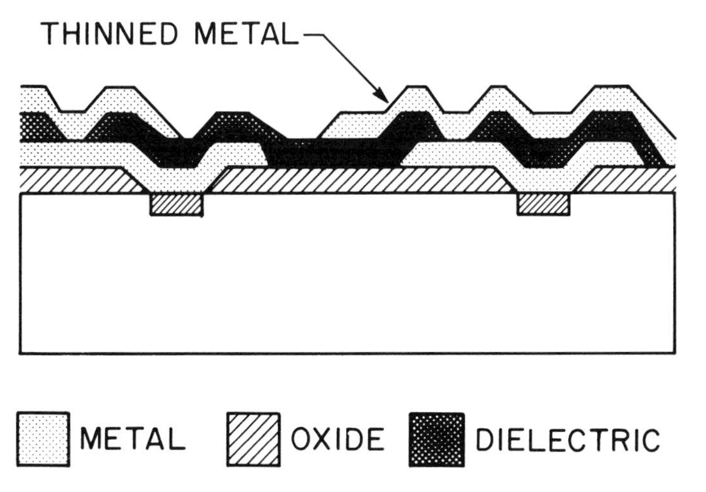


**FIGURE 3-2** Cross section of device topography, showing the thinned conductor path caused by incomplete planarization of the dielectric layer, which can lead to cracking.

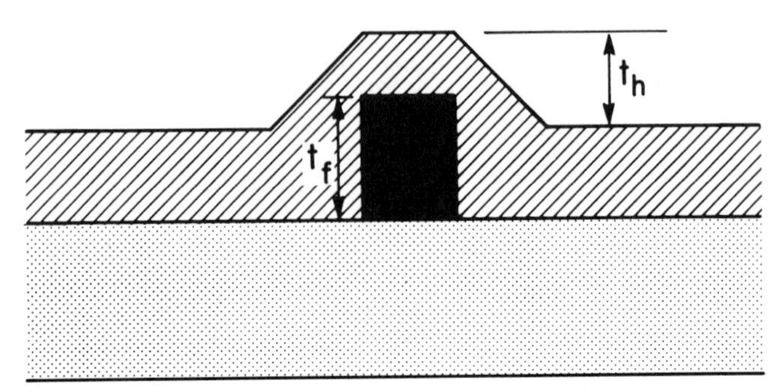


**FIGURE 3-3** The dimensions used in the definition of degree of planarization.

### 3.1.1 Polyimides

Polyimide represents a general class of materials that result from the reaction of dianhydrides and diamines to form an intermediate poly(amic acid) (PAA) which is soluble in polar solvents such as dimethylformamide (DMF). The PAA is then applied, usually by spin coating for wafer processing, and baked to complete the imidization reaction. Spin velocities to achieve planarization are of the order of 5000 revolutions per minute. There are a wide range of both aromatic diamines and aromatic dianhydrides that can be reacted to provide high-stability polyimide structures. The high-temperature properties are derived from the aromatic groups in the starting monomers. The stiff aromatic rings in the polymer backbone raise the temperature of large-scale chain mobility, perceived as a macroscopic softening of the material, to over 400°C. Many polyimides have no high-temperature softening point at all because they thermally degrade before they can be melt processed. In contrast, polystyrene, which has the aromatic ring pendant to the backbone chain, has a glass transition temperature of only 100°C. Polyimides that have excessively rigid backbones, such as those derived from fused aromatic rings, often have poor mechanical properties and are too brittle for most microelectronic applications. Polyimides also have good electrical properties to go along with their excellent thermal properties, as shown in Table 3-1.

The first polyimide introduced for microelectronics was polyimide-isoindroquinazolindione (PIQ) [2]. One of the more common polyimides for use in interlayer insulation is the material derived from the condensation reaction of pyromellitic dianhydride (PMDA) and 4,4′-diaminodiphenyl ether (DAPE), more commonly known as oxydianiline (ODA), to provide the polyimde known as PMDA-ODA. The reaction that forms this polymer is shown in Figure 3-4. Another important class of polyimides are those derived from methyl- or paraphenylene diamine (MPD and PPD). This more rigid diamine requires a more flexible dianhydride to provide acceptable mechanical properties and minimize

**Table 3-1. Representative Properties of Polyimides**

|  |  |
|---|---|
| Density | 1.38 gm/cm$^3$ |
| Coefficient of thermal expansion | $35 \times 10^{-6}$ cm/cm-°C |
| Dielectric strength | 300 kV/mm |
| Dielectric constant (1 MHz) | 3.5 |
| Dissipation factor (1 MHz) | 0.002 |
| Volume resistivity | $1 \times 10^{15}$ Ohm-cm |
| Tensile modulus | 300 kg/mm$^2$ |
| Maximum service temperature | 400°C |

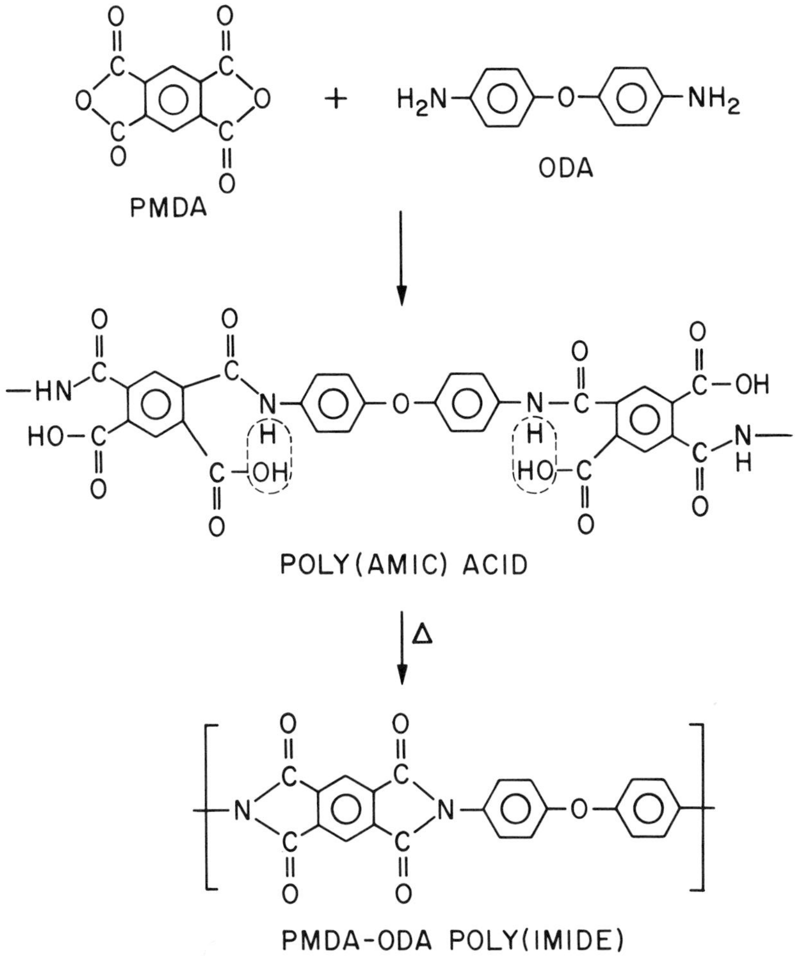

PMDA + ODA

POLY(AMIC) ACID

PMDA-ODA POLY(IMIDE)

**FIGURE 3-4** The synthesis and final chemical structure of PMDA-ODA polyimide.

brittleness. For MPD systems, the less rigid 3,3′, 4,4′-benzophenonetetracarboxylic dianhydride (BTDA) replaces the PMDA forming the polyimide known as BTDA-MPD, illustrated in Figure 3-5.

Commercial offerings of polyimide materials are often copolymers of two or more of the most common materials shown in Figures 3-4 and 3-5. Most are not photodefinable. They require masking and imaging with other resist materials to form and position the interlayer insulators. The excellent solvent and chemical resistance of most polyimides often requires the use of plasma etching to remove polymer in what is known as *bilevel processing*. A silicon-containing material is imaged and then developed with solvent, stopping at the polyimde layer which is relatively insoluble. Plasma etching is then used to develop the polyimide through the silicon containing mask. If the silicon-containing material is not photodefinable itself, then a third layer of photoresist is required for imaging it, creating a *trilevel process*, a version of which is illustrated in Figure 3-6.

Photodefinable polyimides, which were reported in the mid-1980's [3, 4, 5,], can be used to fabricate dielectric layers with far fewer process steps. The chemical structure modification of these photopolymers essentially consists of

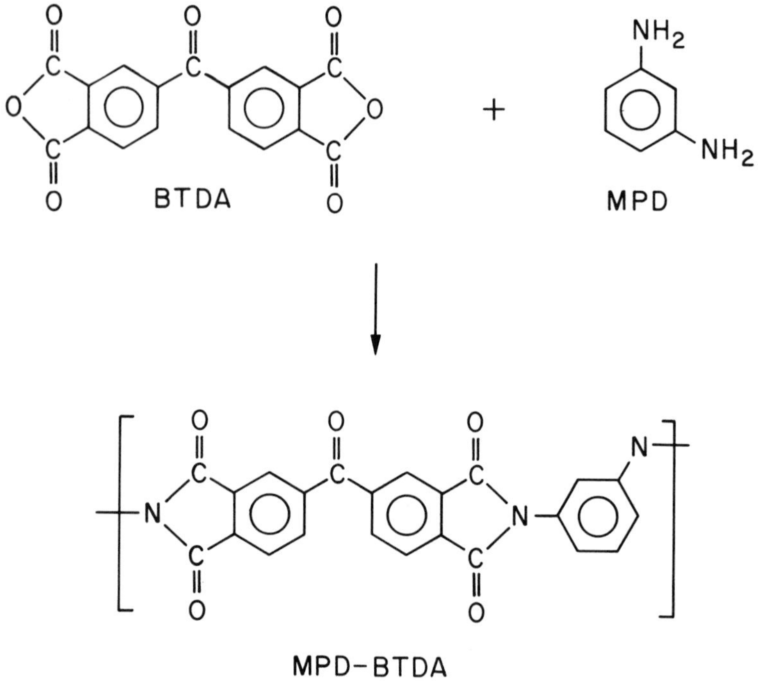

**FIGURE 3-5** The components and the final structure of MPD-BTDA polyimide.

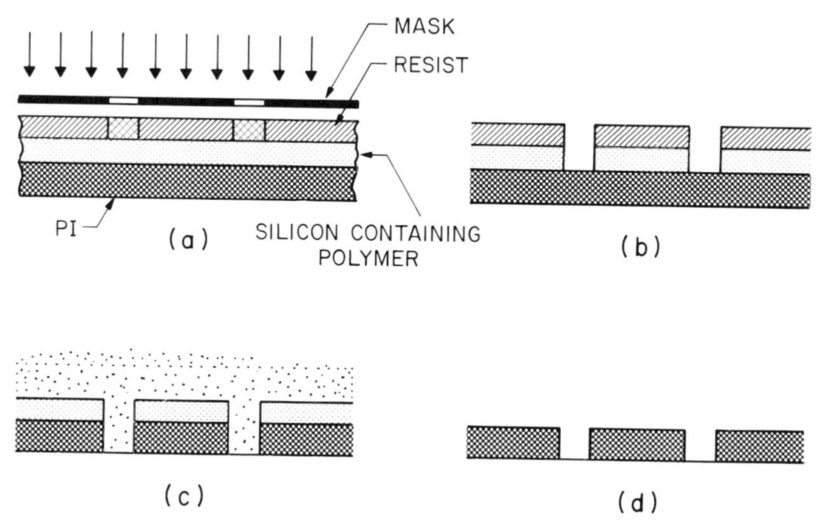


**FIGURE 3-6** Trilevel processing, using both a nonimageable silicon-containing polymer and a photoresist to develop a polyimide pattern; (a) Imaging the resist, (b) developing resist and silicon-containing layer with solvent, (c) removing resist and developing polyimide with plasma etching using the imaged silicon-containing polymer as an etching mask, (d) removing silicon-containing layer with solvent.

adding photocrosslinkable side groups to the polyamic acid, as shown in Figure 3-7 [5]. In most cases, the side group is removed by high-temperature pyrolysis.

The reduction in the number of process steps to form a polymeric interlayer is significant as illustrated in Figure 3-8 [6]. The reduction in steps is greater if trilevel processing is used for etching the nonphotodefinable polyimide. These imageable polyimide materials have not had a large impact on production at the time of this writing. One problem is that these are negative resists which are developed with solvents that can cause swelling problems such as coalescence of lithographic features. Excessive volume contraction of these polymers during crosslinking may also be a problem. Although neither of these problems is insurmountable, their resolution does add process steps, forfeiting some of the benefits of using a photodefinable polyimide in the first place.

## 3.2 CONFORMAL COATINGS ON THE DEVICE

The conformal coating of integrated circuit devices fulfills several important functions, and there are a number of application and performance criteria that it must meet. As a passivation layer, polymer materials are often used as conformal coatings on top of the primary passivation layers of the device, which is typically silicon oxide or silicon nitride. The coating prevents scratches and

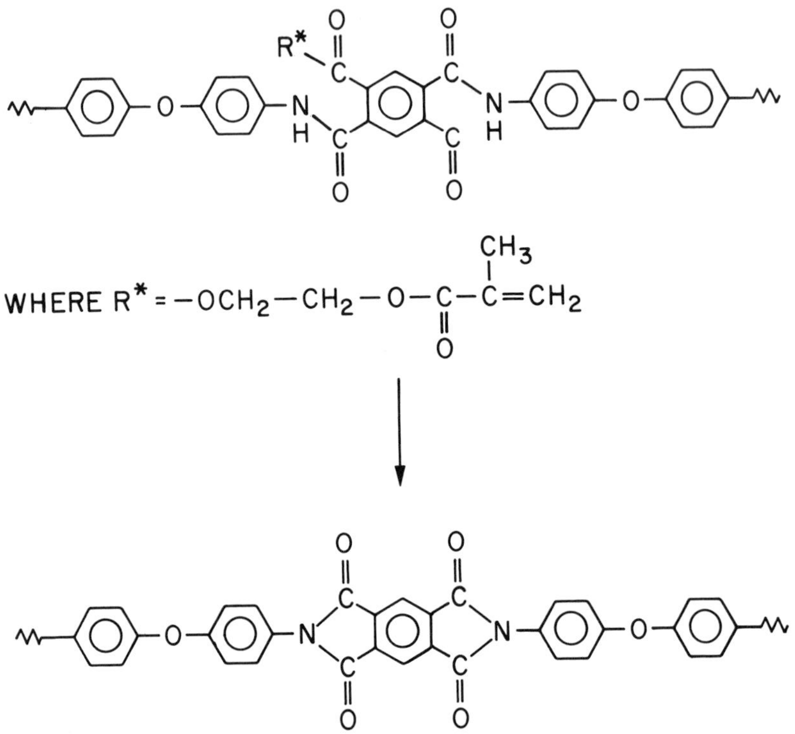

WHERE R* = $-OCH_2-CH_2-O-C(=O)-C(CH_3)=CH_2$

FIGURE 3-7 Chemical structure of photodefinable polyimide. (After Reference 5.)

damage to the exposed metal conductor lines on the device during multiprobe testing, assembly, and packaging. In addition, the chip must be protected from ionic impurities once it leaves the clean-room environment for assembly and packaging, or when the chip is in contact with the plastic molding compound. In devices that are sensitive to alpha particles a conformal coating may be used to block this radiation.

The conformal coating should exert as little thermal shrinkage stress on the device and molding compound as is possible. For this, it should have a coefficient of thermal expansion (CTE) as close as possible to that of silicon, or a low modulus to minimize the shrinkage forces. A low-modulus coating minimizes warping of the silicon in cases where there is a coefficient of thermal expansion mismatch. The stress created by disparity in the coefficient of thermal expansion between the coating and the silicon device is defined by Equation (3-2):

$$\sigma = \int_{T_1}^{T_2} \Delta\alpha E_p \, dT \tag{3-2}$$

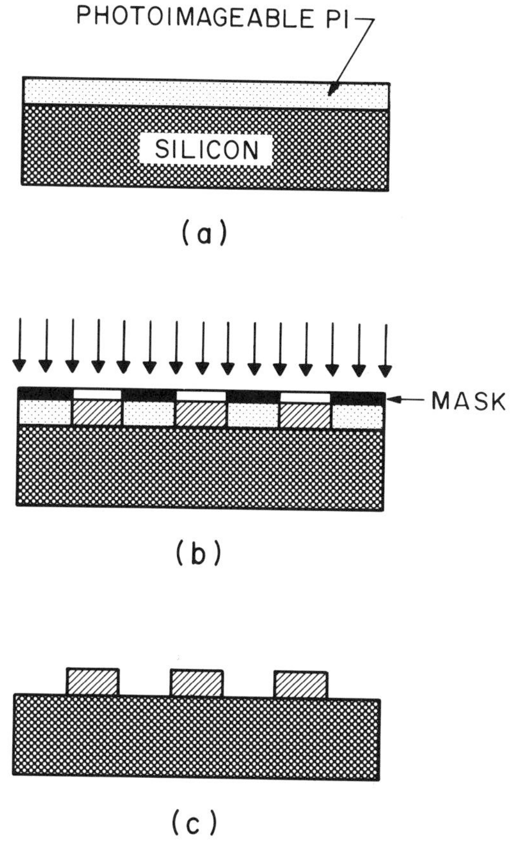


**FIGURE 3-8** The process steps to fabricate a dielectric layer using a photodefinable polyimide. The drying and thermal treatment steps have been omitted for clarity. (After Reference 6.)

Materials that best minimize the integral in this relation cause the least stress and the least deformation to the wafer. Other criteria are that the conformal coating should be formed at or below the annealing temperature of the device which is near 450°C. It should be able to withstand the high temperatures experienced during molding and subsequent temperature cycle testing which exceed 200°C. There are only a few materials that meet these criteria. The most important are polyimides and silicone rubber.

### 3.2.1 Polyimides

Polyimides were discussed at length in the previous section on dielectric layers. Often the same materials used for insulation are used for conformal coatings such as the PMDA-ODA and BTDA-MPD. Polyimide conformal coatings are often applied at the wafer stage. The principal difference between a conformal coating material and a dielectric is the greater need for low thermal shrinkage stresses in the conformal coating because of its continuity over the entire wafer. These stresses can result in cracking of the inorganic passivation layer of the device, aluminum conductor slide, and warpage of the wafer. There have been some material advances reported to alleviate these problems, and there will certainly be others.

From Equation (3-2) it is evident that the thermal shrinkage forces can be lowered by either minimizing the difference in the coefficients of thermal expansion between polyimide and silicon, lowering the elastic modulus of the polyimide, or reducing the processing temperature of the polyimide. Several of these approaches have already been reported. The structure shown in Figure 3-9 is reported [7] to have a thermal expansion coefficient of only 9.6 ppM (between 20° and 50°C) compared to a value of 35 ppM for conventional polyimides, 2.6 ppM for silicon, and 2.8 ppM for $Si_3N_4$. The loss of chain mobility with the elimination of the diphenyl ether linkage appears to be the mechanism for lowering the expansion coefficient. The penalty paid for this reduction in chain mobility is an increase in the tensile modulus to 720 $kg/mm^2$ from a conventional value of 300 $kg/mm^2$. The integral product of modulus and expansion coefficient is still significantly reduced, however.

The tensile elastic modulus can be lowered be incorporating more flexible units in the polymer morphology, either chemically by altering the molecular structure of the polyimide, or physically by the addition of an external plasticizing agent. There has been a report [7] on the development of a chemical approach based on the copolymerization of dimethylsiloxane units into the polyimide backbone chain. This forms a block copolymer with the siloxane block length sized appropriately to provide a microphase separation of the siloxane into a domain morphology. The tensile modulus of the composite is then effectively lowered because the cross sectional area of the hard phase polyimide is reduced. The authors [7] report a tensile modulus for this material of 160 $kg/mm^2$, close to 50% less than conventional polyimide. The soft siloxane

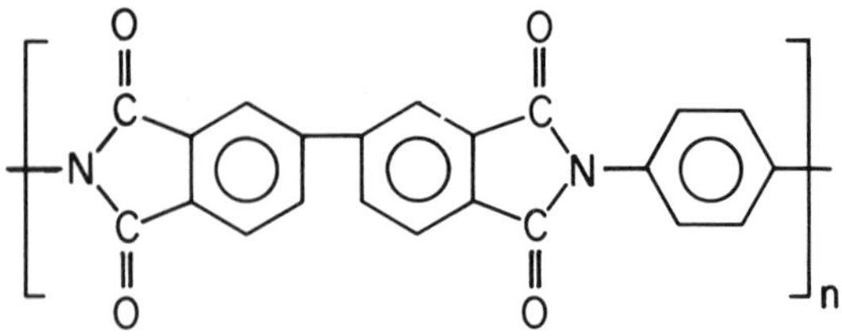

**FIGURE 3-9** Chemical structure of low-stress polyimide. (After Reference 7.)

segments have a higher thermal expansion coefficient, however, so the overall coefficient of thermal expansion for this material is increased. Again, the integral product of modulus and expansion coefficient difference is reduced substantially. Siloxanes may also be copolymerized with polyimides to effect a number of other physical and mechanical property changes such as increased elongation, solubility, and adhesion [8]. Unfortunately, addition of these more flexible chain segments is generally accompanied by a decrease in high-temperature performance [9].

**3.2.1.1 Compatibility of Polyimide with Epoxy Molding Compounds** An important consideration in conformal coating materials is the compatibility, adhesion, and property match between the coating and the epoxy molding compound. In general, the better the adhesion between interfacial layers, the lower will be the maximum stresses developed in temperature cycle tests, and the less the opportunity for delamination and water collection at the interface. Delamination or loss of adhesion between the high modulus coating and molding compound will result in concentrating the thermal shrinkage stresses at the edges and corners of the device, rather than distributing them over the volume of molding compound. Polyimides have fairly good adhesion to epoxy molding compounds and moderate adhesion to the common inorganic passivation layers of the device; in some cases adhesion promoters are required. In addition, the difference in the coefficients of thermal expansion between molding compound and polyimides is not great—approximately 35 ppM/°C for polyimide and 19 ppm/°C for molding compound. The use of very low CTE polyimides with expansion coefficients below that of the molding compound may relieve the thermal shrinkage stresses on the device, but increases the opportunity for delamination between molding compound and conformal coating.

Reliability of semiconductor devices depends on the properties of the polyimide used in dielectric and passivation roles. The effects of incomplete cure [10], leakage current [11], and ionic contamination [12] have been reported. The amount of moisture absorbed into and transported through the polyimide coating also plays an important role in reliability. Water absorption of 4.2% has been reported [13] in 3.6-micron-thick film soaked in water. There is rapid water absorption for the first 150 minutes of exposure, then leveling off after 400 minutes. Water uptake within a plastic package is expected to be much lower than in the neat resin with most estimates in the vicinity of 1% water uptake. Absorbed water will increase both the dielectric constant and the dissipation factor over the value for dry polyimide reported in Table 3-1. The dielectric constant increase is a modest 10–20%, resulting in an actual value of 4.0 for polyimide in equilibrium within the package. The dissipation factor for polyimide is reported to increase by approximately 33% on going from dry conditions to saturation [13].

### 3.2.2. Silicone Rubber

Silicone rubber materials have a long history of use in microelectronics packaging applications. In general, these are low-modulus, rubbery materials that have very good thermal and electrical performance properties. They are often described as RTV (room temperature vulcanization). The description, though common, is actually misleading, because most of the silicones used in the microelectronics industry are thermal cure systems, in contrast to the household silicon caulking materials which are truly RTV. Silicone conformal coatings are more often applied to the individual chips at the assembly stage and not on the entire wafer, although wafer level passivation is also feasible.

Silicones are generally made through a condensation reaction of hydroxysilanes, or silanols, often in the presence of a catalyst. A pure silicone material would have only silicon and oxygen in the main chain with pendant hydrogen, but the commercial materials used in microelectronics packaging are often organosilicone materials that have hydrocarbon groups pendant to the Si—O backbone [14]. Crosslinked silicone materials are obtained through hydrocarbon or oxygen-containing bridges to other Si—O chains as shown in Figure 3-10.

Silicones have very good high temperature properties and good electrical properties as can be seen in the listing of their properties provided in Table 3-2 [14]. They meet most of the criteria for conformal coating materials described above. They show good compatibility with the silicon device, which is not accurately represented by the adhesive strength alone. The exceptionally low modulus and tear strength of silicones will provide only modest adhesive

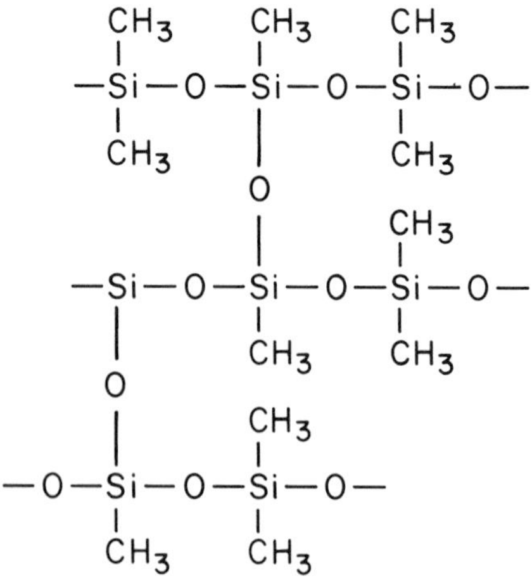

```
        CH3     CH3     CH3
        |       |       |
       —Si—O—Si—O—Si—O—
        |       |       |
        CH3     |       CH3
                O
                |       CH3
                |       |
       —Si—O—Si—O—Si—O—
        |       |       |
        |       CH3     CH3
        O
        |       CH3
        |       |
    —O—Si—O—Si—O—
        |       |
        CH3     CH3
```

**FIGURE 3-10** A crosslinked organosilicone resin typical of the type used in the electronics industry.

**Table 3-2. Representative Properties of Silicones [14]**

|  |  |
|---|---|
| Density | 1.05 $gm/cm^3$ |
| Coefficient of thermal expansion | 200–300 $\times 10^{-6}$ cm/cm-°C |
| Dielectric strength | 550 kV/mm |
| Dielectric constant (1 MHz) | 2.7 |
| Dissipation factor (1 MHz) | 0.001 |
| Volume resistivity | 2 $\times 10^{14}$ Ohm-cm |
| Maximum service temperature | 200°C |

strengths, yet chemical compatibility and chemical bonding between the silicone of the conformal coating and the silicon of the device is thought to be beneficial, particularly as an interfacial moisture inhibitor [15]. In terms of the other criteria of a coating material, the generally poor adhesion between silicones and epoxy molding compound and the low tear strength and very high coefficient of thermal expansion of silicones are all major liabilities. Very-low-modulus materials, known as silicone gels, are gaining acceptance for device coverage to reduce stress-induced deformation to the device features and improve moisture resistance of the package.

The high expansion coefficient of silicones can actually cause a void to form between the coating and the epoxy molding compound, particularly in cases where the coating has enough time prior to encapsulation to heat to the mold temperature of 170°C. Upon cooling to room temperature, the silicone rubber can shrink 10 times as much as the molding compound, which can lead to delamination with the fairly incompatible molding compound. Conversely, the high pressures exerted during the packing stage of the molding process, up to 800 psi, could compress the elastomeric rubber, thereby compensating for most, if not all, of the difference in shrinkage. Hence, depending on process conditions and process variability, the silicone could be under compression or extension, or be delaminated from the molding compound. This variability may be the most serious problem of all. Delamination has the most serious repercussions in terms of device reliability because, first, it allows a void to form which will condense moisture, and second, it concentrates the stresses on the edges of the device rather than distributing them over the entire molded body volume, as illustrated in Figure 3-11.

### 3.2.2.1 Application and Cure of Silicones

Another consideration in the use of silicone rubber materials within the molded body of plastic packages is the application of the material. Unlike polyimides, which are often applied during wafer processing, silicone materials are generally applied at the assembly stage after the device has been mounted on the lead frame and wire bonded. Automated discharge of a precise amount of liquid resin from a syringe is the most common application method. Silicones are often in

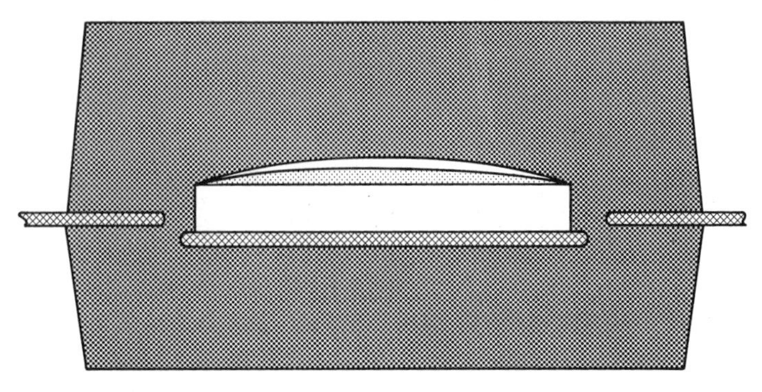

**FIGURE 3-11** Void formation due to excessive silicone rubber shrinkage with attendant thermal shrinkage stress concentration on the edges of the device.

solvent to facilitate dispensing, and the removal of the solvent is an important consideration in itself. The application is relatively straightforward, although there are ample opportunities for problems to arise. One simple problem is related to the amount of material discharged from the syringe; it depends on the applied pressure and the rheology of the silicone resin. If the volume discharged is low, even by less than 10%, the coverage of the device could be incomplete, as shown in Figure 3-12. One consequence of this is the possibility of soft errors where active device area is exposed to the alpha activity of the molding compound. Another concern with incomplete coverage is that some of the bonding

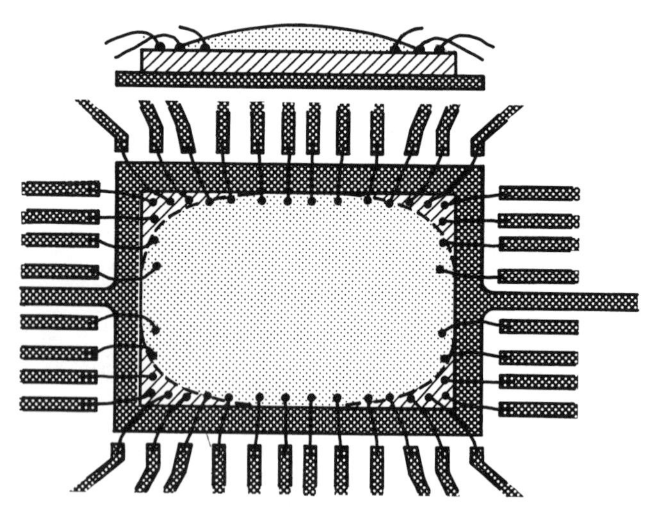

**FIGURE 3-12** Side and top views of paddle support with attached die showing incomplete coverage of the conformal coating material. The ball bonds which straddle the interface between the silicone and the epoxy molding compound are most at risk.

pads and ball bond sites will be exposed, and in direct contact with the molding compound. In the event of delamination between the silicone and the molding compound, moisture may be brought into intimate contact with the ball bond area providing a very high probability of premature corrosion failure. Yet another consequence of incomplete coverage is the effect of thermal shrinkage stresses. Although ball bonds can provide excellent reliability when embedded in molding compound or in silicone rubber, there is some concern that positioning the silicone–molding compound interface at the ball bond site is the more stressful configuration due to interfacial forces and movement. Conversely, excessive volume of silicone will cause runoff onto the die support and possibly off the leadframe itself, fouling the equipment and causing a shutdown.

There is strong motivation to attain the proper amount of coverage on the device by dispensing the precise quantity of liquid. As most of these materials are shear thinning liquids (see Section 2.4.1 on shear rate dependence), the volumetric flow rate does not depend linearly on pressure. Similarly, small differences in resin formulation such as molecular weight and molecular weight distribution can cause large differences in volumetric flow. For this reason, careful quality control and inspection of incoming lots should be performed routinely to minimize coverage problems. Rheological characterization of the silicone precursor material should be performed at the shear rates encountered in the application. For a cylindrical delivery tube and a non-Newtonian fluid represented by a power-law model (see Section 2.4.7), the approximate shear rate can be determined from Equation (2-21), repeated here for convenience as equation (3-3). For typical conformal coating materials and typical applicators, the shear rate range is approximately 200–1000 $\sec^{-1}$.

$$\dot{\gamma}_R = \frac{2(1 + 3n)}{n} \frac{U}{D} \tag{3-3}$$

The testing and qualification should be performed at the same temperature as the application. Although any viscometric method that can attain the determined shear rate is adequate for quality control testing, the most appropriate technique is one that duplicates the shearing and flow encountered in the application most closely, in this case it is the capillary rheometer (see Section 2.4.6). It is preferable to use a capillary of the same diameter as the syringe rather than a larger capillary with a higher-volume flow rate to achieve the desired shear rate, since the shear rate relation provided in Equation (3-3) is only approximate.

### 3.2.3 Encapsulants (Potting and Glob-Top Materials)

Silicone rubber materials are also used as encapsulants in a number of packaging technologies such as potting of premolded devices and glob top coatings for chip on board, flip-chip, hybrid integrated circuits, and multichip modules.

Epoxy and epoxy hybrids are the other materials which are candidates as encapsulants. The major difference between the two is that silicone is rubbery because it is well above its glass transition temperature, whereas epoxies are generally below their glass transitions, providing a hard, rigid coating. There are benefits and drawbacks to each material. The epoxy glob-top coatings have the potential of greater thermal shrinkage stresses because of the mismatch in the coefficients of thermal expansion and the high modulus. With proper loadings of inorganic fillers such as silica, the CTE can be reduced to the order of 40 ppM / °C [16]. This value is almost twice what can be achieved with transfer molding compounds because the glob-top material must be dispensable through a tube whereas the molding compounds can be processed at much higher pressures in a molding tool. Therefore, the maximum filler loading must be kept lower than the 70% by weight that is typical of molding compounds. To their advantage, epoxies have excellent adhesion to silicon and ceramic substrates, and they are effective moisture barriers, although not hermetic. Silicones have a very low modulus which is unable to generate high thermal shrinkage stresses as determined through Equation (3-2), so their very high CTE, of the order of 200 ppM / °C [16], is generally unimportant.

There can be a problem with flip-chip and beam-leaded configurations, where there is often a space of approximately 0.002 inch between the chip and the substrate. The encapsulant material will be either intentionally or inadvertently drawn into this gap during application, as illustrated in Figure 3-13. On thermal cycling, the large degree of expansion of the silicone will move the chip up and down, possibly fatiguing the solder bumps or beam leads until failure. Filled epoxy materials will provide a smaller translational movement, but will generate higher forces. Therefore, the selection of an encapsulant should account for the compliance of the interconnect; gull-shaped leads or beam leads are likely to be more compliant than solder bumps and able to sustain greater deformation, whereas solder bumps will show less compliance. The selection pro-

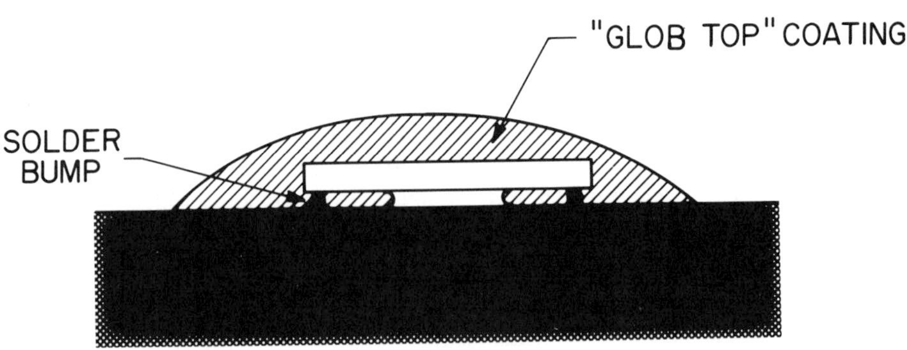


**FIGURE 3-13** Gap between the die and the substrate in flip-chip interconnection. High stresses can be encountered on temperature cycling when the encapsulant material is drawn into this gap.

cess should include the specific properties of the encapsulant material and the precise distances. A simple calculation derived from Equation (3-3) can be then performed to determine both the deformations and forces expected. As with any material specification, actual reliability studies are the safest route.

### 3.2.4 Other Device Coatings and Encapsulants

There are several other materials that can be used for a number of applications in device assembly and packaging. A class of materials that are valued for their low temperature application are the poly-p-xylylenes. The final materials are known commercially as Paralenes®, which are completely linear-chain polymers. The application of these materials is unique in that they require a vapor phase reaction and deposition process. Although the reaction itself is conducted at high temperature, the substrate on which the film is deposited can be at relatively low temperatures below 150°C. They form a closely conforming coating that has good solvent resistance and good dielectric properties.

Another more recent dielectric coating material that can adapt to a number of applications is benzocylobutene (BCB). This material has good dielectric and high-temperature properties, and lower moisture absorption than polyimide. It is a thermoset polymer formed by the polymerization of benzocyclobutene at moderate to high temperatures (250°C) in nitrogen atmosphere. The monomer is applied by conventional spin coating or spraying and then polymerized on the substrate. Planarization properties are also good. The resulting polymer has a glass transition temperature of 370°C and a coefficient of thermal expansion of 34 ppM/°C. Specific end-use information cannot be provided since the material was just introduced at the time of this writing, but potential applications include chip passivation, interlayer dielectric, die attach adhesives, glob-top materials, and even molding compounds

Other dielectric materials include fluoropolymers such as poly(tetrafluoroethylene) (PTFE) and poly(chlorotrifluoroethylene) (PCFE). These materials have excellent electrical and dielectric properties as well as excellent high temperature properties. Their limitation, however, is in applying the material since they are neither polymerizable in situ nor melt processable, and they are largely insoluble in most common solvents. As such, they are mostly used as preformed films and tapes, although they can be applied in the form of aqueous colloidal suspensions.

## 3.3 DIE ATTACH ADHESIVES

Die attach bonding is used to mechanically attach the die to the leadframe. Solder bonding alloys, organic adhesives, and silver-glass composites are used. The die attach solders are usually applied as preforms less than 0.05 mm thick.

Both the back of the die and the leadframe are metallized to promote solder wetting. The die attach solder must be chosen to be compatible with the temperature requirements of the remaining process steps including molding, burn-in, temperature cycle testing, and attachment to the circuit board. A listing of common die attach preform solders is provided in Table 3-3 [17].

Conductive polymer adhesives are displacing solders for die attach in plastic packages because they are less expensive than high-gold-content solders and their application is more easily automated [18]. Solventless conductive epoxies are the material of greatest commercial importance, with about 80% of the overall die attach market [19]. Most formulations are based on silver particles to provide the conductivity. Typical materials contain between 50% and 72% by weight of silver flakes, loadings that fall within the percolation range of electrical conductivity. There may also be coupling agents on the metallic particles to promote wetting and electrical conductivity. In some formulations, the material is nonconductive prior to thermal cure, and conductive afterward. These die attach materials require cures at temperatures of 150–200°C for as long as an hour, constituting a separate process step that requires dedicated equipment. Newer materials that cure in as little as 5–10 minutes have also been introduced. Typical materials have a thermal expansion coefficient of $55 \times 10^{-6}$ cm/cm-°C and a thermal conductivity of 0.008 Watts/cm-°C [20]. Low-stress die attach materials have lower coefficients of thermal expansion to more closely approach the CTE of the silicon die. Impurities in the die attach material can cause reliability problems, particularly if there is outgassing during the cure, so there has been a shift toward much purer materials with extractable ionic contents less than 10 ppM [19]. Polyimides are also used for die attach in high temperature applications. Yet another die attach material option is the silver-filled glass resin adhesives. These materials become wholly inorganic after firing, thus eliminating the outgassing and thermal degradation problems of the organic materials in a form that is still convenient to apply. High-temperature bake cycles of over 400°C for several hours are required to fire the resins. The die and the leadframe still require metallization to assure electrical conduction to the chip.

**Table 3-3. Compositions and Melting Points for Die Attach Solder Preforms [17]**

| *Composition* | *Liquidus Temp., °C* | *Solidus Temp., °C* |
|---|---|---|
| 80% Au/20% Sn | 280 | 280 |
| 92.5% Pb/2.5% Ag/5% In | 300 | |
| 97.5% Pb/1.5% Ag/1% Sn | 309 | 309 |
| 95% Pb/5% Sn | 314 | 310 |
| 88% Au/12% Ge | 356 | 356 |
| 98% Au/2% Si | 800 | 370 |
| 100% Au | 1063 | 1063 |

## 3.4 REINFORCEMENTS OF LEADFRAMES

Polymeric tapes are used to reinforce the leadframes of most larger packages to prevent damage during the assembly and molding operations. The tapes are also useful in reducing damage due to flow-induced stresses of the molding compound. In most flow configurations, the molding compound flows along and through the leadframe, since the gate of the cavity is located only on one side of the frame. The individual leads are taped together to prevent extensive deformation of the frame during this cross flow. Figure 1-14 shows the placement of this reinforcing tape in an assortment of leadframes. The tape must be a high-temperature material that can withstand the molding operations, subsequent temperature cycling, and other reliability tests. Polyimide films, such as the commercial material known as Kapton® are essentially the only material used for taping the leadframes. Kapton is the condensation product of pyromellitic dianhydride with oxydianiline, to provide the polyimide known as PMDA-ODA that was discussed previously and shown in Figure 3-4. The film is made by casting from solution before the imidization step [17]. Substantial orientation of the film is a distinguishing feature of the commercial product.

The size, shape and positioning of the tape on the leadframe can have implications for the molding yield. Tape is often not needed and not feasible on low-pin-count packages such as 16- and 18-pin DIPs. The leads are exceptionally short and sturdy; also, there is not enough space between the edge of the package and the wire bonds to fit the tape. Tape is most often required on the longer DIP packages such as the 40-pin DIP and larger chip carriers with more than 32 leads. The size and shape of the tape are largely determined by the tape manufacturing method, which is to die cut the segments from a continuous thin spool. For this reason, a chip carrier package would be taped with four straight segments rather than a continuous square of tape. Overlapping of the tape is usually avoided because it creates a raised area that can alter the flow path.

## 3.5 DIELECTRIC LAYERS WITHIN THE MOLDED BODY

Dielectric layers are sometimes used within the molded body to provide electrical isolation for power and ground planes incorporated within the package but not part of the device itself. In small outline packages, the leads may pass under the chip, conserving area and eliminating the need for a separate paddle support. In this case, these leads have to be electrically isolated from the underside of the chip with a dielectric layer. Typically these conductive and insulator layers are simply thin sheets of copper and plastic film. The plastic films used most often are the oriented polyimide films discussed above in Section 3.4. An illustration of a 256K DRAM device that utilizes these conducting and dielectric layers is provided in Figure 3-14.

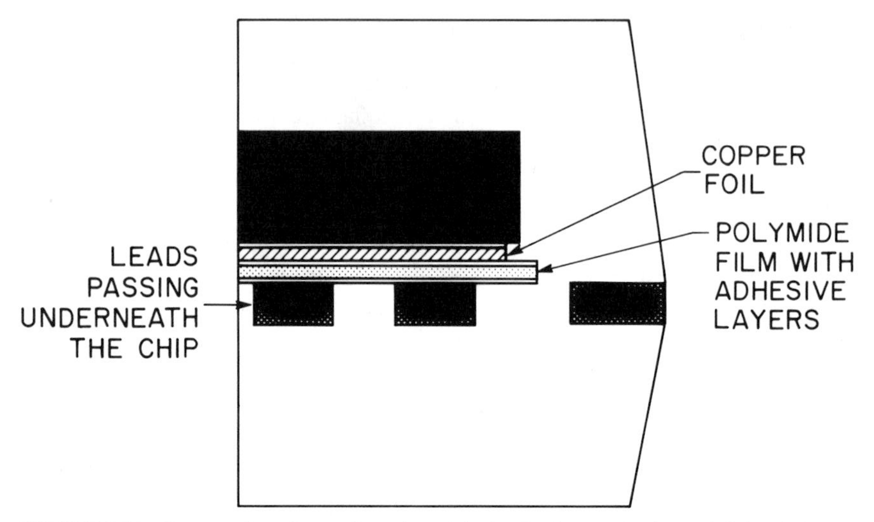


**FIGURE 3-14** Cross section of a plastic packaged device that has conductive and dielectric layers within the molded body but not on the device itself (not drawn to scale).

## 3.6 MOLDING COMPOUNDS

The molding compound is the most important polymer material used in plastic packaging of IC devices. It provides both mechanical and chemical protection to the device at a cost that is significantly below that of metal, ceramic or glass packaging and below that of premolded plastic packaging. Its principal requirements are to offer very high yield, high productivity, and good reliability. Although there are several different types of molding plastics that can be used for IC packaging, the epoxy molding compounds are by far the most commercially important materials. Alternative materials have filled some specialty niches, and their use could grow significantly in the future as the rapid increase in device size and lead count extends epoxy molding materials to their limit.

### 3.6.1 Epoxy Molding Compounds

Postmolded packaging of IC devices requires a low initial viscosity to flow over the leadframe and wire bonds without causing significant deformation, and a high modulus and high heat distortion temperature in the final state to withstand mechanical operations such as trim and form, and solder attachment to a printed circuit board. Thermoset polymers are uniquely suited for these needs because they polymerize to a thermally intractable material from low-viscosity starting resin (see Section 2.1 for a detailed explanation of thermoset materials and properties). Most conventional thermoplastics are largely inappropriate for microelectronics packaging because their melt viscosity is too high to flow over the device without damage, and their softening temperature is too low for solder operations. Hence thermoset polymers have dominated microelectronics packaging since its inception. Epoxy materials are the most important IC packaging material at the present time, but it has not always been the case, and it may not continue to be so given the rapid changes in the technology.

**3.6.1.1 The Formulation of Epoxy Molding Compounds** The formulation for most epoxy molding compounds consists of a complicated and often proprietary mixture of epoxy resin, hardener (or curing agent), catalyst(s), fillers, flame retardants, flexibilizers, coupling agents, mold release agents, and colorants. Each component is described in a subsection below.

**3.6.1.1.1 Epoxy Resin.** Cresol-novolac epoxies are the most common resins used in molding compounds for IC packaging. As explained in Section 3.6.1.2 on the evolution of epoxy molding compounds, novolacs have high-temperature properties superior to the resins based on the diglycidyl ether of bisphenol-A (DGEBA), which are more common in liquid adhesives and potting resins. The chemical structures of both DGEBA and cresol-novolac epoxy are shown in Figures 3-15 and 3-16.

The viscosity of the base resin for the encapsulant grade materials is kept low by limiting the average degree of polymerization of the novolac epoxy to values such as 5, providing $m = 3$ in Figure 3-16. The actual formulation will contain a mixture of degrees of polymerization that averages 5, with the breadth of this distribution being one of the distinguishing features of different commercial formulations. The average molecular weight of the starting resin is therefore approximately 900 for a typical formulation. Longer chain lengths of the starting resin increase molecular weight and viscosity according to the relation developed in Chapter 2, Equations (2-8) and (2-9). Shorter chain lengths provide a tighter crosslink structure that may be more brittle.

DGEBA epoxy has an epoxide functionality of 2 for its average molecular weight of 380, whereas the novolac epoxy has an average epoxide functionality of 5 over its molecular weight of 880. The higher epoxide functionality density of the novolac compared to the DGEBA material imparts a higher crosslink density which increases the glass transition temperature. The mobility of the backbone polymer chain is reduced, requiring higher temperature before it can attain the longer range cooperative motions that manifest themselves as the soft-

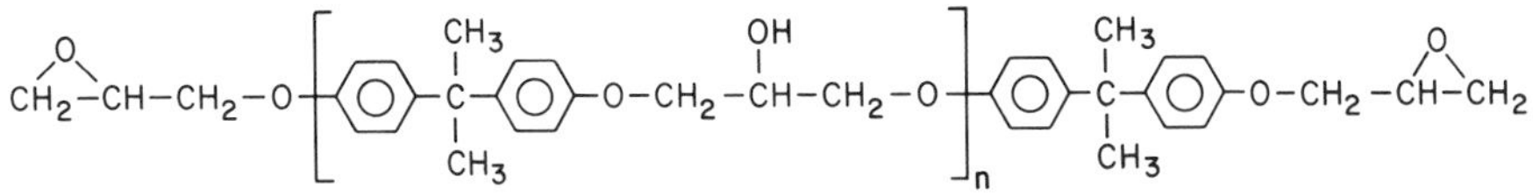

**FIGURE 3-15** The chemical structure of diglycidyl ether of bisphenol-A.

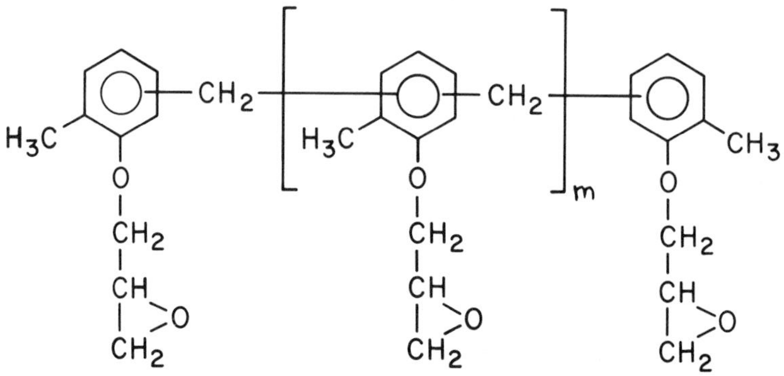

FIGURE 3-16 The chemical structure of polyglycidyl ether of *o*-cresol-formaldehyde novolac epoxy.

ening of the crosslinked network, commonly known as the glass transition temperature or the heat deflection temperature. Too high a crosslink density can excessively reduce the chain mobility and increase the brittleness of the material. For this reason, the balance among molecular weight, functionality and resultant crosslink density is one of the principal tools in formulating materials that have low initial viscosity, high reactivity, sufficient flow time, and good strength and crack resistance.

**3.6.1.1.2. Curing Agents (Hardeners)** Amines, acid anhydrides and phenols can all be used to crosslink epoxy resins. The reactions proceed through different chemistries. Ring opening is the predominant mechanism for amine curatives, whereas crosslinking occurs through pendant hydroxyls for the anhydrides and phenols [21]. The reaction with alicyclic amines is rapid at low temperatures, whereas acid anhydrides and aromatic amines require accelerators because of their lower reaction rates even at elevated temperatures. Phenol novolac and cresol novolac hardeners have become the dominant hardeners for microelectronics packaging because of their excellent moldability, electrical properties, and heat and humidity resistance [21]. The chemical structures of these two hardeners are shown in Figure 3-17 and Figure 3-18.

**3.6.1.1.3. Accelerators** Accelerators are used in most epoxy molding compound formulations to reduce the in-mold cure time and thus improve productivity. The most common agents are amines, imidazoles, organophosphines, ureas and Lewis acids and their organic salts [21, 22, 23]. The important properties of an accelerating agent are its reactivity enhancement and its latency. Faster cure times should not be at the expense of shorter pot life. Amines provide excellent reactivity but provide poor pot life, relatively poor electrical

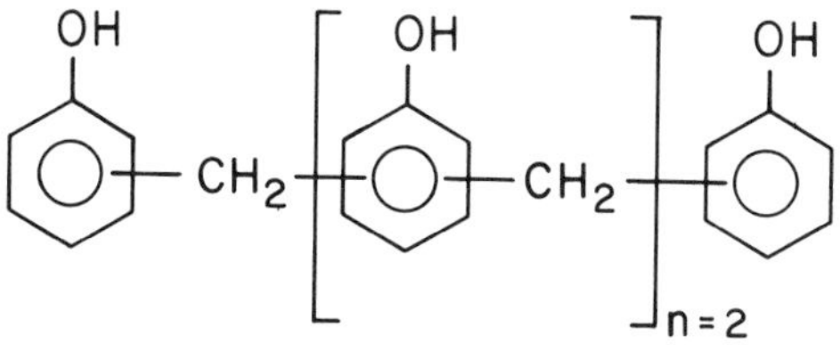

**FIGURE 3-17** The chemical structure of a phenol-novolac hardener.

properties and humidity resistance. Organophosphines provide excellent acceleration, excellent electrical and humidity resistance properties, but poor pot life. Lewis base salts provide moderate acceleration, excellent pot life, and excellent electrical properties and humidity resistance.

**3.6.1.1.4. Fillers and Coupling Agents** Fillers are used to lower the coefficient of thermal expansion and improve the thermal conductivity. They also improve mechanical properties such as the modulus and prevent excessive resin bleed at the parting line of the molding tool. Other criteria of an inorganic filler include good electrical and mechanical properties, chemical stability, heat and moisture resistance, and low alpha particle activity. Yet another important consideration is low abrasiveness. Fillers such as alumina, magnesium oxide, and silicon nitride have significantly higher thermal conductivities than silica, but they are many times more abrasive. Mold wear would be accelerated to the point of impracticality.

Ground, fused silica is the most widely used filler in epoxy molding compounds [24] because it possesses the optimum combination of properties. In particular, it has low enough abrasiveness to provide a reasonable molding tool lifetime. Crystalline silica is used in most applications where high heat dissipation is required. This will become a more important problem in IC packaging as the power generated per chip continues to increase and approach levels that are not easily dissipated through conventional molding compounds.

Most silica filler particles are irregularly shaped flakes. The particle size and size distribution are important considerations in the material compounding. Typical filler sizes are in the range of 10 to 25 microns, but the distribution can be wide with some particles reaching 125 microns (5 mils). Exceptionally large

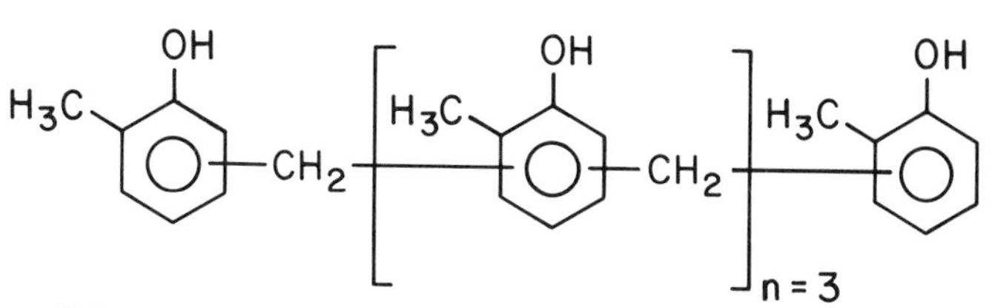

**FIGURE 3-18** The chemical structure of a cresol-novolac hardener.

particles are to be avoided because they can become trapped in the gates which can be as thin as 150 microns (6 mils) or caught between the leads of high pincount packages where the interlead spacing can be as little as 100 microns (4 mils). Gate clogging by filler particles can be suspected if incomplete filling of cavities located randomly over the molding tool is detected. Similarly, trapping filler particles between the leads will result in inordinate lead deformation that will not respond to process parameter changes. Another potential problem with filler particles is that under thermomechanical shrinkage stresses, they can be pressed into the device features, possibly resulting in metallization and/or passivation layer cracks. An illustration of this is shown in Figure 3-19.

Spherical silica particles are also used as inorganic filler in epoxy molding compounds. Most formulations that have spherical particles use them in tandem with flakes to increase the filler loading, thereby further lowering the coefficient of thermal expansion. This is particularly important in the lower stress resins that are formulated to provide very low thermal shrinkage forces. Maximum filler loadings have reached over 75 by weight through the use of these different particle shapes and size distributions. Fillers increase the melt viscosity of the compound, as was discussed in Section 2.4.5. For low filler loadings, the increase of viscosity can be predicted by Einstein's relation provided in Equation 2-12, but for the levels of fillers found in most microelectronic molding compounds, typically 70% by weight (50% by volume), the two parameter Mooney Equation 2-13, which accounts for particle interactions, is more appropriate.

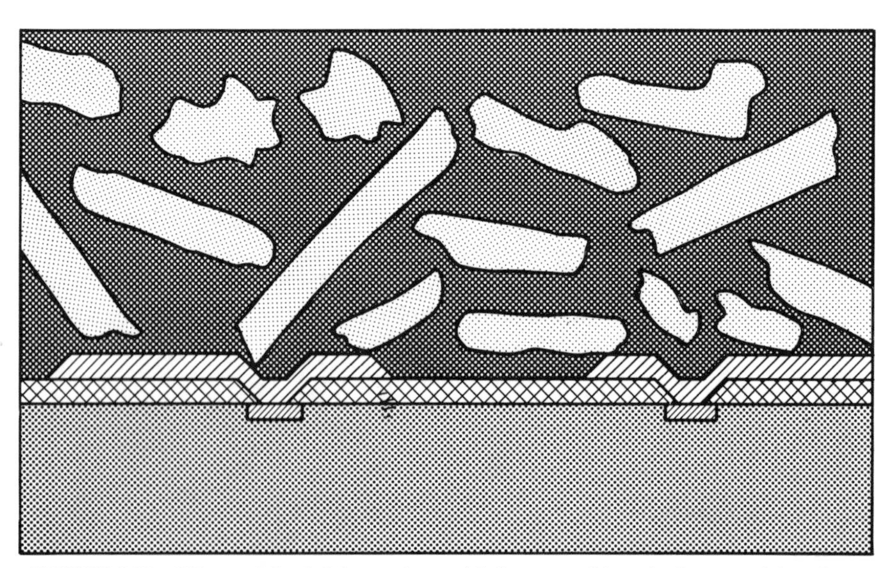

**FIGURE 3-19** Filler particles lodging against and being pressed into the features of the microelectronic device. This drawing is scaled to show the relative sizes of filler particles and device features.

Coupling agents are used on the fillers to promote interfacial adhesion between matrix polymer and inorganic filler that results in improved mechanical properties. These materials have two reactive ends: one attaches to the filler particles while the other reacts with the epoxy. Coupling agents improve humidity and heat resistance because they bind tightly to the silica interface where these effects are influenced. In the same way, they can actually improve adhesion to the device and leadframe, also reducing moisture and high temperature sensitivity at these interfaces. The surface modification of the filler also affects the rheology and flow of the molding compound in ways that are not well understood and therefore difficult to predict. The addition of a coupling agent can either increase or decrease the viscosity of a molding compound, depending on the agent used and the thickness of the coating that is applied. Silanes, titanates, aluminum chelates, and zircoaluminates are all known coupling agents for molding compounds. Of these, silanes are the most common. Figure 3-20 shows a proposed structure of a silanol group reacting with hydroxyls on the filler surface.

**3.6.1.1.5. Flame Retardants** Flame retardancy is now standard for epoxy molding compounds used for IC packaging. Epoxy materials are not inherently flame retardant, so additives must be incorporated into the material for this purpose. There are two principal approaches that are in use. These are the addition of brominated epoxies such as brominated DGEBA epoxy, and the use of antimony trioxide additives, Many formulations use both brominated epoxies and the antimony additives as there appears to be a synergistic effect. The brominated resin is produced through the reaction of brominated bisphenol-A and epichlorohydrin, as shown in Figure 3-21. A typical formulation will have approximately 10% by weight of the brominated resin to 90% of the novolac epoxy. The major concern with the use of the brominated materials is that bromides are serious corrosive agents for microelectronic circuitry, similar to chlorides. Although the bromines attached to the aromatic rings are not mobile or hydrolyzable, there is only limited stability to these covalent bonds which can

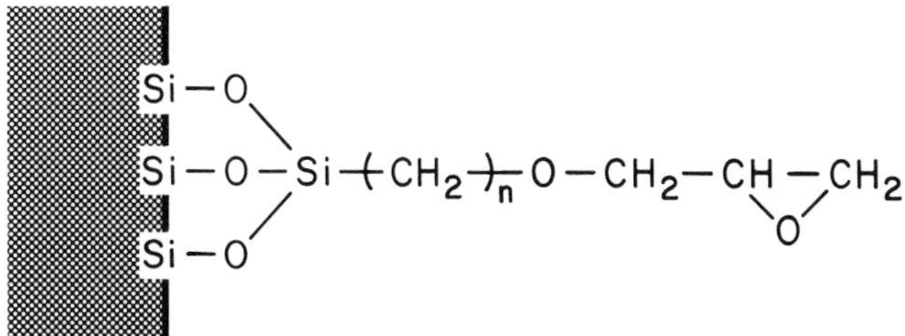

**FIGURE 3-20** A silanol coupling agent attached to the filler surface providing a pendant epoxy group for reaction with the epoxy molding compound.

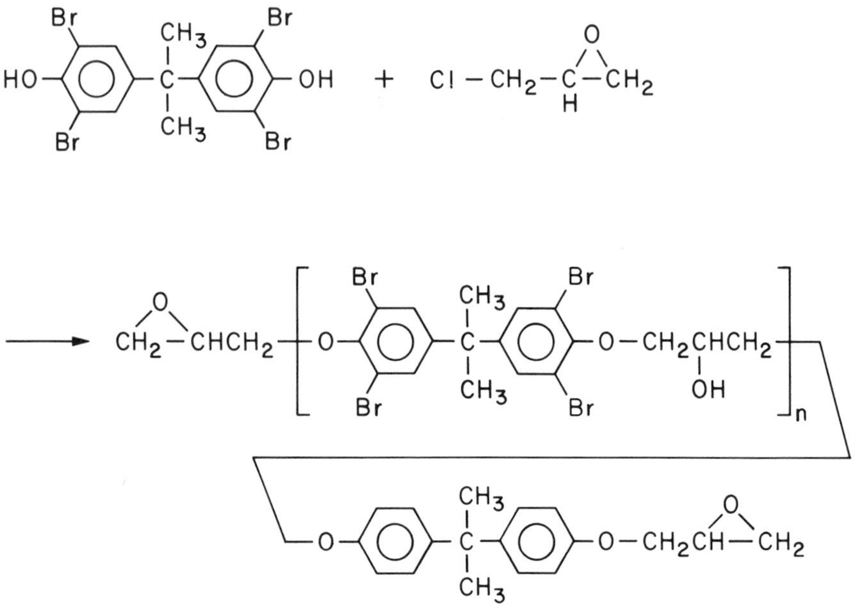

FIGURE 3-21 The precursors and chemical structure of the brominated epoxy used to confer flame retardancy on microelectronic grade epoxy molding compounds.

lead to formation of the highly corrosive bromine ion. Antimony additives are also likely to be corrosive if they are broken down.

**3.6.1.1.6. Stress-Relief Additives** Epoxy materials are inherently brittle due to their high crosslink density which confers the high-temperature performance and solvent resistance that is desirable, but which also decreases polymer chain mobility and relaxation response. Flexibilizing or stress relief agents are added to epoxy molding compounds to inhibit crack propagation as well as to lower the thermomechanical shrinkage stresses that can initiate cracks in the molding compound or in the passivation layer of the device. In general, these materials are used to lower the elastic modulus, improve toughness, and improve flexibility. They may also lower the coefficient of thermal expansion in some cases.

Stress relief agents can be unreactive or reactive. Unreactive (inert) plasticizers include phthalic acid ester or chlorinated biphenyl. Many different types of materials can be used as reactive flexibilizers because of the high reactivity of the epoxy group. If the added agent remains in the epoxy matrix as a single-phase material, then it is more properly termed a *flexibilizer*. Such materials improve the ductility of the epoxy and also lower the tensile modulus. Flexibilizers based on interpenetrating networks are becoming an important aspect of molding compound modification. Conversely, if the material is segregated in a second dispersed phase and there is little or no effect on the epoxy matrix properties, then it is a *stress relief agent*. There are two different types of stress relief agents: those that are initially compatible with the epoxy matrix material and phase separate during cure [25], and those that are already formed into domains and then mechanically dispersed within the epoxy prior to cure [26].

Stress relief agents promote several beneficial effects. They reduce the tensile modulus of the two-phase system by decreasing the cross-sectional area of the hard (epoxy) phase, and they are effective crack propagation inhibitors. A crack propagating through the matrix phase may be stopped or branched upon reaching a rubbery domain. There are several theories on the mechanism of this toughening behavior. Most concern the added energy dissipation mechanisms that are introduced by the presence of the dispersed rubbery phase [27]. McGarry and coworkers [28, 29] attribute the toughening to a perturbation of the stress field in the vicinity of the elastomeric domains, which promotes crazing and microcavitation, both effective energy absorption mechanisms. The continued development of these reactive stress relief agents has been an important technology driver in epoxy molding compounds for IC packaging.

Silicones, acrylonitrile-butadiene rubbers, and poly(butyl acrylate) (PBA) are the major stress relief agents used in epoxy molding compounds. The butadiene rubbers do not have good high-temperature stability so they are not as popular as the silicones and PBA. Silicones have excellent purity and high-temperature properties, but they are relatively incompatible with epoxies, resulting in relatively weak interfacial areas if used without proper modification. The morphology of the segregated phase, the domain size, size distribution, and the integrity of the interfacial region play important roles in the mechanical property improvements that are achieved. Early attempts at elastomer modification of epoxy molding compounds used domain sizes of 5–10 microns, whereas more recent attempts show that domains below 1 micron in diameter are more effective [26]. The domain particles are not formed in situ but are produced separately in an emulsion polymerization; they show fairly uniform size distribution. Elastomer-modified epoxy molding compounds can often provide significant improvements in passivation layer cracking, aluminum line deformation, and package cracking [30].

One issue in elastomer modification of epoxy molding compounds is that the large difference in modulus between the epoxy and the soft elastomer provides a relatively weak interfacial region. Interfacial integrity has been improved through the use of an interaction layer with a modulus intermediate between the epoxy and the modifier. Poly(methylmethacrylate) (PMMA) has been used as the interlayer shell over PBA core particles [30, 31] as depicted in Figure 3-22. These particles are prepared through a seeded emulsion polymerization and then dispersed in the epoxy resin during the compounding stage. The PMMA coating allows the PBA particles ($T_g$ of $-54°C$) to be dried and handled at room tem-

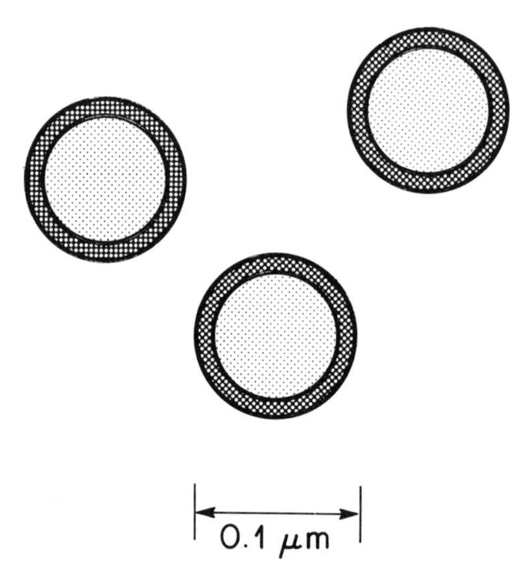


**FIGURE 3-22** Illustration of the shell-core morphology of elastomer-modified epoxy molding compound. (After Reference 30.)

perature. The particles were found to be well dispersed in the epoxy, and the PMMA appeared to have dissolved in the matrix forming an effective interpenetrating network at the interface [26]. Crosslinks in the PBA core were found to reduce the thermal shrinkage of the epoxy, whereas crosslinks in the PMMA shell improved the interaction with the epoxy matrix [31].

**3.6.1.1.7. Coloring Agents** The plastic packages of microelectronic devices are colored for a number of reasons, but the primary one is aesthetic as the filled molding compound without color additives is a translucent yellow. Most plastic packages for silicon integrated circuits are black. Carbon black, the primary coloring agent for black molding compound, is made by the incomplete combustion of hydrocarbons; hence sulfur- and nitrogen-containing impurities will also be present. Also, carbon black is electrically conductive and could cause shorting problems if too high a level were used, or if local high concentrations were created by improper mixing during compounding. Carbon black also degrades the moisture resistance of the molding compound. For these reasons, the level of carbon black additive and the dispersion in the epoxy should both be carefully controlled. The level of carbon black in semiconductor-grade molding compound is typically below 0.5%.

Other colors for epoxy molding compounds are used for different types of devices. Opto-isolators are white to minimize the absorption losses of radiant energy to the packaging material. Red, yellow, blue, and green molding compounds are also used for special applications. These brighter colors are obtained with organic dyes that are not as thermally stable as the inorganic additives, so they could lower the high-temperature properties of the base resin. In any case, there could be a color change with the long-term high-temperature exposure.

**3.6.1.1.8. Mold Release Agents** Epoxy resins are generally excellent adhesives providing good adhesion to the leadframe and device that helps to minimize moisture ingress and corrosion at these interfaces. These same adhesive properties make release from the compounding and molding equipment difficult. For this reason, mold release agents have to be carefully selected and implemented to assure ready release from the process equipment while retaining good adhesion to the components of the package. Often, this is achieved through the softening and functioning temperature of the release agent. The materials are compounded at approximately 100°C and then molded at 175°C. Often, two different release agents are used to provide release at each temperature. Adhesion at room temperature could then be achieved with a release agent that solidifies and loses effectiveness at these lower temperatures.

The common mold release agents for epoxy molding compounds are silicones, hydrocarbon waxes, metallic salts of organic acids, and fluorocarbons. Of these, the waxes are the most common for molding compounds used for microelectronic encapsulation. Materials such as carnauba wax, the hard, high-melting-temperature wax used in some car and furniture waxes, are regularly used. Silicone materials such as silicone oil are not selective enough, since they are soft and even fluid throughout the entire temperature range. They are also notorious for exuding and creeping over long periods of time, which could cause problems at electrical contact points. Organic and metallic salts are corrosive agents.

**3.6.1.2. The Evolution of Epoxy Molding Compounds** The earliest materials used for plastic packaging of microelectronic devices were silicones because of their high-temperature performance and high purity. Epoxy materials offer much better adhesion than silicones, but the common bisphenol-A (BPA) epoxies have lower glass transition temperatures. Also, epoxies can have high ionic impurity levels because their synthesis uses an excess of halogen-containing epichlorohydrin. Hybrid encapsulant materials consisting of epoxies and silicones captured a large segment of the molding compound market in the mid-1970s because they combined some of the high-temperature performance and high purity of silicones with the mechanical properties, adhesion, and solvent resistance of epoxies. In epoxy molding compounds, novolac epoxies displaced BPA materials because of their higher functionality and attendant improvements in heat distortion temperature. Molding compounds have since then moved steadily toward all-epoxy systems as the high-temperature novolac materials were improved and the hydrolyzable impurity levels were driven below 50 ppM. This value does not include bromide in bromobenzene groups used as flame retardant, which is generally resistant to hydrolysis.

*Low Stress Materials* A major advance in epoxy molding compounds was the development of low-stress materials in the early 1980s. Package cracking, passivation layer cracking, and circuit pattern deformation can all result from the disparities in the coefficients of thermal expansion among chip, plastic package, and metal leadframe. This problem was exacerbated as the surface area of memory devices became a significant fraction of the package itself with 64K and 256K DRAMs. The plastic edge thickness became thin and cracking began to occur. The low-stress molding compounds combine several approaches to alleviate this problem. Fillers with different size and shape distributions were used to increase the filler loading to over 72% by weight (50% by volume), thereby reducing the coefficient of thermal expansion by as much as 25% ($21 \times 10^{-6}$ to $16 \times 10^{-6}$ cm/cm-°C). The second advance was the addition of elastomer modifiers in a dispersed domain morphology to improve toughness. Research on the size, size distribution, and interfacial region was a major contributor to the improved strength of these materials.

*Low-Alpha-Emitting Materials* Alpha particles emitted from the silica fillers can cause soft errors in memory devices, and the development of the low-alpha-particle-emitting molding compounds was a major improvement. Although these materials were available from the late 1970s, the silica was from the few known mines with low uranium and thorium contamination. Supplies were scarce and prices were much higher than conventional materials. The introduction in the early 1980s of silica made in the laboratory by chemical vapor deposition (CVD) reduced the price of low-alpha-emitting materials by more than half by 1986. These materials quickly captured the market for memory devices, essentially eliminating the use of conformal coatings in designs where they were used for alpha particle protection.

*Processing Improvements* Processing improvements were also required to keep pace with the increasing demands of higher-pin-count packages. The 40-pin DIP of the early 1980s was one of the first designs to experience flow-induced stress problems. The long cantilever leads and the close proximity of the wire bonds made the package much more susceptible to lead movement, paddle shift, and wire sweep. The forces are proportional to the product of the material viscosity and its velocity. The breakthrough was the development of materials that had significantly lower viscosities than previous generation materials. The introduction of fine-pitch packages and thinner leadframes has continued to push the need for very-low-viscosity molding compounds. Comparisons with published reports [32] show that the viscosity at the molding temperature has been reduced from 1500 poise in 1980 to 300 poise (at 100 $sec^{-1}$) in 1988, while the filler loading and crack resistance of the material have both been increased. At the same time, the cure time for epoxy molding compounds has been reduced from 3–4 minutes in the mid-1970s to 60–90 seconds in the late 1980s, contributing an important productivity increase. A summary of these advances is provided in Figure 3-23 [21].

## 3.7 LEADFRAME MATERIALS

The choice of a leadframe material has important implications on the degree of adhesion of the molding compound to the frame, the thermomechanical shrinkage stresses, the trim and form characteristics, the buckling of the leadframe after molding, and the heat dissipation capacity of the package. All of these characteristics have direct implications for packaging yield and reliability. Several different metallic alloys share the market for leadframes used in plastic packaging. Alloys 194, 155, and 151 are predominantly copper and therefore have a relatively high CTE. (See Table 3-4 for leadframe alloy compositions.)

| | 1970 — 1980 — 1990 |
|---|---|
| DRAM | 16K, 64K, 256K, 1M, 4M |
| LOGIC (# GATES)(CMOS) | 0.1K, 1K, 10K, 100K |
| | FLAME RETARDED |
| | LOW IONIZABLE CHLORINE |
| | LOW STRESS |
| | LOW α- EMISSION |
| | ANHYDRIDE CURE |
| | AMINE CURE NOVOLAC |
| | STRESS MODIFICATION |
| | SYNTHETIC $SiO_2$ FILLER |
| THERMAL STRESS (MPa) | 12, 8, 5, (<2) |
| THERMAL EXPANSION ($x10^{-6}/°K$) | 25, 22, 19, 17, (<14) |
| IONIZABLE CHLORINE (ppm) | 1000, 100, (<100) |
| U/Th CONTENT (ppb) | 100, 10, (<1) |
| FLAME RETARDANCY | UL-94 HB, UL-94 V—O |
| CURE TIME (SECONDS) | 140, 120, 90, 60 |
| MELT VISCOSITY (POISE) (MIN.) | 2000, 800, 500, 300 |
| MOISTURE SENSITIVITY (REL.) | 1, 5, 10, 30 |
| TEMPERATURE CYCLING (REL.) | 1, 10, 100, 500 |

**FIGURE 3-23** The evolution in the properties of epoxy molding compounds, and some immediate needs. (After Reference 21.)

**Table 3-4. Compositions of Leadframe Alloys in Plastic Packaging**

| | |
|---|---|
| *Alloy 194* | |
| Copper | 97.5% |
| Iron | 2.35 |
| Phosphorous | 0.003 |
| Zinc | 0.12 |
| *Alloy 151* | |
| Copper | 99.9 |
| Zirconium | 0.1 |
| *CDA 155* | |
| Copper | 97.8 |
| Silver | 0.034 |
| Phosphorous | 0.058 |
| Magnesium | 0.11 |
| *Alloy 42* | |
| Iron | 58% |
| Nickel | 42 |
| *KLF 125* | |
| Copper | 94.55 |
| Nickel | 3.2 |
| Tin | 1.25 |
| Silicon | 0.7 |

The thermomechanical shrinkage stresses will be higher with these alloys than with others because of the larger disparity between the CTEs of the silicon die and the leadframe. In addition, the adhesion of molding compound to the copper alloys is not as good as it is to other alloys. The principal advantage of the copper-rich alloys is a much higher thermal conductivity. (See Table 2-2 for thermal conductivities of leadframe materials.) Heat conduction along the leadframe is an important heat dissipation mechanism in molded plastic packages, and the better thermal conductivity of the copper-rich alloys makes them a good choice for high-power devices.

Alloy 42 is another common leadframe material that is popular because of its low coefficient of thermal expansion. This alloy contains iron and nickel with just a trace amount of other materials. It has a significantly lower CTE than the copper-rich alloys, providing lower thermal shrinkage stresses. It has better adhesion to epoxy molding compounds, probably because it does not form a passivating oxide film as easily as the copper alloys. Its greater rigidity confers better resistance to damage during assembly and molding operations, and less buckling under the compressive loads applied by the molded body. Its principal drawback is lower thermal conductivity. Packages using Alloy 42 leadframes have less power dissipation capacity than similar package designs using copper alloy leadframes. For this reason, Alloy 42 is being replaced by the copper alloys as higher heat dissipation becomes a more serious consideration. Newer leadframe materials such as KLF 125 (available from other suppliers with different alloy designations) are gaining acceptance because of their high strength advantages in fine-pitch applications where a thinner leadframe is used. Through-hole packages such as DIPs also require high-strength leadframe materials to resist damage upon insertion through the circuit board.

## REFERENCES

1. A. M. Wilson, "Use of Polyimides in VLSI Fabrication," in *Polyimides: Synthesis, Characterization and Applications*, Volume 2, K.L. Mittal, Editor, Plenum Press, New York (1984).
2. M. Tomono, A. Abe, S. Harada, K. Sato, T. Takagi, Y. Oya, and A. Saiki, "Discrete Semiconductor Device Having Polymer Resin as Insulator and Method for Making Same," U.S. Patent 4,017,886, (April 12, 1977).
3. N. Yoda and H. Hiramoto, "New Photosensitive High Temperature Polymers for Electronic Applications," *J. Macromolecular. Sci. Chem.*, **A21** (13–14), 1641–1663 (1984).
4. F. Katoaka, F. Shoji, I. Obara, H. Yokono, and T. Isogai, "Characteristics of Highly Photoreactive Polyimide Precursor," Extended Abstracts of First Technical Conference on Polyimide, New York, (Nov. 1982).
5. O. Rohde, M. Reideker, A. Schaffner, and J. Bateman, "Recent Advances in Photoimageable Polyimides," *Advances in Resist Technology and Processing II*, SPIE Volume 539, 75 (1985)
6. M. Ahne, H. Kruger, E. Pammer, and R. Rubner, "Polyimide Patterns Made Directly From Photopolymers," in *Polyimides: Synthesis, Characterization and Applications*, Volume 2, K. L. Mittal, Editor, Plenum Press, New York (1984).
7. T. Takeda and A. Tokoh, "Low-Stress Polyimide Resin for IC," Sumitomo Plastics Bulletin, Sumitomo Bakelite Company, Ltd. Basic Research Laboratory, Yokohama, Kanagawa 245 Japan.
8. A. Berger, "Modified Polyimides by Silicon Block Incorporation," in *Polyimides: Synthesis, Characterization and Applications*, Volume 1, K. L. Mittal, Editor, 67, Plenum Press, New York (1984).
9. M. Fryd, "Structure - Tg Relationship in Polyimides," in *Polyimides: Synthesis, Characterization and Applications*, Volume 1, K. L. Mittal, Editor, 377, Plenum Press, New York (1984).
10. A. Gregoritsch, *Reliability Physics*, 14th Annual Proceedings, 228 (1976).
11. K. Mukai, A. Saiki, K. Yamanaka, S. Harada, and S. Shoji, *IEEE J. Solid State Circuits*, **SC-13** (4), 462 (1978).
12. G. Brown, *Reliability Physics*, 19th Annual Proceedings, p. 282 (1981).
13. G. Samuelson and S. Lytle, "Reliability of Polyimide in Semiconductor Devices," in *Polyimides: Synthesis, Characterization and Applications*, Volume 2, K. L. Mittal, Editor, 751, Plenum Press, New York (1984).
14. W. T. Shugg, *Handbook of Electrical and Electronic Insulating Materials*, Van Nostrand Reinhold, New York (1986).
15. M. L. White, J. W. Serpeillo, K. M. Striny, and W. Rosenzweig, "The Use of Silicone RTV Rubber for Alpha Particle Protection on Silicon Integrated Circuits," *IEEE Proceedings*, IRPS, 43 (1981).
16. T. Gabrykewicz, D. Sengupta, T. Thuruthumaly, and L. Frazee, *Proceedings of the 1986 Intl. Symp. on Microelectronics*, 707-13, Intl. Soc. Hybrid Microelectronics (1986).

17. C. E. T. White and J. Slattery, "An Update on Preforms," *Circuits Manufacturing*, 78 (March 1978).
18. C. A. Steidel, "Assembly Techniques and Packaging," in *VLSI Technology*, S. M. Sze, Editor, McGraw Hill, New York (1983).
19. K. M. Kearny, "Trends in Die Bonding Materials," *Semiconductor International*, 84 (June 1988).
20. Peter J. Planting, "An Approach for Evaluating Epoxy Adhesives for Use in Hybrid Microelectronics," *IEEE Trans. Parts, Hybrids, Packag.*, **11**, 305 (1975).
21. N. Kinjo, M. Ogata, K. Nishi, and A. Kaneda, "Epoxy Molding Compounds as Encapsulation Materials for Microelectronic Devices," in *Advances in Polymer Science 88*, Springer-Verlag, Berlin, Heidelberg (1989).
22. W. C. Mih, American Chemical Society Symposium Series No. 242, p. 273 (1984).
23. J. D. B. Smith, *J. Appl. Polym. Sci*, **26**, 979 (1981).
24. O. Nakagawa, I. Sasaki, H. Hamamura, and T. Banjo, *Electronic Materials*, **13**, 231 (1984).
25. L. T. Manzione, J. K. Gillham, and C. A. McPherson, "Rubber Modified Epoxies, Transitions and Morphology," *J. Appl. Poly Sci.*, **26**, 889 (1981).
26. Y. Nakamura, H. Tabata, H. Suzuki, K. Iko, M. Okubo, and T. Matsumoto, *J. Appl. Polym. Sci.*, **32**, 4865 (1986).
27. L. T. Manzione, J. K. Gillham, and C. A. McPherson, "Rubber Modified Epoxies, Morphology and Mechanical Properties," *J. Appl. Polym. Sci.*, **26**, 907 (1981).
28. J. N. Sultan and F. J. McGarry, M. I. T. Research Report R69-59 (October 1969).
29. F. J. McGarry, *Proc. Royal Soc. London*, **A319**, 58 (1970).
30. Y. Nakamura, S. Uenishi, T. Kunishi, K. Miki, H. Tabata, K. Kuwada, H. Suzuki, and T. Matsumoto, "New Profile of Ultra Low Stress Resin Encapsulants for Large Chip Semiconductor Devices," *Proc. of the 37th Electronic Components Conference* (1987).
31. Y. Nakamura, H. Tabata, H. Suzuki, K. Iko, M. Okubo and T. Matsumoto, "Internal Stress of Epoxy Resin Modified with Acrylic Core-Shell Particles Containing Functional Groups Prepared by Emulsion Polymerization,' *J. Appl. Polym. Sci.*, **33**, 885 (1987).
32. L. L. Blyler, H. E. Blair, P. Hubbauer, S. Matsuoka, D. S. Pearson, G. W. Poelzing, and R. C. Progelhof, "A New Approach to Capillary Viscometry of Thermoset Transfer Molding Compounds," *Polym. Eng. Sci.* **26**(20), 1399 (1986).

# 4 Testing and Evaluation of Thermoset Molding Compounds

The productivity and reliability of a plastic molded package will be influenced more by the thermoset molding compound than by any other material. This chapter reviews the analysis, testing, and characterization that can be done to assure that optimum molding compounds are selected.

A considerable amount of technical information can be obtained from the material supplier's bulletins. This information source often includes recommendations on the process conditions, the cure time, the flame retardancy rating, and the flow behavior. There is also much information on the physical properties of the molding compound such as the coefficient of thermal expansion, the mechanical modulus and mechanical strength. This information is generally accurate and reliable, and in most cases there is no need to reproduce the tests to confirm the values. The major concern in basing material selection decisions on this information alone is not with regard to the accuracy of the information, but rather the type and usefulness of the information supplied in choosing among materials that may show little if any difference in these key specification parameters. The intent of this chapter is to describe how to make the best use of the suppliers' information, and then to go beyond it to implement tests that are more discriminating in choosing among molding compounds. The chapter is divided into segments of suppliers' information, productivity analysis, and reliability analysis.

## 4.1 USING THE MATERIAL SUPPLIERS' INFORMATION

A large amount of information is available from the material supplier. When used knowledgeably and prudently, it can significantly reduce the amount of effort that has to be expended in selecting materials. Information provided by essentially all material suppliers includes a listing of physical properties and reliability performance data such as is shown in Table 4-1. These properties are described in detail in the following sections.

### 4.1.1 Coefficient of Thermal Expansion

The coefficient of thermal expansion (CTE) is an important parameter that is measured and reported by all suppliers. The technique of measurement is usu-

Table 4.1. Typical Specification of Molding Compound Properties

| | |
|---|---|
| Coefficient of thermal expansion | $16 \times 10^{-6}$°C$^{-1}$ |
| Glass transition temperature | 155°C |
| Flexural modulus | 1400 kg/mm$^2$ |
| Flexural strength | 15 kg/mm$^2$ |
| Thermal stress | 0.45 kg/mm$^2$ |
| Thermal conductivity | $16 \times 10^{-4}$ cal/cm-sec°C |
| Volume resistivity | $7 \times 10^{16}$ ohm-cm |
| Hydrolyzable ionics | <20 ppM |
| Uranium content | 0.4 ppB |
| Pressure cooker test | 3000 hr (median life) |
| Pressure cooker test (bias) | 500 hr (median life) |
| Thermal shock | 2000 cycles (median life) |
| Spiral flow length | 30 inches (76 cm) |
| Gel time | 23 seconds |
| Hot hardness | 85 (Shore D) |

ally thermal mechanical analysis, known as TMA. In TMA, a small block of the fully cured molding compound is positioned in a temperature chamber and the end of a quartz rod is rested on the sample. The movement of the rod is then precisely measured with a magnetic field transducer as the sample grows in size with increasing temperature. The coefficient of thermal expansion is the resulting slope of the plot of displacement versus temperature shown in Figure 4-1. The measurement of CTE is specified in the American Society of Testing Materials (ASTM) test procedure D 696 [1]. Standardized test procedures specific to the semiconductor industry have been published by Semiconductor Equipment and Materials International (SEMI) [2]. The SEMI standard for measuring CTE of molding compounds is G 13-82. Two distinct slopes are obtained for epoxy molding compounds, a lesser slope below the glass transition temperature denoted $\alpha_1$, and a greater slope above $T_g$ denoted $\alpha_2$. The value most often reported is the lower $\alpha_1$ number. This test is easy to perform if the equipment is available, and it is one of the parameters that the customer may want to verify, if only because of its importance to package reliability and the ease of testing. The rate of heating during measurement is an important parameter of the test because it can shift observed transition temperatures, yet this information is not always clearly stated when the CTE is reported. Nonetheless, the reported CTE can be confidently compared with values from other suppliers.

### 4.1.2 Glass Transition Temperature

Polymeric materials can exist in several different physical states that control their thermomechanical response. Crystalline and amorphous states differentiate the ordering of the polymer chains. Polymer crystals are similar to crystals of

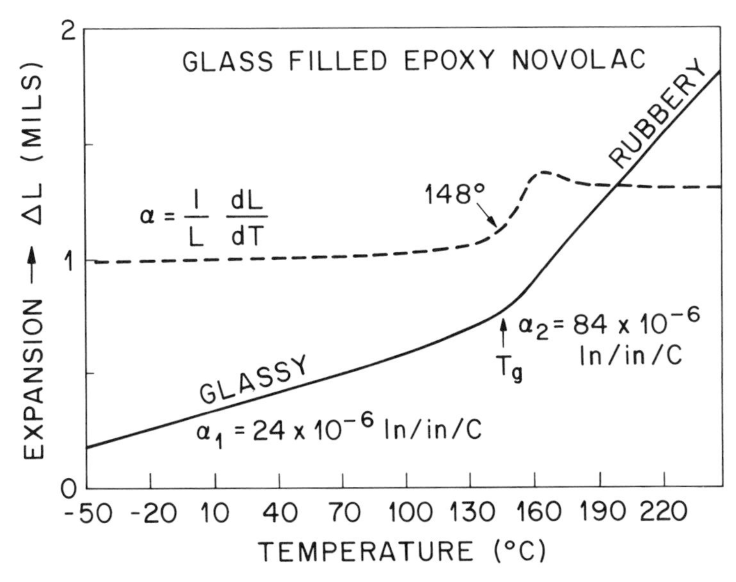


**FIGURE 4-1** A plot of expansion versus temperature for an epoxy molding compound. The dashed line is the temperature derivative of expansion, which is defined as the coefficient of thermal expansion. Note the higher CTE above the glass transition temperature.

other materials in that the chain atoms are arranged in a highly ordered array. Amorphous polymers can be either glassy or rubbery depending on the temperature. The glass transition temperature, denoted $T_g$, separates the glassy and rubbery regions of material response. It is usually reported in any listing of material properties. $T_g$ is actually a manifestation of the total viscoelastic response of a polymer material to an applied strain. The transition point therefore depends on the rate of strain, the degree of strain, and the heating rate. Slower heating rates and slow deformation rates both translate to lower apparent transition temperatures. More important than the transition temperature alone is the total mechanical response, which will be discussed in Section 4.3.1. The transition temperature reported in most suppliers' literature is determined through TMA analysis by noting the break in the displacement versus temperature plot. Some instruments can take the derivative of the displacement curve, highlighting the discontinuity, and making it easier to assign the transition temperature as is shown in Figure 4-1. In general, the glass transition temperature can be confidently compared with values reported for other molding compounds since the differences due to testing technique will not be great. One important reason to perform a check of this parameter is to assure that your own process conditions are correct. Improper molding and post-cure conditions could lower the $T_g$ below the value reported in the data sheet. The SEMI standardized test for

$T_g$ is G 13-82, the same for CTE since it uses the same instrument and test procedure to measure CTE and $T_g$ from volume expansion data.

### 4.1.3 Flexural and Tensile Properties

The flexural properties of molding compounds including the flexural strength and flexural modulus are derived from a standardized ASTM test (ASTM D 790-71, D 732-85) that is both reliable and can be compared to reported values from other materials and material suppliers. The tensile modulus, tensile strength, and percent elongation are also reported by some vendors. These are derived from ASTM D 638 and D 2990-77. Section 4.3.1 describes these mechanical tests in greater detail.

### 4.1.4 Thermal Stress

This parameter is usually derived from a calculation or through an experimental measurement. The equation used is typically the stress calculation shown below:

$$\sigma = \int_{T}^{T_g} \alpha_1(T) E_p(T)\, dT \qquad (4\text{-}1)$$

It is important to realize that Equation (4-1) is actually one of several simplifications of a more complicated expression that includes the modulus of the silicon and its expansion coefficient. See Section 8.2.1 for a more thorough discussion of Equation (4-1). Also, different limits of integration may be used by different suppliers, and the integration may be extended into the rubbery region with the inclusion of $\alpha_2$ and $E_p$ in the rubbery region. Therefore, it is effectively impossible to compare this calculated thermal stress parameter with values from different suppliers. In addition, comparing it to the strength of the material to predict package cracking is useless. The stress parameter is only a crude approximation of the stress level in the material and it does not account for stress concentration points or other geometric and interfacial features that influence cracking.

In some cases, the thermal stress parameter may be determined through a photoelastic experiment. The most common experiment is the ASTM F-100 test in which the molding compound is molded around a glass cylinder. The molding compound experiences thermal shrinkage in cooling from the molding temperature down to the test temperature, stressing the inner glass cylinder, shown in Figure 4-2. The deformation of the glass is then read by counting the fringe pattern produced under polarized light through crossed polarizers. The specifics of the experiment would have to be fully disclosed to enable the comparison of these data with those from other suppliers.

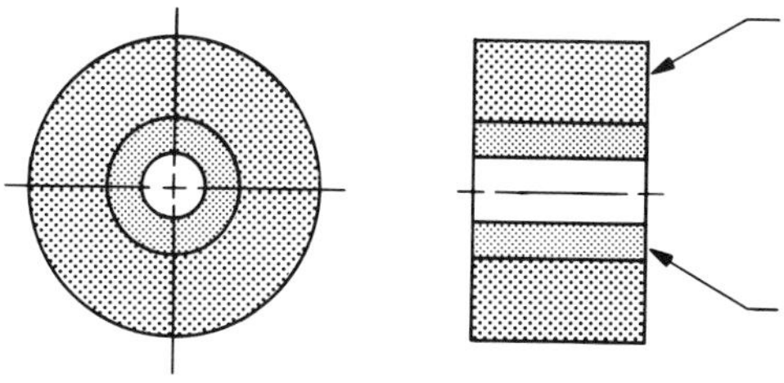

OUTSIDE CYLINDER- MOLDING COMPOUND

INSIDE CYLINDER (GLASS)

**FIGURE 4-2** Specimen for experimental measurement of thermal stress from ASTM F-100. (Copyright ASTM. Reprinted with permission.)

### 4.1.5 Thermal Conductivity

The thermal conductivity is determined experimentally through commercial instruments known as thermal conductance testers. The conductivity of a material is a difficult parameter to measure, and different instruments can provide results that are significantly different. It should be noted that there will probably be greater error with this parameter than with most others reported in the material data sheet.

### 4.1.6 Volume Resistivity

This is a straightforward electrical test (ASTM D 3320-82). Reported values are reliable and can be compared to information from other vendors.

### 4.1.7 Hydrolyzable Ionics

The aluminum conductors and bonding pads on the device will corrode under either acidic or basic conditions. Although aluminum oxide passivates the surface and inhibits further corrosion, the oxide itself can be damaged by ionics and expose the aluminum to corrosion. Aluminum lines under the passivation layer of the device are fairly well protected from ionic corrosion. The gold ball bond-aluminum bond pad interface is particularly vulnerable in the presence of water, however. It is not protected by the passivation layer, and the two metals plus the electrolyte provide the essentials of an electrochemical cell leading to rapid corrosive failure of the ball bond. It is for this reason that ionic impurities must be driven to very low levels to ensure reliability. The principle source of ionic impurities in molding compounds is the epoxy resin itself since it is made from a chlorine-containing precursor, epichlorohydrin. Other ionics can be introduced by decomposition of the flame retardant agents as described in Section 3.6.1.1.5. Early epoxy molding compounds had ionic impurity levels near 1000 ppM. Careful refining of resin synthesis and purification have resulted in reducing hydrolyzable ionics to well below 100 ppM and below 20 ppM in many instances. Clearly, lower hydrolyzable ionic impurity concentrations are highly desirable and are an important selection criteria for molding compounds. The only caution in comparing suppliers' data for this parameter is to assure that the extraction conditions are similar. Milder extraction conditions such as lower extraction temperatures, larger particle sizes, shorter extraction times and less agitation contribute to lower reported impurity levels.

### 4.1.8 Uranium Content

Soft errors can be induced in memory devices from alpha particles emanating from the radioactive impurities in the molding compound and other components of the package [3]. Alpha particles have very low permeability in solids, so the problem is most severe where the molding compound is in direct contact with active surface of the device. The contribution of the molding compound can be quantified through its uranium content. It is also expressed as alpha particle flux, since both uranium and thorium contaminants contribute alpha particles. The principal source of the radiation is silica filler particles obtained from natural mines. Low-alpha molding compounds use synthetic silica, which has essentially no uranium and thorium contaminants.

### 4.1.9 Pressure Cooker Tests

The pressure cooker is a reliability test reported by many material suppliers. The conditions of the test are usually saturated steam at two atmospheres pressure, corresponding to a temperature of 121°C. The devices are subjected to these conditions until failure is induced. Other conditions are also used by different vendors, and in general the conditions for most reliability tests have become harsher as improvements in molding compounds and passivation layers increase the number of hours without failure to long periods that become impractical. An important parameter in using this information is the type of integrated circuit used for the test. Some circuits are inherently more prone to corrosion failures induced by pressure cooker tests than others. Typically, a vendor will use the same device they have used in evaluating their previous generation molding compounds so that they can show the improvements of new grades. Different vendors will use different devices making it impossible to make a sensible comparison of these reliability data.

### 4.1.10 Temperature Cycling

This is another common reliability test of molding compound performance. In temperature cycling, the packaged device is repeatedly cycled from low to high temperature with periodic inspections for package cracking, device cracking or device failure. Many purchasers of IC products require the devices to pass the U.S. military specification (Mil. Std. 38510 Group D) for temperature cycling between −65 and +135°C. The temperature endpoints, the rate of temperature change and the type of heat transfer are important characteristics of this test. In general, higher rates of temperature change and high convective heat transfer coefficients, as by liquid immersion, are more severe test conditions that will decrease the number of cycles to failure. The temperature endpoints of the cycling test are becoming more severe as improvements in molding compounds increase the number of cycles without failure to several thousand. Several suppliers now report that temperature cycle limits of liquid nitrogen (−196°C) to molten solder (+215°C) are needed to limit the test to a reasonable time period.

### 4.1.11 High Temperature, Humidity, Bias (HTHB)

HTHB is an older reliability test where the devices are subjected to elevated temperatures and high humidity, usually under bias, for long periods of time to induce failure. Typically these conditions are 85°C and 85% relative humidity, commonly known as 85/85 testing. The reliability is measured by the number of hours of exposure required to produce failure in a specified fraction of the initial population, typically 50%, which is then called the half-life. This test is not as popular as it once was simply because it is not severe enough to discriminate among the improved molding compounds. Excessively long half-lives can be achieved rendering the test tedious and inconclusive. Similar to the pressure cooker test discussed above, it is difficult to compare HTHB data from different vendors who use different integrated circuit devices and test conditions.

### 4.1.12 Spiral Flow Length

The spiral flow length is a standardized test of thermoset molding compounds that is extensively used in the semiconductor industry and reported by all material vendors. It is not only used to compare different materials, but also as a quality control tool written into the specifications for molding compounds. The test is based on ASTM D 3123 or SEMI G 11-88 which specify the design of the molding tool used to measure the flow length, as in Figure 4-3. The test consists of loading the molding compound into the heated transfer pot of the press and transferring it through a spiral coil of semicircular cross section until the flow ceases. It is actually a measure of three phenomena: fusion under pressure, melt viscosity, and gelation rate. The pressure and temperature for the test are recommended in the test procedure; for ASTM D 3123 they are 1000 psi on the material (6.90 MPa) and 150°C for the mold temperature, but these

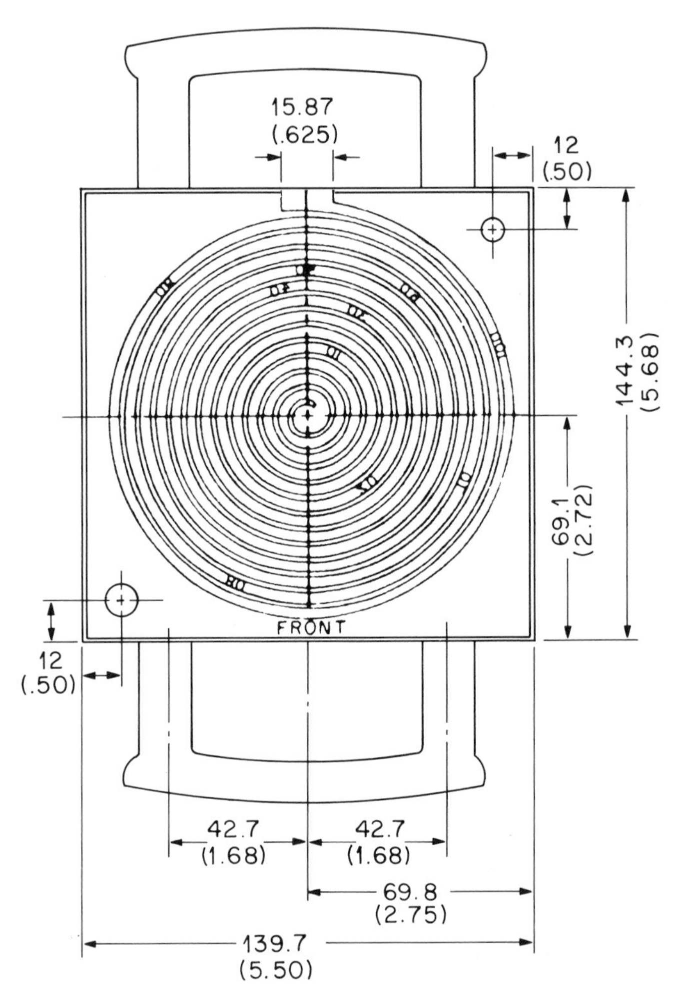


**FIGURE 4-3** Molding tool used for the spiral flow length test from ASTM D 3123. (Copyright ASTM. Reprinted with permission.)

conditions are altered in many cases. SEMI Standard G11-88 recommends using the specified mold temperature for the material, typically 170–175°C. The ASTM procedure specifies no preheating of the molding compound charge, whereas the preforms are preheated in most implementations of this test in the semiconductor industry. The rate of transfer is also specified in the test procedure, but it is actually a dependent variable of the pressure applied to the material.

The shortcoming of the spiral flow test is that it is a one condition measurement which is unable to resolve the viscous and kinetic contributions to the flow length. In essence, two materials can show the same spiral flow length even though one has twice the viscosity and twice the gel time as the other. The longer flow time allowed by the slower-curing material offsets the slower rate of flow caused by the higher viscosity. One remedy to this problem is to include a "ram-follower" device on the transfer press used to conduct the spiral flow studies. The ram-follower provides a flow profile which is a trace of ram displacement versus time. The flow profile for the two materials described above would be completely different with the longer curing material showing half the slope and twice the flow time, even though the flow length into the spiral coil would be the same. A drawing of this device and its output are shown in Figure 4-4. An improved quality control criteria would be based on specified tolerances on spiral flow length and flow profile, rather than on flow length alone. The ram follower is specified in SEMI G 11-88 for the spiral flow measurement.

### 4.1.13 Gel Time

The gel time is often reported by material suppliers as a way of comparing the expected productivity of a molding compound. In general, shorter gel times

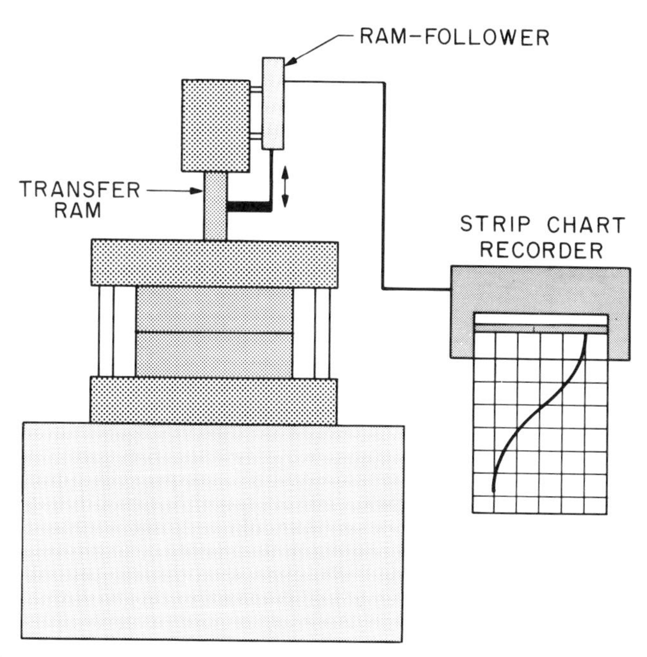


**FIGURE 4-4** Ram-follower device attached to a transfer molding press and its function and output in a spiral flow test.

mean faster polymerization rates and shorter in-mold cure times. The disadvantage of shorter gel times is shorter flow times for filling the mold which translate to higher velocities and higher flow-induced stress damage to the devices. The gel time of a thermoset molding compound is usually measured with a gel plate, although SEMI standard G11-88 recommends using the spiral flow test as a means to evaluate gel time. The gel plate approach uses a simple hot plate with precise temperature control and measuring apparatus. A small amount of the molding compound powder, about a few grams, is applied to the hot plate which has been set to the prescribed temperature, usually 170°C for molding compounds. The specimen softens on the hot plate to a thick fluid consistency. With a wooden stick or other suitable instrument, the specimen is probed periodically to determine if it has gelled. Gelation is noted by a significant increase in the viscosity of the material and an inability to smear it into a thin coating. This endpoint is subjective, hence the procedure should be repeated at least three times to arrive at a representative average. Reported gel times are accurate to no more than 10% for these highly filled molding compounds.

## 4.2 PRODUCTIVITY EVALUATION

There are a number of productivity parameters that can be assessed from the thermoset molding compound without an extensive molding trial in the factory and statistical analysis of the results. These specific parameters can be measured in carefully controlled experiments in a relatively short time, thereby allowing the survey of several molding compound candidates. The one or two materials that then appear as the strongest candidates from the laboratory studies can be used in more extensive production molding trials.

Yield of the plastic molding operation can be assessed through the application of several standard techniques. The basis of this assessment is the realization that the yield and productivity losses in molding are caused by several different effects. The primary cause is flow-induced stress. There are also other molding problems that can be assessed by evaluation of the molding compound such as the propensity for incomplete fill of the molding tool, commonly known as a *short shot*. A short shot essentially ruins all of the devices in the mold cavity because partially filled leadframes cannot be trimmed and formed or molded again, and even cavities that appear to have filled have not been properly packed. Packing is the application of high pressure after all the cavities have filled. Also, problems such as resin bleed, which is molding compound contamination of the leadframe outside of the molded body, can cause downstream yield losses at the trim and form and solder dipping stations, not to mention the extra time and expense of adjusting the deflashing apparatus to remove excessive resin. The following sections describe the fundamentals and the techniques for evaluating molding compounds.

### 4.2.1 Assessing Flow-Induced Stress Potential

Flow-induced stresses are generated by the viscous molding compound flowing along and through the leadframe and wire bonds of the device. When a fluid flows in this way, pressure and viscous drag forces are generated on all of the wetted surfaces. In addition, there are also surface tension forces generated as the free surface of the molding compound contacts lead fingers and wire bonds creating new surface area and causing surface forces. An earlier estimation found that these surface tension forces are small compared to the pressure and viscous drag forces [4].

The pressure forces generated during flow over the device can be scaled through the equation that relates the pressure driving the flow to the viscosity of the fluid and its volumetric flow rate. For non-Newtonian materials in flow between parallel plates, this relation is [5]:

$$Q = \frac{nWB^2}{2(1 + 2n)} \left\{ \frac{B \Delta P}{2KL} \right\}^{1/n} \tag{4-2}$$

In this equation, the magnitude of the viscosity is incorporated in the coefficient $K$, whereas its shear rate dependence is in the power law index $n$. $B$ is the thickness, and $L$ is the length over which the pressure drop is experienced. Equation (4-2) is for infinite parallel plates with the result provided on a basis of per unit width $W$. The volumetric flow rate $Q$ can easily be converted to the velocity through the cavity, $U$, by dividing by the cross-sectional area of the flow. Rearranging Equation (4-2) to determine the dependence of pressure on viscosity and velocity yields:

$$\Delta P = 2KLQ^n \left\{ \frac{nW}{2(1 + 2n)} \right\}^{1/n} B^{2/n-1} \tag{4-3}$$

This indicates that the pressure forces generated by the flow are proportional to the product of the viscosity multiplied by the velocity raised to the power-law index of the material as shown in Equation (4-4). The power-law indices of epoxy molding compounds are usually in the range of 0.3–0.8. For scaling purposes it suffices to use 0.5 as a representative value:

$$\text{Flow-Induced Pressure Forces} \propto \eta U^n \approx \eta U^{.5} \tag{4-4}$$

The dependence of the viscous drag forces on the viscosity and velocity is more difficult to determine, since the drag coefficient of a power law fluid is not explicitly defined in terms of the viscosity and velocity, but is instead dependent on the specific geometry and the degree of non-Newtonian behavior. For Newtonian materials, however, the drag coefficient for the very slow flow, known as *creeping flow*, experienced in the packaging mold cavity is defined as [6]:

$$C_D = \frac{8\pi}{Re(2.002 - \ln Re)} \tag{4-5}$$

where the Reynolds Number includes both the viscosity and velocity:

$$Re = \frac{\rho UL}{\eta} \tag{4-6}$$

The drag force is directly proportional to the drag coefficient and the square of the velocity:

$$F_D = \frac{1}{2} C_D \rho U^2 A_w \tag{4-7}$$

In this equation, $A_w$ is the surface area of the wire bond or lead finger that the drag force is being exerted upon. From these relations, the dependence of drag force for Newtonian materials can be extracted:

$$\text{Viscous Drag Force} \propto \frac{\eta U}{\ln \frac{U}{\eta}} \tag{4-8}$$

Although the dependence on viscosity and velocity is not as straightforward for viscous drag force as it is for pressure, and the dependencies for non-Newtonian fluids are still more complicated, it is apparent that the drag force again scales to the product of the viscosity and velocity with greater dependence on the viscosity. The velocity is much more process dependent than material dependent, although some fast-cure materials will require faster fill times. The conclusion with regard to material selection is that molding compounds with lower viscosity at the conditions at which they flow over the chips will produce lower flow-induced forces.

**4.2.1.1 Rheometry** The use of rheological measurements is an effective way to predict a molding compound's potential for flow-induced stresses. A viscometric measurement determines a viscosity at a specified shear rate. The molding compound is a non-Newtonian material whose viscosity is shear rate dependent so the use of a viscometric technique is required to properly characterize these materials. Before applying these techniques, it is necessary to determine the shear rates that the molding compound experiences during different phases of the molding process. Shear rates are calculated through the velocity gradient. In most cases, the no-slip boundary condition at a solid surface can be applied, although these highly filled molding compounds probably exhibit some wall slip. The velocity gradient is then computed as the difference between the maximum velocity minus the zero velocity at the wall divided by the distance between the maximum velocity position, usually the centerline of the flow, and the wall:

$$\frac{dv}{dz} = \frac{U_{\text{max}} - U_{\text{wall}}}{\Delta z} \tag{4-9}$$

Figure 4-5 shows the shear rates calculated in different sections of a molding tool for a conventional mold design and standard molding conditions. These shear rates are not applicable to all molding tools as the exact shear rates depend on the specific runner, gate and cavity dimensions, but they do indicate the general ranges: hundreds of reciprocal seconds in the runner, thousands through the gate, and tens in the cavity.

Rheological tests can also be used to determine the temperature sensitivity of the molding compound viscosity. As discussed in Section 2.4.4, there are two accepted ways of characterizing this temperature behavior: the WLF relation or an Arrhenius flow activation energy. The flow activation energy form will be used for this discussion, yielding a viscosity relation which includes both the shear thinning and temperature sensitivity parameters in the form shown below. In Equation (4-10) only the initial viscosity is considered.

$$\eta = \eta_{\text{inf}} \exp\left(\frac{E_\eta}{RT}\right) \dot{\gamma}^{n-1} \tag{4-10}$$

The viscosity also has a time dependence which is discussed later in this section.

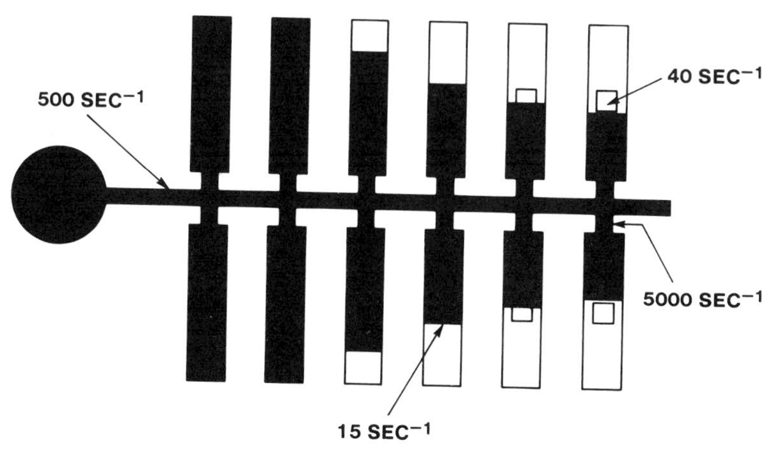


**FIGURE 4-5** The shear rates in reciprocal seconds calculated for the different sections of a packaging mold. Shear rates are specific to the actual feature geometries which in this case are a 0.45 × .50 cm runner, 0.025 × 0.70 cm gates, and a 1.3 × 0.45 × 5.0 cm cavity. The fill time is 17 seconds.

*Rheological Selection Criteria.* Shear Dependence of Viscosity. Selection criteria should be based on the specific problems expected. For fragile devices where wire sweep and/or paddle shift are concerns, the test program should try to identify molding compounds that have the lowest viscosity at the low shear rates and high temperatures encountered in the cavities. Likewise, in molds where complete filling is a concern as it would be in a 300-cavity mold or the mold for a very large package, the test program should seek molding compounds that have lower viscosity at the very high shear rates encountered through the gates which largely determine the fill character of a mold. The pressure drop in the runner is rarely an important consideration in package molding, hence the viscosity at shear rates of hundreds of reciprocal seconds is probably the least important. Different instruments may be required to characterize the viscosity over this wide range of shear rates.

Temperature Dependence of Viscosity The temperature sensitivity of the molding compound is also a yield influencing parameter. There is no clear preference of whether high or low flow activation energy is desirable because it depends on the particular mold geometry. In general, a molding tool that combines both long and short flow lengths leading to the cavities, such as is shown in Figure 4-6, would not perform well with a molding compound that had a high flow activation energy, that is, a material whose viscosity is highly sensitive to temperature. Cavities off the medium to long flow lengths would see molding compound that had normal high temperatures and lower viscosities, whereas cavities off the short flow lengths would see material that had relatively low temperature and much higher viscosities due to the high flow activation energy. A material whose viscosity was more temperature insensitive would probably perform better in this suboptimal tool design. Nearly all molding tools, other than multiplunger molding, will have different flow lengths to the cavities providing different molding compound temperatures in different parts of the mold. It is, therefore, important to assess the temperature dependence of the molding compound viscosity to determine the differences in viscosity that can occur due to these flow length and temperature differences.

Time Dependence The third selection criterion is the time dependence of the viscosity. This effect was discussed in detail in Section 2.4.2. The molecular weight and hence the viscosity will increase with time. In transfer molding, the viscosity decrease with increasing temperature will overwhelm the small increase in molecular weight induced viscosity increase at the early stages of filling, but ultimately the molecular weight and the viscosity go to infinity at gelation. The time-dependent viscosity increase can influence flow-induced stresses at the later stages of mold filling and especially in cavities more distant from the material source (see Figure 4-6). The selection criteria for time de-

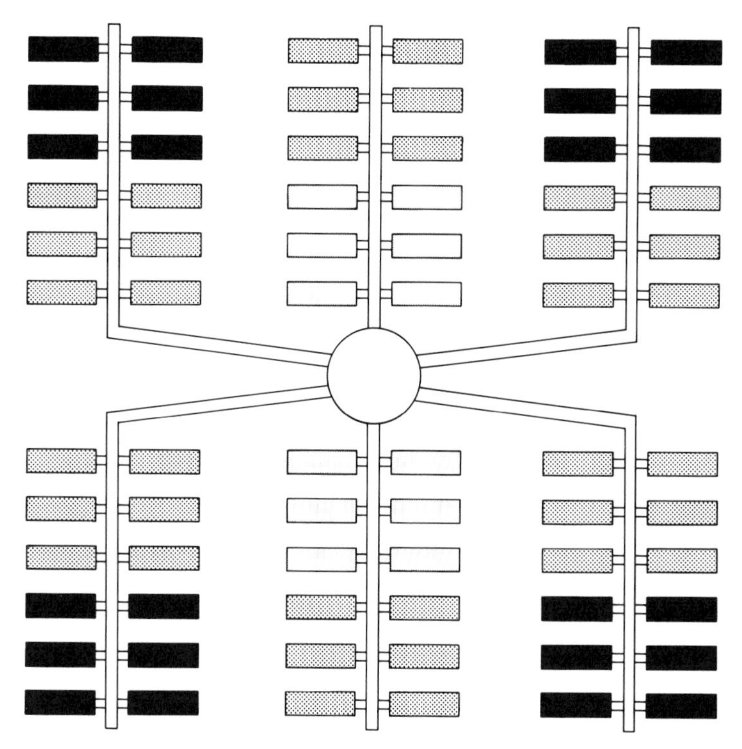

**FIGURE 4-6** Cavity and runner layout of a package molding tool that has significantly different runner lengths to the cavities. Those cavities with the light shading will show relatively low temperatures and high viscosities, those with medium shading would be expected to show moderate temperatures and moderate viscosities, those with the dark shading would show high temperatures and high viscosities due to advanced cure.

pendent viscosity are clear: molds with longer flow lengths and longer flow times should use molding compounds with longer gel times at the relevant mold temperatures. The nominal average temperature of filling of 150–160°C is a good range for these gel time measurements. The chemical kinetics of the polymerization reaction are highly temperature dependent, so one could be misled in that a molding compound with a shorter gel time at 175°C could actually have a longer gel time at 150°C. Molding compounds with shorter gel times can provide higher productivity through shorter cure times in most instances, but there must be sufficient flow time to fill the mold at moderate velocities. The reaction kinetics are also concentration dependent so that a material could show a shorter time to reach the conversion to gel, typically less than 25% conversion, but a longer time to reach the conversion to eject the parts, typically 80-90% conversion. For this reason, a thorough kinetic study should be made to conduct this level of cure time analysis.

Rheological Test Methods The following paragraphs describe the rheological test methods that can be used to evaluate materials according to the criteria established above.

Spiral Flow Length The spiral flow length test described in Section 4.1.12 is one of the techniques that can be used to evaluate viscosity, although the test is not viscometric, meaning it is not a measure of the viscosity. In general, materials with longer spiral flow lengths have lower viscosity, but it is important to bear in mind that the molding compound is a non-Newtonian material whose viscosity is shear rate dependent. The shear rate of the typical spiral flow test is several hundred reciprocal seconds, hence it covers the shear rate range of the runners. This shear rate region probably has the least impact on yield and productivity. Materials that have lower viscosity at medium shear rates may or may not have lower viscosity at higher or lower shear rates. For these reasons, the spiral flow test is of limited value in assessing yield loss potential due to flow included stresses.

Spiral flow testing does have some value in assessing the gel time or maximum flow time of thermoset molding compounds, although the results have to be used with caution since it is only a one point test. Longer spiral flow lengths are preferable in molds with longer flow lengths and flow times.

Capillary Rheometer The theory of the capillary rheometer was discussed in Section 2.4.6. Although the flow field, shear rate, and shear stress are not uniform, it is acceptable to compute point values of shear rate and shear stress as representative parameters (See Equation 2-15). Characterization of epoxy molding compounds for plastic packaging using a capillary rheometer has been reported [7]. The shear rate ranges for this instrument, shown in Figure 4-7, were 200–1200 $sec^{-1}$. This range covers the shear rates encountered in the runners, but misses both the lower shear rate ranges of molding compound flowing in the cavities, and the much higher shear rates of molding compound flowing through the gates. Capillary instruments that could reach the lower shear rate ranges of 10–50 $sec^{-1}$ experienced in the cavities are required to evaluate flow-induced stress potential. There are some commercial capillary instruments that have this capability, but the user should be certain that the extrusion die has sufficient length over diameter ratio to minimize end effects otherwise the extensional component of the flow will be enhanced, obscuring the shear viscosity which will ultimately damage the devices. Conversely, capillary instruments with $L/D$ ratios of more than $5/1$ show predominantly shear viscosity and are of limited value in assessing the extensional contributions to the pressure drop through the gate. For this high deformation rate, shorter capillaries may be more representative of the flow field and the rheology.

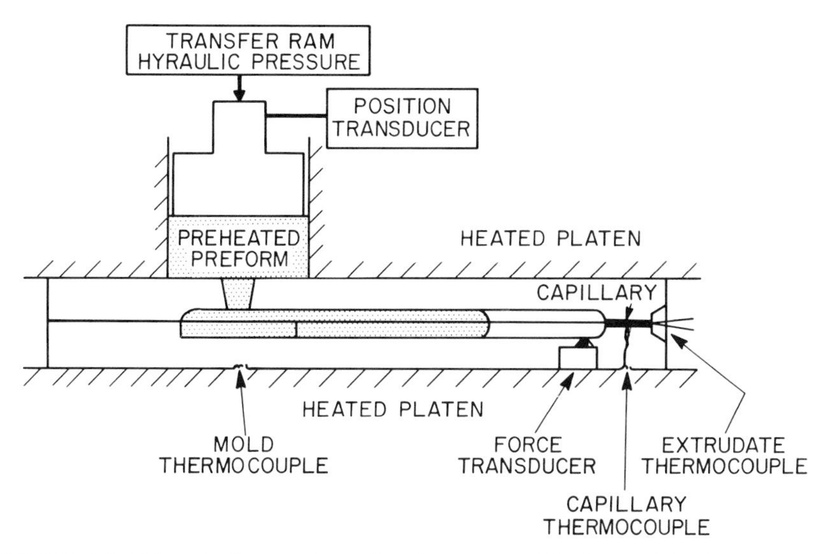


**FIGURE 4-7** Schematic diagram of a capillary rheometer used for characterizing thermoset molding compounds [7].

Cone-and-Plate Rheometry The theory and practice of cone and plate rheometry were described in detail in Section 2.4.6. Although this viscometric method is theoretically capable of covering the entire shear rate range encountered in plastic package molding, practical considerations such as slippage of the highly filled molded compound on the polished surfaces often limit the technique to low shear rates below 50 $sec^{-1}$. This range, however, is well suited to assess the potential of flow-induced stress damage to the devices. In practice, molding compounds can be tested with a cone and plate apparatus either in powder form or by pressing the powder into preforms as thick as a coin. The plates should be preheated and maintained at a temperature near the temperature at which they would contact the leadframes. For typical mold temperatures of 175°C and initial preheat temperatures of 95°C, the contact temperature is approximately 150–160°C. Lower temperatures can often be encountered with shorter flow lengths ($<8$ cm), and temperatures approaching the mold temperature can also be found at the end of long flow lengths ($>40$ cm). Shear rate sweeps over the entire range of rates up to the point of slipping should be conducted to assess the shear rate dependency of the molding compound, even though a power-law index cannot be accurately estimated from this limited shear rate range. A temperature sweep at several shear rates should also be conducted to estimate the temperature dependence of the viscosity. A separate sample and run is required for each temperature and shear rate data point because the rapid curing of the molding compound at these elevated temperatures precludes changing the shear rate and temperature during a run.

**4.2.1.2 Molding Trial Evaluation of Flow-Induced Stresses** Flow induced stresses can be effectively evaluated in a molding trial using wire bonded devices. The material candidates do not need to be processed at identical process conditions, but instead should be processed at their individual optimum conditions. X-Ray analysis of the molded packages and cross sectioning of the molded bodies through the paddle support are the primary methods of evaluation for these trials. It is often preferable to exaggerate the effect of flow-induced stresses by using long wire bond spans that are more susceptible to flow-induced deformation. This can be accomplished by using a small silicon device on an oversized paddle. Wire bond spans greater than 0.100 inch are known to enhance flow-induced deformation and can be used to create a worst case scenario. There has been a report on characterizing the degree of wire sweep based on the observed deformation of the wire bond loop [8]. Similarly, the paddle support with the greatest area can be used to enhance the displacement caused by pressure imbalances of the flow fronts moving over the chip and paddle. These and other features are illustrated in Figure 4-8.

Careful interpretation of the results is necessary to draw any firm conclusions in discriminating among the material candidates subjected to such molding trials. Wire sweep can be held to undetectable levels in many instances, so one should not accept small degrees of deformation as normal. The greater numbers and the greater inconsistencies of production manufacturing virtually assure that production parts will have more wire sweep problems than what was expected from a limited experiment.

### 4.2.2 Assessing Molding Compound Production Rate

The productivity of plastic package molding depends on the rate of chemical conversion of the molding compound to the fully cured plastic of the final pack-

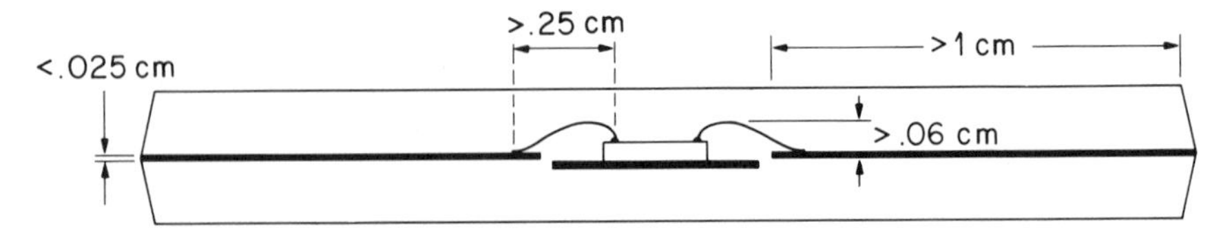


**FIGURE 4-8** Cross section of an integrated circuit device with features intended to enhance the effects of flow-induced stress for the purpose of a molding trial. These features include a long lead span, a thin leadframe, a small chip attached to an oversized paddle support to provide extra long wire bond spans, and a high wire loop height.

age. The rate of this conversion will vary with different molding compounds, and as such it is one of the parameters that could be used as a selection criteria. The cure time required before the parts can be ejected from the mold varies over a range of 1 minute to almost 4 minutes. This range is clearly enough to make a significant difference in production rate since the cure time consumes approximately 70% of the overall cycle time. There are, however, other considerations that prevent one from selecting the molding compound that has the shortest recommended cure time. The most important of these is the flow time. Molding compounds that have shorter cure times will generally have shorter flow times prior to gelation. Materials with exceptionally short cure times of less than 1.5 minutes are usually formulated for special equipment such as multiplunger machines which have very short flow lengths, requiring flow times under 10 seconds. They are typically incapable of flowing for the 18 seconds required to fill most conventional packaging molds. One of the most important needs in the plastic packaging industry is a molding compound that could flow for 20–30 seconds, but then cure to an ejectable conversion is less than one additional minute. Understanding the chemical kinetics of polymerization and how they depend on temperature and degree of conversion can help in selecting materials that can provide higher production rates.

**4.2.2.1 Polymerization Rate Analysis** The rate of the conversion of the polymerization reaction is controlled by the chemical kinetics. A kinetic rate relation expresses the molecular dynamics of the reaction including the reactive species, the rate of their collisions, the effectiveness of these collisions in causing reaction, and the temperature dependencies of these processes. The order of the reaction is the number of chemical species that must collide to effect the reaction. As the concentration of the reactive species is depleted due to consumption, the reaction slows. Higher-order reactions will show a more severe drop off in reaction rate as the species are depleted. The novolac epoxy reaction is difficult to characterize with this type of molecular dynamic kinetics. It is not a single reaction but actually several competing reactions among three or four reactive species. Also, the chain segments that form are complicated and difficult to predict. For these reasons, empirically derived kinetic expressions where both the form of the expression and the rate constants are fit from experimental data are usually preferred. Several different empirical forms have been offered to fit conversion data for the epoxy molding compounds used for IC packaging. One of the more popular forms is [9, 10, 11]:

$$\frac{dX}{dt} = (k_1 + k_2X^m)(1 - X)^n \qquad (4\text{-}11)$$

This relation has four fitting parameters to express the conversion of epoxide groups, $X$, as a function of time during the reaction: $m$ and $n$ are pseudo-reaction orders, $k_1$ and $k_2$ are the rate constants which have associated temperature dependencies. For a model epoxy molding compound with cure characteristics similar to the materials used in IC packaging, Hale et al. [11] report the following parameters for Equation (4-11):

$$m = 3.33 \qquad n = 7.88$$

$$k_1 = \exp(12.672 - 7560/T)$$

$$k_2 = \exp(21.835 - 8659/T) \tag{4-12}$$

Figure 4-9 [12] is a plot of the conversion of epoxide groups as a function of time at four different isothermal cure temperatures. Note the dropoff in reaction rate, derived from the slope of the conversion versus time plot, as the extent of reaction approaches 100%. Figure 4-10 [12] shows a plot of the corresponding glass transition temperature as a function of degree of cure. This plot shows that there is very little difference in the $T_g$ for materials that arrived at the same degree of cure through different isothermal cure temperatures.

The conversion data required to develop a kinetic expression for actual molding compounds can be obtained through dynamic scanning calorimetry (DSC) measurements [13]. There are many other techniques for measuring the conversion rate data for epoxy systems including infrared spectroscopy (IR) and nuclear magnetic resonance (NMR), but these methods are largely inappropriate for opaque solids with higher filler loadings. The DSC technique works well with the filled molding compounds. It involves measuring the heat of reaction

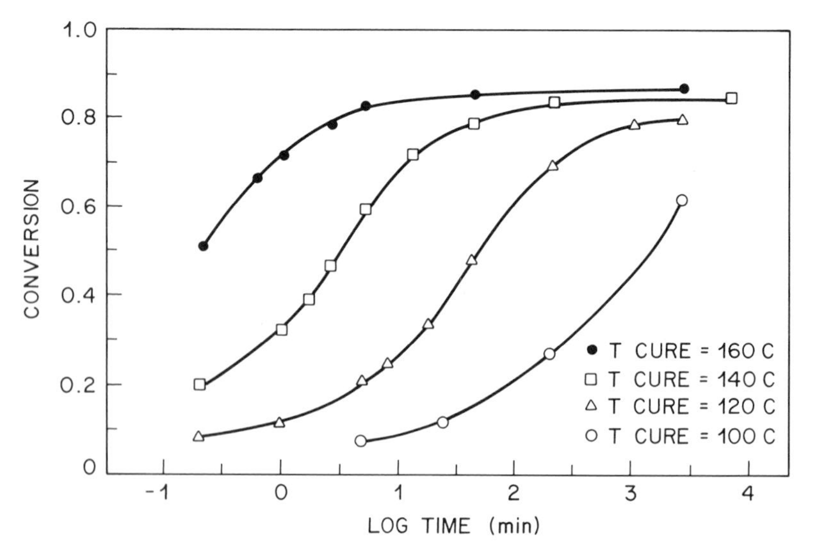


**FIGURE 4-9** A plot of conversion versus time for an epoxy molding compound during cure [12]. (Reprinted with Permission of Society of Plastics Engineers)

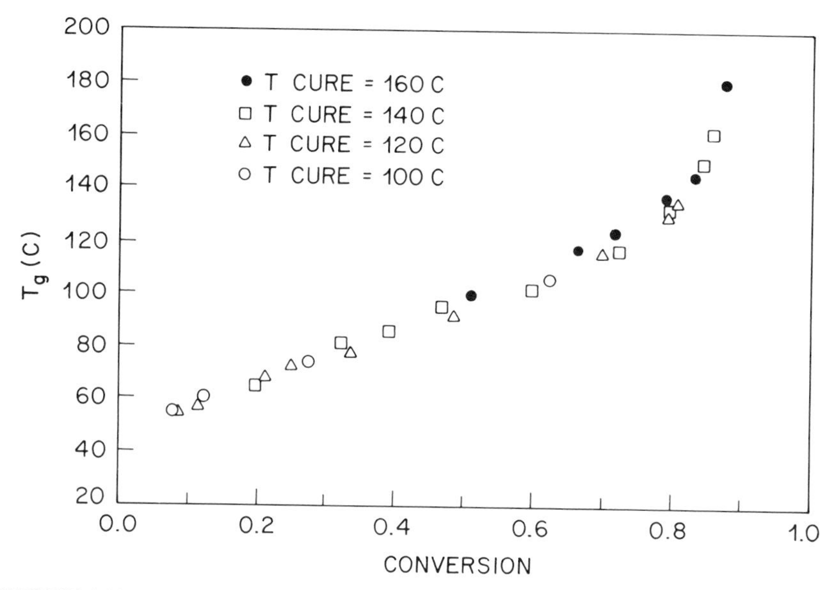


**FIGURE 4-10** A plot of glass transition temperature versus time for an epoxy molding compound during cure. Note the rapid increase in $T_g$ as the reaction approaches full chemical conversion [12]. (Reprinted with Permission of Society of Plastics Engineers.)

for complete conversion and then proceeding with the assumption that the fraction of the total heat liberated for an incomplete reaction corresponds to the same fraction of complete chemical conversion as expressed simply in Equation (4-13). Measuring the heat liberated during isothermal cure then provides the data on degree of conversion versus time which are needed to determine a kinetic rate expression for the molding compound:

$$\frac{\Delta H}{\Delta H_{rxn}} = \frac{X}{100\%} \tag{4-13}$$

Conducting a chemical kinetic analysis of the molding compound is fairly involved, requiring personnel who are skilled in the use of calorimetric equipment and have some understanding of chemical reaction engineering. An extensive analysis of kinetics is probably not needed for material selection since secondary effects of the cure kinetics, such as gel time, mechanical properties, and glass transition temperature can be used to effectively compare materials. More in-depth analysis of cure kinetics may be required for more sophisticated process modeling that will be discussed in Chapters 6 and 7.

**4.2.2.2 High-Temperature Hardness** High-temperature hardness, also known as *hot hardness* or *Green strength*, is the stiffness of the material at the end of the recommended cure cycle. This is a productivity-influencing parameter because this hardness may not be sufficient to eject the parts from a particular molding tool. If the molded body is too soft, the part can stick in the mold and deform the molded shape, effectively ruining the package and all other packages on the leadframe strip because the strip cannot be trimmed and formed. This ability to eject depends on both molding compound and mold characteristics. Mold characteristics include the draft angle of the vertical surfaces, the surface finish of the tool, and the number and size of ejector pins. Some materials are inherently harder at the time and temperature of ejection and will have less likelihood of experiencing this problem. Additional in-mold curing can resolve the problem in some instances by further advancing the cure and hardening the material. Other materials simply have low mechanical modulus above the glass transition temperature even as they approach 100% conversion. Most material suppliers can provide information on high temperature hardness. It is also worthwhile to check for hangup due to low hardness in a molding trial. The hot hardness can be determined in situ at a molding trial by testing for hardness within 10 seconds of opening the mold. Hardness values of approximately 80 on the Shore D scale are normal.

**4.2.2.3 Postcure** Nearly all epoxy molding compounds require a post-cure treatment to complete the polymerization reaction and bring it near 100% chemical conversion. Four hours at 170–175°C is typical for this treatment. Some molding compounds may specify shorter periods of postcure, but this would be unusual, whereas eight-hour postcures are not uncommon. There is actually little motivation for selecting a molding compound with a shorter postcure treatment because postcure is conducted batchwise in large cure ovens. Little capital equipment or labor is tied up in the procedure. The greatest improvement would be to eliminate postcure entirely. As discussed in Section 4.2.2.1, the polymerization kinetics provide for a rapid falloff in reaction rate at the later stages of conversion, making elimination of postcure unlikely.

### 4.2.3 Evaluating Molding Problem Potential

There are a large number of molding issues that affect package molding productivity and some of these also effect subsequent package reliability. Unlike wire sweep and paddle shift problems which directly lower yields, these concerns are indirectly related to molding productivity and yield. Most of these considerations are practical and many times intermittent. They manifest themselves primarily on the production floor, making them more difficult to identify in shorter qualification and preproduction runs.

*Inability to Fill the Mold* Commonly known as a short shot, the inability to consistently fill and pack the molding tool is an important problem in plastic package molding. Incomplete filling or packing in just a few cavities effectively ruins all of the devices loaded onto the tool because the full packing pressure is essential to develop the proper material density and minimize the inherent porosity of the molding compound that would lead to premature moisture-induced corrosion failures. A photomicrograph of the molding compound with and without packing pressure is shown in Figure 4-11. In that up to 300 silicon devices can be packaged in a single molding cycle, it is obvious that just a few short shots per shift can significantly reduce molding yields. Insufficient packing pressure can lower reliability of devices that appear to be properly molded. Incomplete filling is more common on molding tools with large numbers of cavities, greater than 150, and also on packages with large molded body volumes such as the large chip carriers and PQFPs. Quad packages are particularly prone to incomplete fill problems because the four-sided leads restrict them to corner gating on many molding tools, limiting the size of the gate and inhibiting the flow of material into the cavity.

*Resin Bleed and Flash* These are molding problems where the molding compound bleeds out of the cavity and onto the leadframe at the parting line of the mold. Resin bleed usually infers that the resin-rich part of the molding compound is the component that oozes through the parting line, straining out the filler particles which remain in the cavity, whereas flash is a more substantial escape of resin and filler at the parting line. These problems are as much processing and mold design related as material related since essentially all molding compounds will show flash under improper molding conditions. Imperfect parallelism between the two halves of the molding tool will lead to nonuniform clamping pressure and flash. Molding with a tool that is too large for the press, or turning the clamp pressure down below the minimum for the wetted, pressurized area causes low clamping force per unit area and subsequent flash. Resin bleed is thought to be more material-dependent, since some materials do exhibit this problem more than others. It probably correlates to the viscosity of the resin and the size of the filler particles; materials with low resin viscosity and larger filler particles would be expected to have the more serious resin bleed. Excessive packing pressure, particularly at the time soon after the cavity fills, can cause resin bleed, as well as improper clamping pressure on the molding tool in the press. Typically, lower viscosity molding compounds and those with long spiral flow lengths will have greater potential for resin bleed and flash, so the advantages of these materials in reducing flow-induced stresses must be weighed against their potential for this molding problem.

There is a standardized test that can be applied for assessing a material's potential for resin bleed and flash: SEMI G 45-88. This is a transfer mold experiment based on a molding tool. The tool has one or more thin rectangular flow channels that extend from a common source of material. Materials that provide the greatest flow length into the thinnest sections are likely to show the

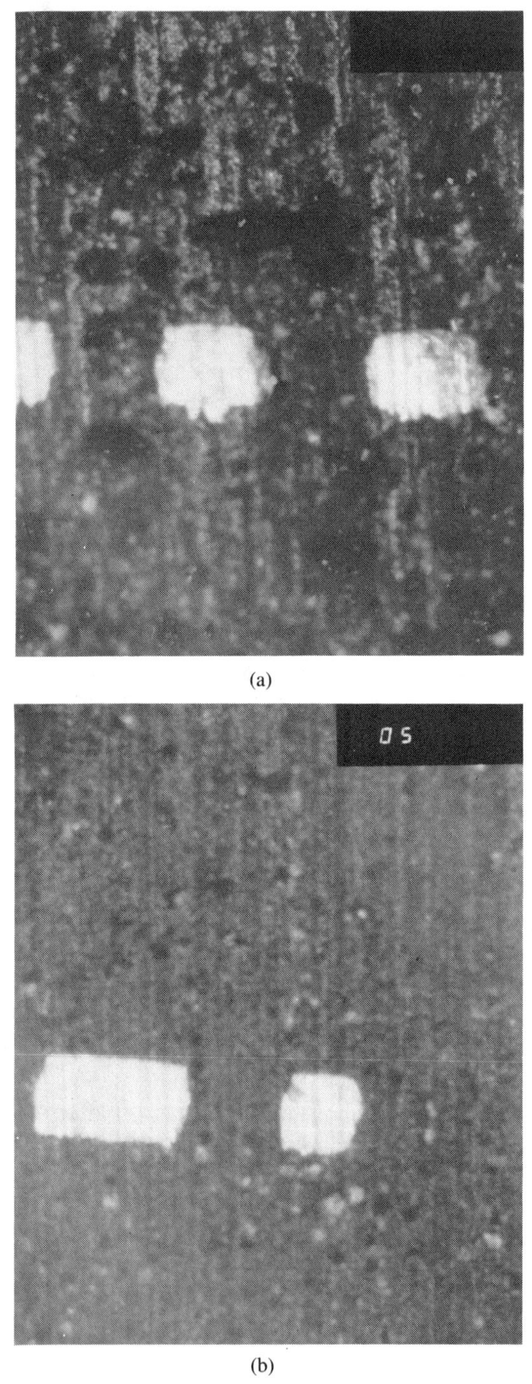


(b)

**FIGURE 4-11** Photomicrographs of cross sections of molded plastic packages that had (a) insufficient packing pressure and (b) full packing pressure. The magnification is 50×, the metal lead fingers have a thickness of 0.025 cm (0.010 in.).

greatest potential for flash and resin bleed simply because they are able to flow into the smallest crevices. Transfer molding compounds used for plastic packaging are known to flow into spaces as thin as 0.0005 inch with the resin portion being able to bleed out further into crevices less than half that thickness. Custom molds are often used for this determination instead of the standardized test mold.

*Mold Sticking* Molding compound sticking to the wetted areas of the molding tool is a problem since it requires extra operator time for cleaning which lowers productivity. Another serious consequence of mold sticking is that a higher occurrence of sticking debris raises the chances of some of it being overlooked resulting in loss of yield from clogged gates and runners and possible tool damage from clamping on the debris. Although the surface of the molding tool plays a major role in sticking (surface treatments for the tool are discussed in Section 5.5.4.2), some molding compounds have a higher propensity to stick to the mold surface than others. The introduction of new molding compounds with improved adhesion to the leadframe and silicon die has exacerbated this problem since these same improvements lead to greater adhesion to the molding tool as well. It appears that by using a mixture of mold release agents that soften and work at different temperatures, compounds can be made that have good adhesion to the leadframe and die while still releasing well from the tool. Notwithstanding these new developments, it is still a general trend that molding compounds with better adhesion will have more problems with sticking in the mold. That is not to infer that molding compounds with the best adhesion, a highly desirable trait, will provide sticking problems. Most applications of molding compounds with the greatest adhesion do not show sticking problems when the tool design and process conditions are correct.

*Mold Staining* Molding compounds can stain the molding tool and the accumulation of stains requires periodic cleaning with a cleaning compound. The cleaning procedure halts production for as much as several hours, so excessive need for cleaning can hurt productivity. The cleaning compounds are also strong chemical agents that can be irritating, providing strong motivation to minimize their use. Staining can be process condition dependent, but some molding compounds do show this trait more than others. Determination of mold staining potential is best done in a molding trial that is long enough for this cumulative effect to manifest itself.

**4.2.3.1 Molding Trial for Assessing Molding Problem Potential** A molding trial is an effective way to evaluate a material's potential for molding problems such as incomplete fill, resin bleed, mold sticking, and mold staining. The test should be long enough to allow all of the inconsistencies of the molding process to manifest themselves. A very thorough molding trial should include two or more lots of molding compound, normal and extraordinary time intervals after removal from the freezer, normal and extraordinary tenures in storage, and both humid and dry ambient conditions (if no material conditioning is used). More limited trials with only a single lot and an average time interval out of the freezer will also provide much information. The number of packages molded should also be large enough to provide confident sampling of all stochastic phenomena such as clogging of a gate due to gel particles or oversized filler particles.

Analyzing the results of a molding trial for productivity issues should include the percentage of cycles that show incomplete mold filling, excessive resin bleed, or excessive mold sticking. In a trial that included hundreds of molding cycles, no incomplete fills should be found, especially if the mold design has already been proven. Even a low occurrence of short shots is a harbinger of an underlying problem in the material and/or mold design. The analysis should also include very close scrutiny of the molding compound density. Dozens of molded packages should be sectioned for porosity appraisal. Examination of molded body cross sections under a light microscope, glass lens, or naked eye, can indicate in what cavities, if any, more porous material is located. Typically, these cavities of lower density material will be clustered in certain arms of the mold or in certain sections closer or further away from the transfer pot. Mold mapping of the low-density sites should be conducted to interpret the results properly. Maps of other molding problems should also be drawn up to aid in interpreting the results since molding problems are often related. Resin bleed is sometimes associated with overpacking the cavity, a condition indicated by deformation of the leadframe. Figure 4-12 shows a mold map displaying an exaggerated variety of material-related problems.

**4.2.3.2 Rheological Testing for Evaluating Molding Problem Potential** Molding trials can be time consuming and expensive, especially if more than one molding compound has to be evaluated. It is essential to screen molding compound candidates through laboratory experiments and scrutiny of the suppliers' literature prior to embarking on molding trials for a limited number of final candidates. Several of the tests that can be performed are rheological tests. In particular, the potential for incomplete filling of the mold can be assessed with judicious rheological experiments.

The mold-filling behavior is dominated by the flow of the molding compound through the gates of the tool. These pressure drops are much greater than the pressure drops encountered on flow in the runners or flow in the cavities. Incomplete fills are often the result of the pressure drop through the gates being excessive compared to the pressure applied at the transfer pot. Molding compounds that have lower viscosities at the high deformation rates encountered in

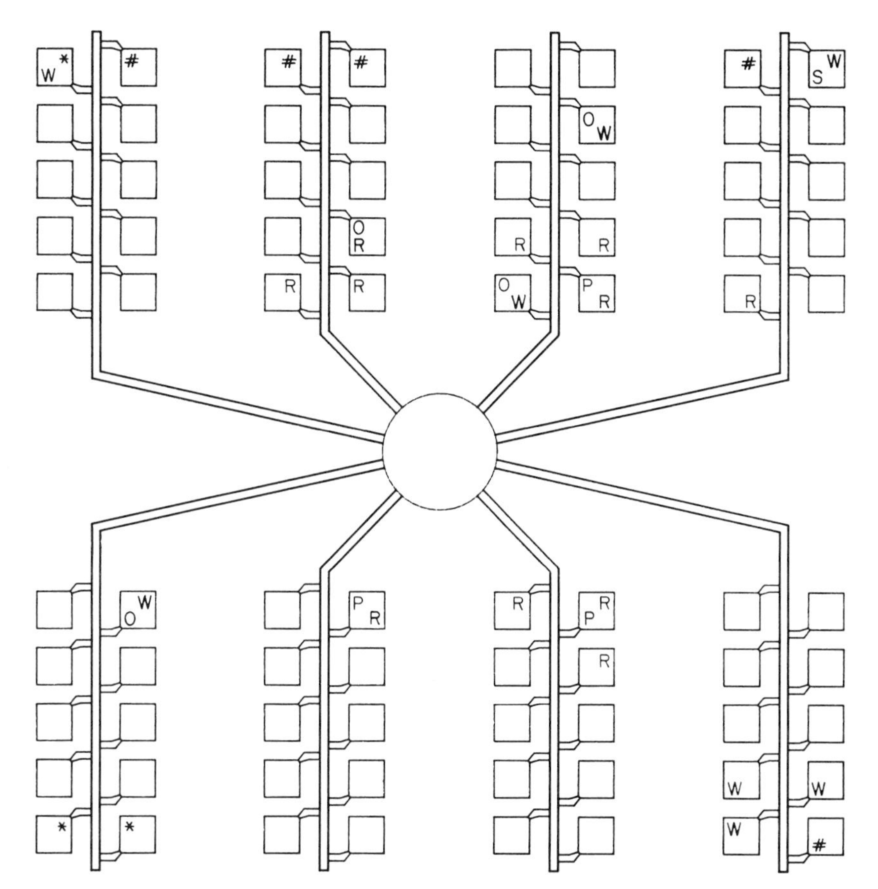


**FIGURE 4-12** Mold mapping of typical molding problems. The map legend is: O—void, *—underfill, #—low density, P—overpacking, R—resin bleed, W—wire sweep, S—paddle shift. The frequency of defects is exaggerated for illustration purposes.

the gates are less likely to cause incomplete fill problems. The difficulty is in identifying the nature and the magnitude of the deformation rate, and then devising rheological experiments to characterize the molding compounds under these same conditions. One problem is that the flow field through the gate is not simple shear flow, hence it is not completely characterized by a single shear rate or even a spectrum of shear rates. The flow more closely approximates flow through a sudden contraction or a converging channel, flow fields that have both shear and extensional components. Extensional deformation [14] is straining that extends the polymer material along the pull axis, similar to pulling taffy. The coefficient relating stress to strain rate in this type of deformation is the extensional viscosity, $\lambda$, similar to the more common shear viscosity relating stress to strain rate of shear flows. The stresses are the tensile forces applied to the ends of the specimen, whereas the strain is the ratio of the length divided by the initial length. The strain rate is the instantaneous strain increment divided by the time increment.

The analysis of flow fields that combine both shear and extensional flows is difficult in simple geometries, nearly intractable in the complicated geometries of most mold gates. The analysis of Cogswell for flow in a convergent wedge channel [14, 15] provides a useful starting point for this analysis. A flow field where the streamlines converge can be interpreted as an extensional deformation superimposed on a simple shear flow with the assumption that these flows can be treated separately. Appendix A contains the complete analysis of Cogswell for converging wedge flow [15] as it was adapted to treat flow through molding tool gates [16]. This analysis allows the characterization of the molding compound rheology under the deformation mode and deformation rate experienced in an actual molding tool. By conducting experimental measurements of pressure drop versus volumetric flow rate through a converging channel, or an actual convergent gate (see Figure 4-13), the viscosity constants for the material are obtained. The rheological constants can then be used to predict how the molding compound would behave in molds that have different gate geometries. This is an important selection criterion, since it is uncommon, at least inefficient, to qualify many different molding compounds for the various molding tools in use. Typically, one or two materials are qualified for a large number of different package codes molded in a large number of molding tools. Therefore, molding compounds that show good mold filling behavior over the widest range of gate sizes and shapes have an advantage over materials that will work well with only certain gate geometries.

The Cogswell analysis was applied to study the pressure drop through convergent gates typical of IC packaging molds [16]. Pressure drop data through actual gate geometries, modeled as flow through a convergent channel, were used to derive the shear and extensional components of the flow. Good agreement was obtained in that the power-law behavior derived from the pressure drop measurements matched the behavior determined separately by combined cone-and-plate and capillary techniques, both in magnitude and shear rate dependency [16]. These experiments appear to confirm the value of this analysis in conducting rheological studies to characterize and compare the mold filling behavior of different molding compounds.

## 4.3 RELIABILITY EVALUATION

Measurable properties of the molding compound have a direct impact on package reliability. Unlike the properties that influence productivity and yield, the reliability-related characteristics mostly concern ultimate properties of the fully

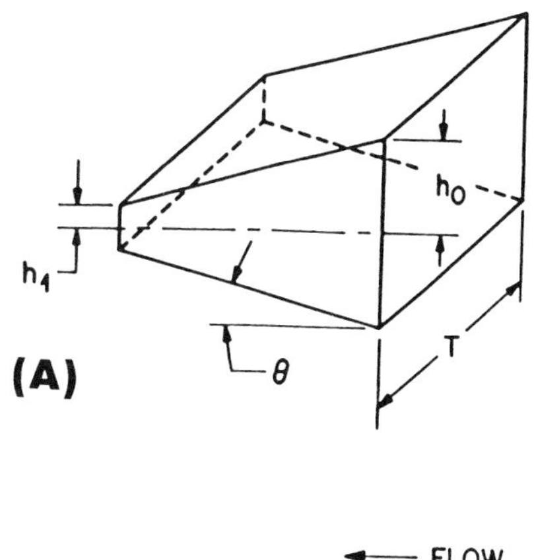


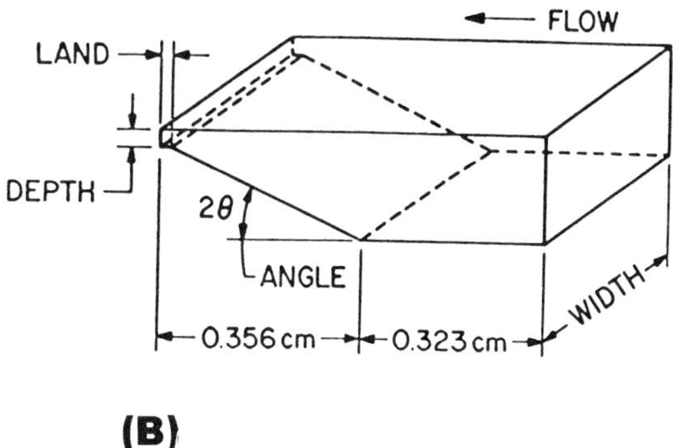


**FIGURE 4-13** Drawings comparing an actual gate geometry (bottom) with a symmetric convergent channel (top) that is more amenable to mathematical analysis (see Appendix 4A for treatment of the convergent channel).

cured molding compound. The role of mechanical properties in contributing to thermal induced stresses was introduced in Equation (4-1). Adhesion also has a major effect on package performance and reliability. Neither mechanical properties nor adhesion assessment lends itself to simple measurements, however, since the actual failure phenomena are more complicated than the laboratory tests that are used. For this reason, actual device reliability testing has assumed a major role in selecting molding compounds.

### 4.3.1 Mechanical Property Testing

Equation (4-1), repeated below for convenience as Equation (4-14), is the starting point for assessing the reliability potential of a molding compound from its mechanical properties:

$$\sigma = \int_{T}^{T_g} \alpha_1 E_p \, dT \tag{4-14}$$

This relation states that stress is proportional to the product of the tensile modulus of the molding compound and the strain, which in this case is the product of the coefficient of thermal expansion of the molding compound and the temperature excursion. There appears to be a growing body of evidence [17], mostly published in molding compound suppliers bulletins, which shows that lowering the stress parameter defined in Equation (4-14) leads to improved device reliability. The stress parameter is reduced in molding compounds that have any or all of the following: lower modulus, lower coefficient of thermal expansion, or lower anchor temperature (ostensibly the glass transition temperature). The coefficient of thermal expansion and the glass transition temperature are straightforward measurements that have been discussed previously; they are often provided in the material supplier's literature. The flexural or tensile modulus at room temperature is also provided by most suppliers, but this parameter requires more careful consideration before being used directly in stress calculations. The modulus is temperature dependent and Equation (4-14) is an integration over temperature requiring modulus versus temperature data [17]. These data will often have to be determined experimentally if a comprehensive comparison of different molding compounds is needed.

There are several test methods available for measuring the modulus of the molding compound as a function of temperature. A mechanical tensile test of the type illustrated in Figure 4-14 is a standard, well established technique for obtaining the tensile modulus. The specimens are "dog-bone" shapes that are

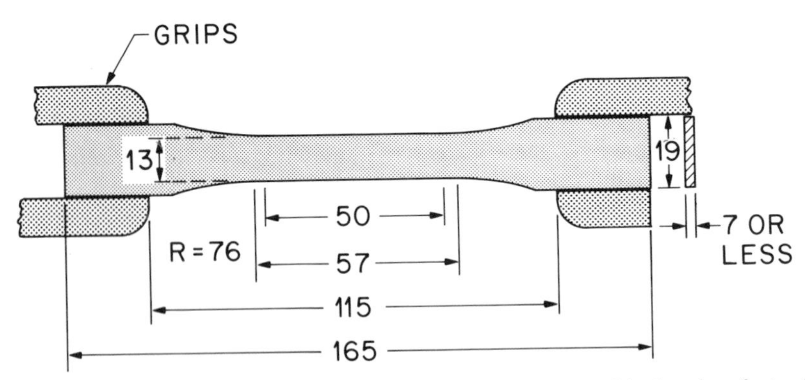


**FIGURE 4-14** Drawing of a mechanical tensile test specimen mounted in the grips of a tensile test machine after ASTM D 638. All dimensions are in mm. For high-modulus epoxy molding compounds, thicknesses less than 7 mm are recommended. The 50 mm dimension is the gage length contained within the 57 mm narrow section. (Copyright ASTM. Reprinted with permission.)

produced either in a special molding tool or cut from a sheet of cured molding compound when it is at a temperature near or above its glass transition temperature. A special cutting die can be bought or machined for this purpose. Both specimen preparation techniques have drawbacks that should be considered. Molded dog-bone specimens may show mechanical properties that are influenced by the flow of the material into the mold, a flow pattern that is different than what would be experienced during flow into the device cavity. Conversely, cut specimens can have notches or other surface defects on the cut edges that can lead to premature failure. Molding is the preferred preparation method. Most tensile test instruments can be equipped with a temperature controlled chamber to conduct measurements as a function of temperature, providing the data needed for Equation (4-14).

The flexural modulus, also denoted by $E$ since it is derived from the same type of tensile and compressive deformation as the tensile modulus, is often used in the stress calculation. This parameter can be measured in a standardized test (ASTM D 790-71) that is illustrated in Figure 4-15.

In most instances, the shear modulus can be estimated from the tensile modulus if the Poisson's ratio for the material is known (see Equation (2-25) for the relation between the shear and tensile moduli). The stresses encountered in molded plastic packages are actually a complex mixture of tensile and shear stresses, therefore a more appropriate analysis would be to measure and compute stress values with both types of deformation. An example of shear modulus versus temperature for an epoxy molding compound is provided in Figure 4-16.

### 4.3.2 Adhesion Property Evaluation

Adhesion plays a critical role in package performance. There are several distinct failure mechanisms that are directly attributable to loss of adhesion. One of the

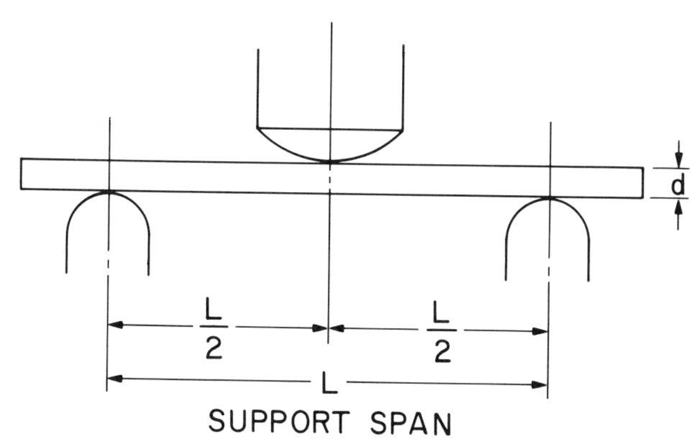


**FIGURE 4-15** Illustration of flexural modulus test from ASTM D 790-71. (Copyright ASTM. Reprinted with permission.)

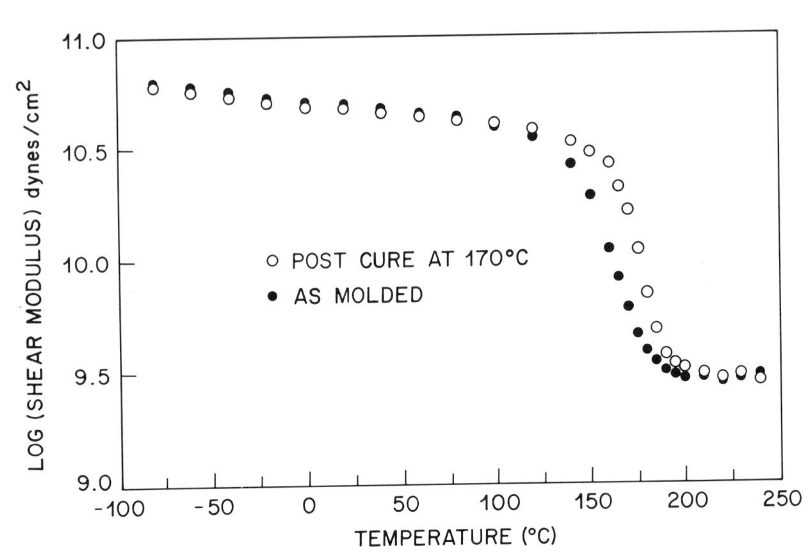


**FIGURE 4-16** Shear modulus versus temperature data for an epoxy molding compound. Data show the effect of postcure on the thermomechanical properties. Frequency is 1 rad/sec.

principal means of rapid moisture ingress is loss of adhesion or delamination at the interface between the molding compound and the lead. Although moisture can move through the molded body without the aid of a crevice, the free space at a delaminated surface allows liquid water to accumulate. Liquid water combined with extractable ionics from the molding compound provide the essential components for an electrochemical cell, leading to accelerated corrosion. Bonding at all interfaces, particularly those on the lead fingers, wire bonds, and silicon device prevents water clustering and rapid corrosion.

The second type of failure associated with adhesion is the increase in stress due to delamination along shear surfaces. Adhesion to the shear surfaces distributes the stresses and deformation over the entire bonded area as well as throughout the molded body volume. Reducing the area of the bonded interface concentrates the stresses in a smaller area thus increasing the force per unit area. Similarly, deformation and stress dissipation in the molded body are reduced, concentrating the stresses at the line between bonded and delaminated areas. Experimental and numerical studies on stresses in molded plastic packages support these conclusions on the important role of adhesion in reducing thermomechanical stresses and improving device reliability [17]. Figure 4-17 is a plot of the predicted principal stress plotted against the length of delamination between the die and the epoxy molding compound at the point of maximum stress which is located just inside the adhered zone [17]. Section 8.2 presents a more comprehensive discussion of the role of adhesion in device reliability.

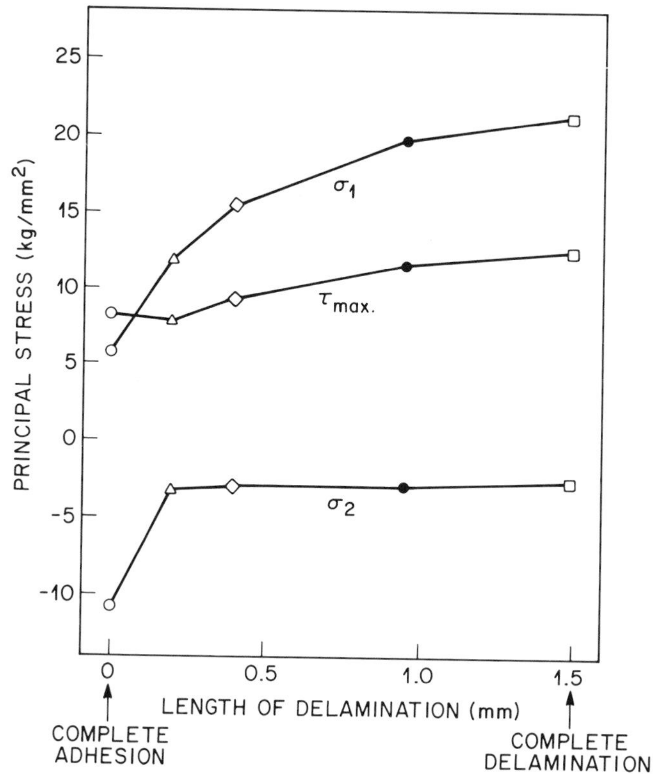


**FIGURE 4-17** The principal stress on the surface of a silicon chip plotted against the length of the delamination. The delamination moves inward from the outer edges of the chip toward the center, hence the distance shown on the horizontal axis is the distance from the edge of the chip. The stresses are computed for the maximum stress point which is a line just inside the adhered zone [17].

Molding compounds that have better adhesion to the leads and silicon die will better resist delamination and the problems associated with it such as accelerated corrosion, package cracking, chip cracking, and aluminum conductor deformation. It is worthwhile to conduct tests that can discriminate among the molding compound candidates in their ability to adhere to chips and the lead-frame materials that will be used. Adhesion is highly surface dependent rather than an intrinsic characteristic of a material, therefore the adhesion to specific surfaces must be measured. Plating and other surface finishes strongly influence adhesion, so these features must be included in the test program.

One method of assessing adhesion is to mold plugs of the molding compound onto metal inserts made of the same materials and with the same surface finish for which you want to measure the adhesion as is shown in Figure 4-18. These

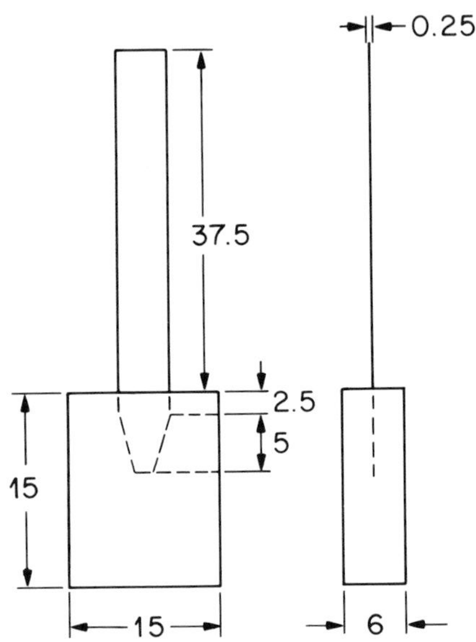


**FIGURE 4-18** Drawing of the molded body and the molded-in metal tab for a leadframe adhesion test method based on a molded plug. Dimensions in mm. (Courtesy of Hitachi Chemical Company, Ltd.)

specimens can then be tested for adhesive strength in a conventional tensile test experiment where the metal insert is pulled from the molded body. The force required to pull the insert from the molding compound is a good indication of the adhesive strength. One configuration for the test method is shown in Figure 4-19. In this example, three plugs are molded in one cycle in a three-cavity molding tool. The molded specimens are then separated and tested individually in a conventional tensile tester for which a special fixture has been made to hold the molded body. The metal tab extending from the molded body is held in the other set of grips without any special hardware.

The effect of the molding conditions on the adhesive properties is another point to consider in designing a program to measure adhesion. The specimens should be molded at the same process conditions that they experience in an actual production mold to provide the most confidence in extending the results to actual product. These conditions include the same mold and preheat temperatures, cure time, and postcure conditions. More subtle differences could be attributed to the temperature of the molding compound when it contacts the lead surface and its precise flow time and flow length at that point. To achieve this level of similarity, it is best to work with an actual production mold. An adhesion test could be developed through the design of a custom leadframe that incorporates an adhesive test specimen. These leadframes are then used in the actual molding tool in which product will be made. An example of this type of leadframe adhesion test is shown in Figure 4-20.

FIGURE 4-19 Molding tool and molded specimens for the leadframe adhesion test method.

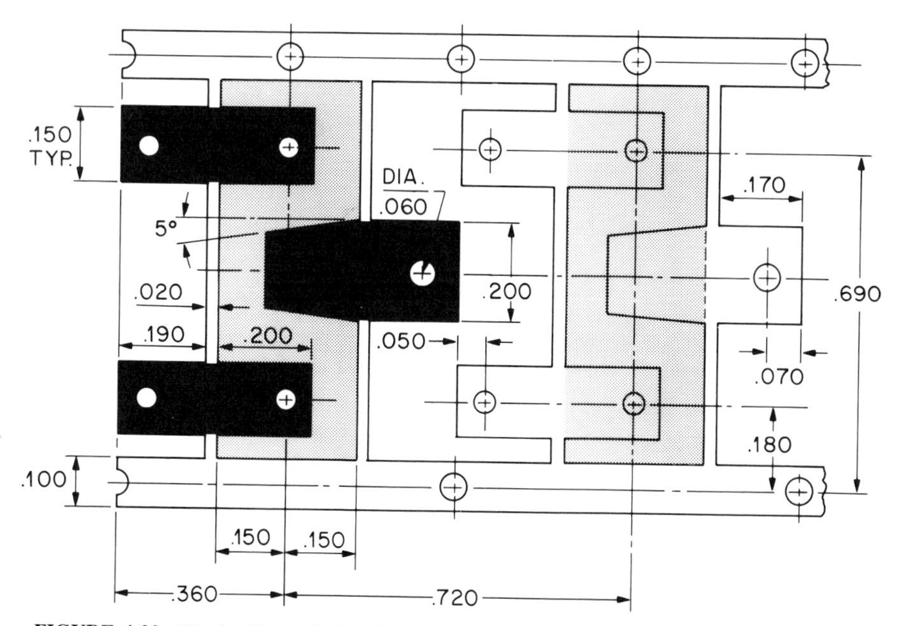


FIGURE 4-20 The leadframe design for an adhesion test. These leadframes can be molded in production tooling to minimize differences between test and production molding conditions. Dimensions in inches. (Courtesy of Plaskon Electronic Materials, Incorporated, a Rohm and Haas Company.)

Another important aspect of adhesion testing is to determine the adhesive bond strength to other materials within the molded package such as the silicon die, its passivation layers, polyimide or silicone overcoats, and polyimide film dielectric layers. The adhesion tests described above are not amenable to these materials because they do not have sufficient strength to be subjected to tensile testing. For these materials it is preferable to mold the epoxy compound against a flat sample of the material to be tested and then measure the force required to pull this configuration apart. The sample preparation and testing for this method are illustrated in Figure 4-21.

Careful interpretation of the results from adhesion studies based on tensile testing is required. As a failure test, the data usually show significant scatter. A minimum of 20 specimens is needed to arrive at a representative average adhesive strength. Also, the shape of the stress-strain curve obtained may be significantly different. In some cases, the stress will build uniformly at constant modulus until adhesive failure between the insert and the molded body occurs. This is often accompanied by a snapping sound. The high-stress point, which is the strength of the adhesive bond, may then be followed by a constant-stress region of the curve where the insert is simply pulled out of the molded body under a compressive loading, a pinch hold. In other cases, the failure point may be absent or difficult to discern. The insert is pulled from the molded body at constant rate almost from the onset of the stress. These data are also related to

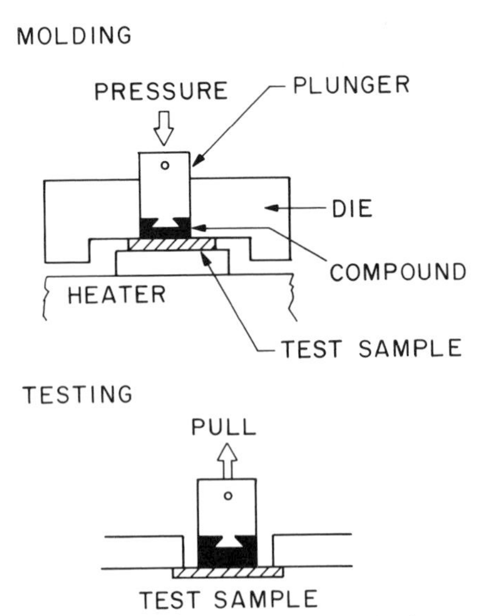


**FIGURE 4-21** Configuration of molding and testing for an adhesion test method that can be used for most substrates such as silicon, polyimide, and leadframe metals. (Courtesy of Hitachi Chemical Company, Ltd.)

adhesion, but they do not describe a true adhesive strength. Experience with a large number of specimens is the best route to design and interpret adhesion experiments.

### 4.3.3 Cracking Potential Evaluation

The potential for cracking is a major selection criteria, particularly for package designs where package cracking is an issue. Small outline and memory device packages usually have a relatively small amount of molding compound surrounding a relatively large silicon device, making them more susceptible to package cracking failures than other larger volume packages. There are no standard procedures in the semiconductor industry for evaluating the potential for molding compound cracking, but there are several tests that are both realistic and discriminating. One of these is the common Izod impact test (ASTM D-256 A). For impact testing, a specimen is molded of the material to be tested into a rectangular solid. Typical dimensions of the test specimen for molded thermosets are shown in Figure 4-22. A 45° notch has to be machined into the specimen with the appropriate machine tool. The specimen is mounted as a vertical cantilever beam in the test apparatus as is shown in Figure 4-22. A weighted pendulum is released and strikes the specimen on the notched surface, breaking it off in the typical case. The energy absorbed in the fracture is reflected in the angular travel that the pendulum falls short of reaching its initial height. Specimens with greater fracture energy, that is, the energy required to break the specimen, will show proportionally less pendulum travel after impact. A variation of this method is to mount the specimen horizontally and strike it on the side opposite the notch (ASTM D-256 B). The shortcomings of these

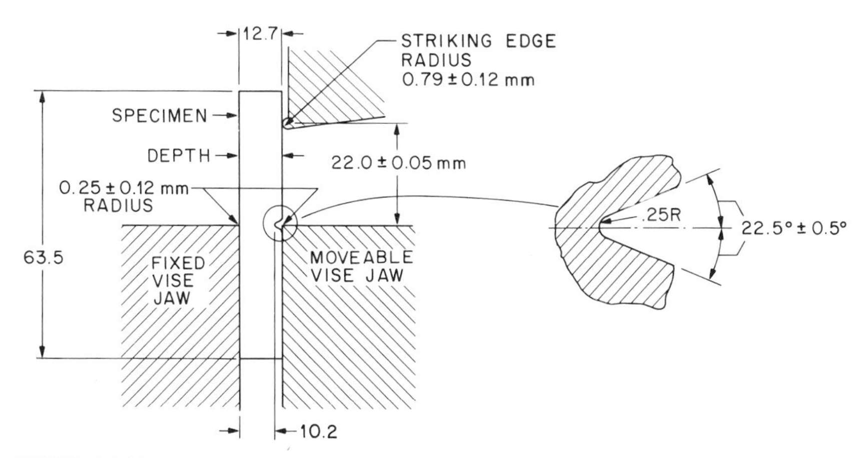


**FIGURE 4-22** Izod impact test from ASTM D 256. (Copyright ASTM. Reprinted with permission.)

techniques for epoxy molding compounds are that the rate of strain is significantly greater than the strain rate in a molded package experiencing thermal shrinkage stresses, even under extreme conditions of very rapid temperature quenching. Therefore the test does not probe the region of viscoelastic behavior where the true differences in properties may lie. It is useful, however, for evaluating the potential for a material to chip or crack on impacts that occur during trim and form operations or when the device is dropped, both being high-strain-rate deformations.

A fracture test that more closely models the actual strain history of the package for thermomechanically induced failure is a three-point flexural bending test similar to the flexural modulus experiment that was described previously. In this test a specimen is again molded under standard conditions. A small notch can be either molded into the specimen or machined in afterwards. The experimental apparatus is as shown in Figure 4-23. The rate of strain should approximate the deformation rate experienced during a temperature cycling test or during actual on-off operation of the device. In temperature cycle testing with liquid to liquid heat transfer, the cooling rate can be of the order of 200°C/sec providing a strain rate of approximately 20%/minute. On-off operation with free cooling in air provides a much lower strain rate of 0.1%/minute for a temperature rate of 1°C/min. The test is conducted on a modified tensile tester with a compression head, a transducer that converts the compressive force on the fulcrum to an electrical signal. The data are similar to conventional flexural test data with the area under the stress-strain curve being proportional to the

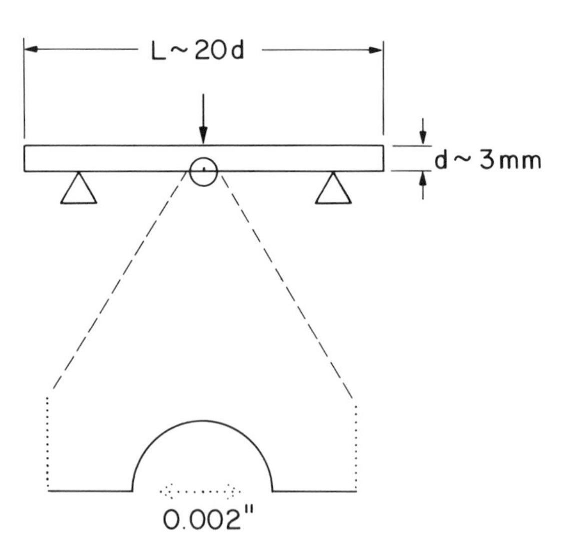


**FIGURE 4-23** Notched flexural test apparatus, showing a detail of the notch region.

energy to break. The flexural energy to break should also be tested at different temperatures if possible to evaluate the temperature dependence and to determine where the material is most brittle. The low-temperature data is of particular importance because the molded body experiences the greatest stress at lower temperatures far removed from the molding temperature. This flexural strength and the corresponding energy to break are important discriminating factors in selecting epoxy molding compounds. Materials that have significantly higher energy to break at the strain rates and temperatures of importance are likely to provide better crack resistance in molded plastic packages.

### 4.3.4 Experimental Reliability Evaluation

A manufacturer's reputation for quality depends on the reliability of their products. It is therefore essential to conduct reliability evaluation of molding compounds as thoroughly and as carefully as possible. Although molding compounds can be screened for reliability parameters such as adhesion and propensity for cracking, experimental evaluation of reliability is an essential aspect of the overall material selection process, especially since the differences between the measurable parameters are narrowing, yet large differences in material performance remain. At the same time, reliability testing is getting more difficult and time consuming because the combination of the significant improvements in molding compound performance and the improvements in passivation layer technology have increased the reliability of plastic packages to the point where failures are not easily induced in relevant tests. There are three principal reliability tests that are in use in the semiconductor industry: high temperature, humidity, bias (HTHB), also known as 85/85 testing; temperature cycling (TC); and pressure cooker tests (PCT). These three tests were introduced in Sections 4.1.9, 4.1.10, and 4.1.11. Commercial instruments are available for each of these tests, or custom equipment can be fabricated if there are unusual requirements.

The test program has to be designed carefully to derive results that can be used with confidence. Essentially all of the reliability tests are carried out until some segment of the initial population has failed, usually 50%, providing a half-life. Other testing criteria may require no failures below a specified number of cycles or hours. The sample size should be large enough to accommodate the difference in device and assembly features that could lead to premature failure. Since the chips and their assemblies are produced in processes that are exceptionally uniform, and the devices have already been thoroughly tested prior to assembly, the sample size does not have to be extraordinary to achieve trustworthy results. Most tests use between 50 and 100 specimens. In cycling tests or other tests where the devices are not testable during the procedure, it is necessary to test device function at regular intervals and eliminate failed specimens. This interval between testing depends on the expected lifetime of the devices in the particular test. The expected lifetime should be divided into 10–20 test increments. It is also useful to test very early in the process to assess the initial dropout rate. The complete results can be presented in a variety of ways. One way is to plot the number or percent of live devices remaining against the elapsed time of testing or number of cycles. Log or linear plots can be used as appropriate. Another important way of presenting reliability data is through a Weibull plot, discussed in Section 8.1.1, which is a plot of the natural log of the natural log of the percentage failed versus the natural log of the number of cycles or time. This plot makes it easy to determine what fraction of devices are functioning after a selected time or number of cycles, and it is effective in identifying anomalous failure trends because the curve shows a rapid rise when major failure events occur. Other approaches to failure rate analysis are presented in Section 8.1.

The HTHB test is probably the least discriminating since it is the mildest. Most current molding compounds and devices can withstand thousands of hours of HTHB without reaching the half-life. HTHB is being supplanted by the pressure cooker test (PCT), known also as the steam bomb test. Both HTHB and PCT measure a packaged device's resistance to corrosion-induced failure. The PCT test is much more effective in promoting corrosion failure because the moisture is at a higher pressure and temperature, significantly improving its penetrating capabilities. PCT testing is relatively insensitive to the package design, making it an excellent material selection test because it isolates the properties of the molding compound. It is a discriminating test of moisture permeability, moisture uptake, and adhesion. Loss of adhesion at an interface will allow clustered water to form there in the saturated steam conditions of the test chamber, resulting in premature corrosion failure.

Temperature cycle testing is much more effective in evaluating the propensity for package and/or chip cracking, so it should be included in the selection criteria when the design is susceptible to these types of failures. The test is highly design dependent, so results for two different package designs should not be compared. The temperature extremes of the test should be set to provide a large though not extraordinary number of cycles to reach half-life. Half-lives of the order of several hundred cycles are most appropriate, although numbers reaching several thousands are not uncommon. Some temperature extremes are set according to the actual exposure limits expected, such as outdoor equipment in the United States having to sustain temperature cycles between $-40$ and $+135$°F. In this mode, the temperature cycle test is used more as a reliability criterion than a reliability assessment. Failures in temperature cycle testing are either package cracks or electrical defaults. Electrical failure usually follows external package cracking by a small number of cycles.

Removal of the molding compound after the device has failed is an integral part of analyzing the results, since it elucidates the failure mechanism. This decapsulation process is difficult since the molding compound is designed to be inert and solvent resistant. The specific procedure will depend on the molding compound in use and should be recommended by the material supplier. In general, these procedures require hot acids or solvents so they should be conducted by trained personnel with appropriate safety precautions. Decapsulation also allows the determination of the extent of device feature deformation caused by the shrinkage of the molding compound. Not all failed packages need to be decapsulated: a very small percentage sampling is usually sufficient. One approach is to sample devices that failed at different times over the test duration. In many cases, there are different failure mechanisms at early and late stages of the test.

## 4.4 THE EFFECT OF WATER ON MOLDING COMPOUND PROPERTIES

Water has a significant influence on the kinetics, rheology, and ultimate properties of the molding material. Epoxy molding compounds absorb water to a level of approximately 0.5% by weight when exposed to 100% relative humidity at room temperature. The water uptake of the preforms tracks almost linearly with relative humidity as is evident in the plot presented in Figure 4-24. These relatively high levels of moisture content have significant effects on material rheology, cure kinetics, and ultimate mechanical properties. Although the effects of water are largely the result of characteristics that are common to all epoxy molding compounds, the additives and curing agents play an important role in making the effect of water on the properties of some molding compounds different than others. The effects of water therefore become a selection criteria. Some of these effects are discussed in the following sections.

### 4.4.1 Effect on Rheology and Processing

Moisture has a profound effect on the viscosity of the epoxy molding compounds used in microelectronics packaging. These materials are not unique in this regard because the effect of water on glass-filled transfer molding compounds has been known for decades. The general effect is that humid conditioning significantly reduces the molding compound viscosity in the molten state. The percent decrease can be as much as 40% compared to bone dry conditions [7]. The effect appears to be an increasing function of moisture uptake in most materials in that higher moisture content provides lower material viscosity over the span of 0–100% of saturation. The exact form of the functional dependence of viscosity on percent saturation has not been documented, and it is likely to be different for different materials. Similarly, the effect of moisture on the shear

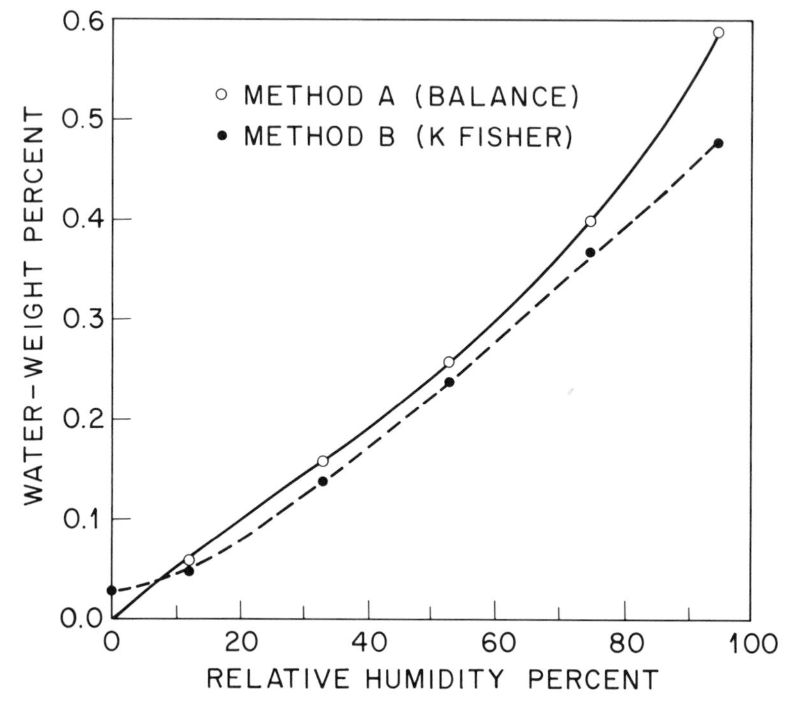


**FIGURE 4-24** Plot of water uptake of epoxy molding compound preforms as a function of relative humidity of the conditioning atmosphere after three weeks of exposure at room temperature. K. Fisher is a hydrolysis technique for measuring water content. (Data of D. J. Boyle, J. T. Ryan and H. E. Bair, AT&T Bell Laboratories.)

thinning behavior is not well understood, although preliminary results indicate that the viscosity is simply lowered with little change in shear rate dependence [7]. If the data were fit to a power-law model as discussed in Section 2.4.1, the power-law index would be unchanged after humid conditioning. Figure 4-25 [7] shows the effect of humid conditioning on viscosity and power-law index.

The viscosity reduction on moisture exposure is not entirely unwelcome. The higher viscosities of bone dry material can cause mold filling problems and increase the flow-induced stresses on the devices. Conversely, excessive moiscure uptake can cause other problems such as greater resin bleed and the possibility of more voids. One concern is that materials that are exceptionally sensitive to viscosity reduction by moisture will be more unpredictable in the factory where, in the absence of conditioning in special chambers, the relative humidity can vary from 20% to 80%.

Moisture sensitivity can be evaluated by testing the viscosity and shear rate dependence of molding compound that has been conditioned in controlled atmospheres. Spanning the range of relative humidities from 0% to 80% in 20%

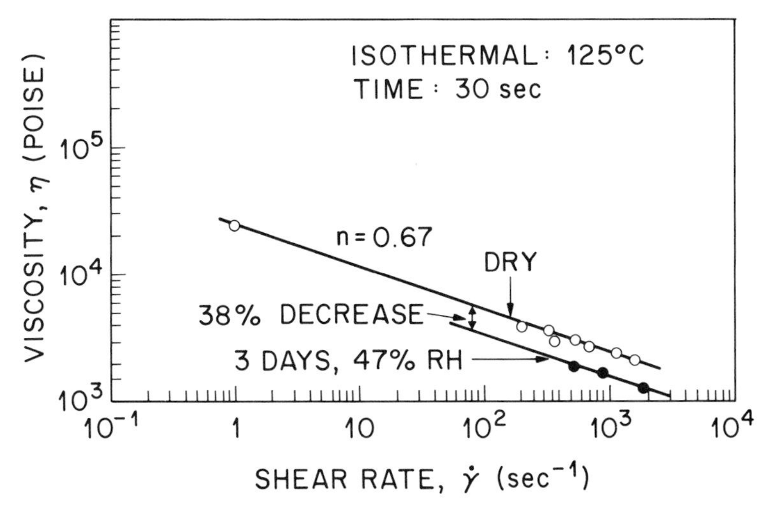


**FIGURE 4-25** Viscosity of an epoxy molding compound plotted against shear rate, showing the effect of conditioning in dry atmosphere and 47% relative humidity [7].

intervals is sufficient for this analysis. These humidity levels can be achieved through the purchase of a humidity chamber or through the use of special saturated salt solutions that provide a range of water vapor pressures. Table 4-2 lists the saturated salt solutions and the humidity levels that they provide [18]. The humidity chamber is much more convenient and reliable, albeit far more expensive. Equilibration times for moisture uptake are between 24 and 48 hours depending on the size of the preform and its packing density. The viscosity of the material should be measured in any of the techniques described in Sections 4.2.1.1 within 20 minutes after removal from the humidity chamber.

## 4.4.2 Effect on Kinetics and Ultimate Properties

Moisture also has a strong impact on the kinetics and ultimate properties of epoxy molding compounds. The ultimate glass transition temperature of the

**Table 4-2. Relative Humidity over Saturated Salt Solutions [18]**

| *Salt* | *Humidity, % at 25°C* |
|---|---|
| Phosphorous pentoxide | 0.0 |
| Lithium chloride | 11.3 |
| Magnesium chloride | 32.7 |
| Sodium dichromate | 53.7 |
| Sodium chloride | 75.1 |
| Potassium sulfate | 97.0 |

cured material can be lowered by humid conditioning of the preform material. This indicates that the crosslink network that forms in the presence of water is different than the one formed under dry conditions. This crosslink structure cannot be altered without thermal degradation of the polymer. Moisture can also have an effect on the rate of the polymerization reaction since it can bind to catalyst sites and decrease catalytic effectiveness. The specific effects of moisture on kinetics and ultimate properties depend on the specific catalyst and chemistries used, so there will be differences among molding compounds in this regard.

Moisture also has a profound effect on the occurrence of cracking during reflow soldering used for the attachment of surface mount packages. Moisture absorbed after molding undergoes a volume expansion as it vaporizes on exposure of the package to the solder temperature. This causes the package to expand, inducing package cracking in the expansion or contraction. Resistance to this cracking depends on a combination of low moisture uptake, high strength in the rubbery state above $T_g$, and good adhesion to die and paddle support. Molding compounds will show different degrees of resistance to this problem, and these differences are primary selection criteria for surface mount packages. A more complete discussion of this problem is provided in Section 8.2.2.1.

## 4.5 MATERIAL SELECTION SUMMARY

Molding compound selection is one of the most important aspects of plastic packaging. This chapter has reviewed some of the elements of this selection process including how to use the material supplier's information and how to conduct an evaluation of molding compound productivity, reliability, and moisture sensitivity. Different aspects of the evaluation will be more important for certain package designs than others. For this reason, the criteria for selecting molding compounds cannot be universal, or even applied across the entire product line. Although it is preferable that the number of different molding compounds in production is minimized, it is very often the case that some materials are simply better suited for some applications than others, and the difference is enough to warrant qualifying several molding compounds.

Prior to starting a program to qualify new materials, it is important to determine the most critical features of the design or designs in question. If the design is one where there will be a thin edge of molding compound surrounding the device, then the thermomechanical stress factor provided in Equation (4-1) becomes more important. Conversely, if the part is large with a thin lead frame, such as the newer fine-pitch packages, or if the wire bond span is long, then flow-induced stress considerations become prominent. In this case, molding compounds which have low viscosity at low shear rates and high temperatures are stronger candidates. If molding yields and crack resistance are both satisfactory, then secondary considerations such as resin bleed, mold staining, high temperature hardness, ionic content, cure time and moisture sensitivity become deciding factors.

There are no standard procedures for qualifying a molding compound for production. Different manufacturers follow different procedures but they will usually involve specifications of the following: (1) spiral flow length, (2) glass transition temperature, (3) purity (in terms of hydrolyzable ionics), (4) cure time, (5) reliability tests, and (6) a molding trial.

An often neglected parameter specification is the viscosity of the molding compound at certain shear rates. Viscosity at low shear rates provides an indication of flow-induced stress on the devices during encapsulation, whereas the viscosity or other pressure drop measurement through a converging channel at high shear/elongation rates is an indicator of mold-filling problem potential. Other productivity and quality improvement criteria include a DSC scan of the molding compound to verify the heat of reaction is within an established window. In cases where the crack resistance and shrinkage stresses of the molding compound are also an issue, as they are becoming on nearly all new package codes, then the thermomechanical stress parameter and the crack energy should also be considered. A more comprehensive listing of molding compound selection criteria is therefore:

1. Spiral flow length
2. Glass transition temperature ($T_g$)
3. Purity (hydrolyzable ionics)
4. Cure time
5. Reliability tests including PCT and TC tests.
6. Molding trial to assess mold filling behavior, hot hardness, resin bleed, and mold staining
7. Shear viscosity at low shear rates
8. Flow resistance in convergent die at high shear/elongation rate
9. Consistency of heat of reaction in DSC measurement
10. Thermomechanical stress parameter ($\sigma$), or experimental stress determination
11. Adhesive strength to leadframe materials
12. Crack sensitivity from notched flexural test
13. Moisture sensitivity of viscosity and $T_g$
14. Alpha particle emission (if applicable)
15. Occurrence of moisture induced cracking on reflow soldering (for surface mount technology)

Specific qualification criteria will depend on the individual package design, device type, molding tool technology, and quality criteria. The above list suggests that there are parameters other than the conventional ones of spiral flow length and glass transition temperature that are important.

## REFERENCES

1. American Society of Testing Materials, Specifications, 1916 Race Street, Philadelphia, PA 19103.
2. Semiconductor Equipment and Materials International, 1988 SEMI Test Methods for Packaging, 801 East Middlefield Road, Mountain View, CA 94043.
3. T. C. May and M. H. Woods, "Alpha Particle Induced Soft Errors in Dynamic Memories," *IEEE Trans. Electron. Devices,* **26**, 2 (1979).
4. L. T. Manzione, "Plastic Packaging of Microelectronic Devices," Society of Plastics Engineers, ANTEC Papers, Volume 29 (1983).
5. S. Middleman, *Fundamentals of Polymer Processing*, McGraw Hill Book Company, New York, (1977).
6. H. Schlichting, *Boundary-Layer Theory*, Sixth Edition, McGraw Hill Book Company, New York, (1968).
7. L. L. Blyler, Jr., H. E. Bair, P. Hubbauer, S. Matsuoka, D. S. Pearson, G. W. Poelzing, and R. C. Progelhof, "A New Approach to Capillary Viscometry of Thermoset Transfer Molding Compounds," *Polym. Eng. Sci.*, **26**(20), 1399 (1986).
8. L. T. Nguyen, "Wire Bond Behavior During Molding Operations of Electronic Packages," *Polym. Eng. Sci.*, **28**(14), 926 (1988).
9. S. Sourour and M. R. Kamal, *Thermochimica Acta*, **14**, 41 (1976).
10. M. E. Ryan and A. Dutta, *Polymer*, **20**, 203 (1979).
11. A. Hale, M. Garcia, C. W. Macosko and L. T. Manzione, "Spiral Flow Modelling of a Filled Epoxy-Novolac Molding Compound," Soc. Plastics Engineers, ANTEC Papers, p. 796 (1989).
12. A. Hale, H. E. Bair, and C. W. Macosko, "The Variation of Glass Transition as a Function of the Degree of Cure in an Epoxy-Novolac System," Soc. Plastics Engineers, ANTEC Papers, Volume 33, p. 1116 (1987).
13. A. Hale, "Epoxies Used in the Encapsulation of Integrated Circuits: Rheology, Glass Transition and Reactive Processing," Ph.D. Thesis, University of Minnesota, Department of Chemical Engineering (1988).
14. F. N. Cogswell, "Measuring the Extensional Rheology of Polymer Melts," *Trans. Soc. Rheol.*, **16**(3), 383–403 (1972).
15. F. N. Cogswell, *Polym. Eng. Sci.*, **12**(1), 64 (1972).
16. L. T. Manzione, J. S. Osinski, G. W. Poelzing, D. L. Crouthamel, and W. G. Thierfelder, "A Semi-Empirical Algorithm for Flow Balancing in Multi-Cavity Transfer Molding," *Polym. Eng. Sci.*, **29**(11), 749 (1989).
17. S. Oizumi, N. Imamura, H. Tabata, and H. Suzuki, "Stress Analysis of Si-Chip and Plastic Encapsulant Interface," Nitto Technical Reports, Sept. 1987, p. 51 (ISSN 0285-2462).
18. J. F. Young, "Humidity Control With Laboratory Salt Solutions—A Review," *Appl. Chem.*, **17**(Sept. 1967).

## APPENDIX 4A
## PRESSURE DROP THROUGH A CONVERGENT CHANNEL*

This analysis separates the viscometric effects from the geometric effects in the pressure drop/flow rate correlation of flow through a molding tool gate allowing

*This analysis is derived from: L. T. Manzione, J. S. Osinski, G. W. Poelzing, D. L. Crouthamel, and W. G. Thierfelder, "A Semi-Empirical Algorithm for Flow Balancing in Multi-Cavity Transfer Molding," *Polym. Eng. Sci.*, **29**(11), 749 (1989).

a priori calculation of flow resistance for any specific geometry, or the calculation of gate dimensions to achieve a given resistance. This is done by combining simplified algebraic expressions for convergent flow with experimental data. Material constants for the power-law fluid used in the experiments also result from this analysis. Resistance to flow is then expressed as a function of gate angle, gate width, gate depth, gate length, and material power-law index.

Typical analytical representations of isothermal, steady, creeping, converging flow can describe this flow phenomenon for an incompressible Newtonian fluid, or for a non-Newtonian fluid of known behavior, but a full solution of three dimensional flow generally requires elaborate numerical techniques. Experimental results, however, are often cast into the simple relationship between pressure drop across and flow rate through a mold gate according to Equation (4A1):

$$\Delta P = KQ^n \tag{4A1}$$

where $K$ is an empirical constant and $n$ is the power law index for a fluid described by

$$\eta = C\dot{\gamma}^{n-1} \tag{4A2}$$

Equation (4A2) relates fluid viscosity $\eta$ to shear rate $\dot{\gamma}$ through parameter $C$, a constant at a given temperature for a given material.

It will be shown by a simplified analysis that the constant $K$ in Equation (4A1) can be resolved into a geometric and material contribution

$$K = BC \tag{4A3}$$

Where $B$ is a function of convergent gate geometry and power-law index, and $C$ is as in Equation (4A2). Full utility of the experimental pressure drop versus flow rate is then possible, since all data from different gate geometries can be related by a single expression.

Cogswell [A1] has presented a simple analytical solution to non-Newtonian convergent flow according to the top diagram in Figure 4-13. According to this analysis, the total pressure drop for a power-law fluid undergoing wedge flow is given by the sum of pressure drops due to entrance losses ($P_0$), losses due to shear ($P_s$), and losses due to extensional flow ($P_e$):

$$\Delta P_{\text{tot}} = P_0 + P_s + P_e \tag{4A4a}$$

$$P_0 = \frac{4}{3n+1}\dot{\gamma}_0(\eta_0\lambda_0)^{1/2} \tag{4A4b}$$

$$P_s = \frac{\sigma_{s1}}{2n \tan\theta}\left[1 - \left(\frac{h_1}{h_0}\right)^{2n}\right] \tag{4A4c}$$

$$P_e = \frac{\sigma_{e1}}{2}\left[1 - \left(\frac{h_1}{h_0}\right)^2\right] \tag{4A4d}$$

where $\sigma_{si} = C\dot{\gamma}_i^n$, $\dot{\gamma}_i = 3Q/2TH_i^2$, $\sigma_{ei} = \sigma_{si}/\tan\theta$, $\lambda_i = 3\sigma_{ei}/4\dot{\epsilon}_i$, and $\dot{\epsilon}_i = (\dot{\gamma}_i/3)\tan\theta$ (see Figure 4-13). Note that $\theta$ is the half-angle of convergence and $h$ the half-height. Substituting and simplifying, we obtain the compact form:

$$\Delta P_{\text{tot}} = BCQ^n \tag{4A5}$$

where $B$ is a unique geometric constant for each gate geometry and is described by

$$B = A_1\left(\frac{3}{2Th_1^2}\right)^n + A_2\left(\frac{3}{2Th_0^2}\right)^n \tag{4A6}$$

and

$$A_1 = \frac{1}{2\tan\theta}\left[\frac{1 - (h_1/h_0)^{2n}}{n} + 1 - \left(\frac{h_1}{h_0}\right)^2\right] \tag{4A7a}$$

$$A_2 = \frac{6}{(3n + 1)\tan\theta}. \tag{4A7b}$$

$B$ can be termed the hydraulic resistance of a particular gate. The geometry of the top diagram in Figure 4-13 is not identical to the gate geometry shown in the bottom diagram of Figure 4-13, but the approximation can be made. Most of the pressure drop through the gate pictured in the bottom diagram of Figure 4-13 will occur in the convergent section, and is therefore nearly described by the above equations derived from the geometry in the top diagram of Figure 4-13.

Equation (4A5) can be linearized by taking logarithms (the subscript has been dropped):

$$\log\left(\frac{\Delta P}{B}\right) = n\log Q + \log C \tag{4A8}$$

A plot of $\log(\Delta P/B)$ vs. $\log Q$ should then yield a straight line of slope $n$, intercept $\log C$. Since the power-law index $n$ must be known to calculate the $B$ value for each gate, an iterative scheme is necessary to solve for $n$ and $C$ from pressure-drop data. An initial guess for $n$ can be used in the first calculation, and an improved $n$ is then obtained from the slope of the data according to Equation (4A8). This process is repeated until convergence is reached in approximately four iterations. The analysis can be used to predict the pressure drop through the gate for a specific gate geometry and material rheology, or it can be used to determine the molding compound rheology at the relevant deformation rate on flow through a gate from pressure-drop data.

## Experimental Correlation

The $\Delta P$ vs. $Q$ data for the various gates can be subjected to the above analysis. Negligible curing and constant temperature of the thermoset molding compound must be assumed when making use of Equation (4A8), since the constant $C$ is actually a function of cure state and material temperature. This is a reasonable assumption for a first approximation. Before applying Equation (4A8), however, some consideration of the geometric differences between the experimental case (Figure 4-13, bottom) and the model case (Figure 4-13, top) is necessary. According to the pressure drop versus flow rate data reported in Section 6.4.3.1, the presence of a 0.018 cm land on a gate increases the pressure drop by approximately 7.5% in comparison to the same gate with no land. In addition, the rectangular entrance region on an actual gate slightly increases the pressure drop over the model case as well. Simple calculations for flow in a rectangular slit can be used to approximate the magnitude of this contribution, although it is important to note that the short channel length of this entrance region suggests that the equations describing fully developed flow are not completely applicable. Nevertheless, it appears that the contribution of the rectangular entrance region is in each case approximately 2–3% of the total pressure drop. To adjust experimental data, therefore, the pressure drop for gates with a land length of approximately 0.018 cm land should be reduced by 10% to account for the entrance region and land, and the pressure drop through a gate with no land should be reduced by 3% to account for the entrance region alone.

## REFERENCE

A1. F. N. Cogswell, "Converging Flow of Polymer Melts in Extrusion Dies," *Polym. Eng. Sci.* **12**, 65 (1972).

# 5 Equipment and Facilities for Plastic Package Molding

Equipment choices and configuration have a considerable effect on the productivity and yield of plastic packaging operations. The capital expenses associated with this equipment are large, and often they are multiplied several times over by the number of molding stations installed or upgraded at a time. The equipment costs are small, however, compared to the losses associated with poor molding yields and low productivity over the lifetime of the equipment, particularly when more sophisticated and expensive devices are packaged. Assembly equipment was reviewed briefly in Chapter 1 in the overall process description. This chapter concentrates on the molding equipment: its design, construction, function and optimization.

## 5.1 MOLDING COMPOUND STORAGE FACILITIES

Thermoset molding compounds are usually packaged in plastic pails holding approximately 20 kilograms of material. The pails are shipped in refrigerated containers or surrounded with dry ice, particularly if they are shipped from overseas and their transit time is more than a few days. The shipping container will usually have a temperature recorder that you may be able to read or that you will have to return to the supplier for reading. This is an important facet of material quality control and should not be overlooked since material that is subjected to normal room temperature for even a few days will experience significant advancement of the polymerization, and could behave differently in sensitive molding tools. The characteristics of advancement are reduced spiral flow length, reduced exotherm in thermal analysis, higher viscosity, and reduced flow length and flow time in the molding tool. Increased flow-induced stresses are the consequence of these physical parameter changes.

The molding compound should be stored at freezer temperature of $-18$ to $-7$°C (0 to 20°F) as soon as it arrives at the facility. Several large walk-in freezers will be required for this purpose at production molding sites since thousands of kilograms of molding compound have to be stored. Conventional size freezers (2.5 $m^3$) are adequate for test and laboratory facilities where only a few hundred pounds of material need to be stored. A large walk-in freezer may have $30,000 worth of molding compound in it at a time, so it may be worthwhile investing in power loss alarms and auxiliary power supply to the freezers to protect this investment, especially when power losses are common. No special conditioning is required for the atmosphere inside the freezer. The low temperature assures that the humidity level is low, possibly below optimum. Molding compounds can be stored for months at these conditions with no appreciable loss of properties. Much longer storage times, even at freezer conditions, are discouraged. The user should have a system for keeping track of the age of the inventory to assure that excessively aged material is discarded rather than molded.

## 5.2 MOLDING COMPOUND CONDITIONING FACILITIES

The effects of moisture on the viscosity, kinetics, and ultimate mechanical properties of the epoxy molding compound were described in Section 4.4. These property changes can have a significant impact on the productivity, yield and reliability of the plastic package. In view of this important influence, it may be worthwhile to provide special conditioning for the molding compound prior to molding. This is especially important in production facilities that do not conduct the molding operations in clean rooms, and where large swings in temperature and humidity are common. In most cases, humidity levels below 20% and above 50% provide adverse effects on molding compound behavior. Below 20%, the molding compound viscosity is generally too high, whereas above 50% the deleterious effects on glass transition temperature and thermomechanical properties become significant. Material removed directly from a shipping container stored in a freezer is likely to have a moisture content near or below the acceptable range. If left standing in a room at relatively high humidity, the cold preforms will rapidly absorb moisture, and condensation on the preforms is also likely.

Preform conditioning is much easier in molding operations that are done in controlled environment rooms, although the molding operation cannot be conducted in a true clean room because of the hydraulic oil and preform dust associated with the process. In such cases, the preforms are conditioned in the ambient room atmosphere. The humidity level should be adjustable and maintainable at the desired level that works best with the molding compounds in use. Fortunately, this humidity range of 20–50% corresponds closely to the comfort range for people. Conditioning in ambient conditions in a controlled environment room has the advantage of being free of the humidity disturbances that occur when conditioning is done in a separate enclosure. Repeated opening of the enclosure keeps disturbing the equilibrium in the chamber, making it almost impossible to maintain a constant relative humidity level.

In the absence of a controlled environment room, a conditioning chamber is the next best option for preform conditioning. This is fairly difficult, however, because of the large number of preforms involved and the need to access the chamber often, thereby disrupting the humidity equilibration. The most suitable equipment for conditioning preforms is a large humidity controlled chamber or storage room. A reasonable equilibration time at a humidity level is approximately 48 hours, therefore the chamber has to be large enough to accommodate a two day supply of preforms per machine. A full size transfer molding machine operating three shifts per day will consume approximately 1000 preforms per day, requiring storage capacity of 2000 preforms per machine. The preforms can be stored in their plastic shipping containers with the covers removed. Each machine will require the removal and use of one shipping container every four hours. The moisture uptake incurred during this four hour period on the production floor is not significant, but longer exposures should be avoided.

The 2000 preforms per molding machine will require a storage volume of approximately 1 $m^3$ (30 $ft^3$). For a moderate size molding operation of 10 presses, this translates to 10 or so smaller chambers or one walk-in humidity chamber. The room-sized chamber has a problem because it services all the molding machines. It will have to be opened repeatedly, which will disrupt the humidity level—a significant problem, since most humidity control systems are slow to recover. A smaller, chamber-sized unit that serves only one molding machine will have to be opened only twice in an eight-hour shift. In this case, however, a fairly large (refrigerator-sized) piece of equipment has to be accommodated in the vicinity of the molding press. The logistics of preform conditioning also have to be carefully observed. The operators have to be certain they use the preforms that are fully conditioned, rather than mixing them with preforms that were just placed under conditioning. Because of these difficulties and the significant space requirements, many manufacturers find that the logistics of preform conditioning are too difficult to warrant the benefits it provides. Most molding compound vendors ship their products with a moisture content that correlates with conditioning in 30–40% relative humidity. This is within the optimum range for molding compound performance. If the preforms are used within several hours of humid exposure, there should be no problem with using them without any conditioning.

## 5.3 PREHEATERS

Preheating is the first significant process step in plastic package molding. The preforms have to be brought to a consistent temperature of 90–95°C. This is accomplished through a dielectric heating of the material using radio frequency radiation, as in a microwave oven. The preheater has to have sufficient power to heat the mass of molding compound to the desired temperature in less than 40 seconds, preferably 25–35 seconds. Underpowered units that require longer heating times will begin to advance the polymerization since the reaction accelerates at elevated temperatures, particularly after it is heated above the glass transition temperature. A production scale preheater needs to heat up to 1.5 lb of molding compound. The typical power rating required for this size of charge is approximately 5000 Watts.

An important feature of a dielectric preheater is the control mechanism of the heating. Many older units use a simple timer for this function. The timer is set by conducting experiments on the preheat temperature versus increasing preheat time with the intended charge size. The time corresponding to the proper temperature is the timer setting. The problem with this type of control is that the heating rate is dependent on the moisture content of the preforms. In the case where the preforms are not conditioned, the temperature obtained after a set preheat time can vary by as much as 5°C. Figure 5-1 shows a plot of preheat

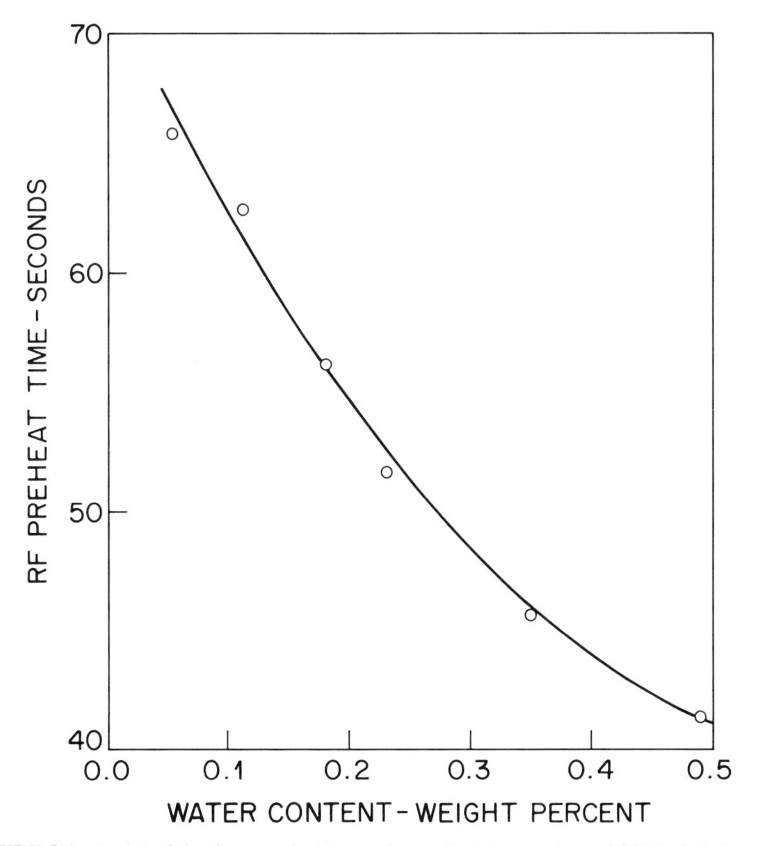


**FIGURE 5-1** A plot of the time required to reach a preheat temperature of 90°C plotted against the moisture content of the preforms.

time to reach a specified preheat temperature versus moisture content of the preforms where this significant effect on preheating is clearly evident. Preheat variations of this extent can significantly affect the process characteristics of the molding compound. Both low and high preheat temperatures can cause problems. Low temperatures will increase the viscosity, increasing the flow-induced stresses on the devices and altering the mold filling profile. High temperatures will overly accelerate the cure rate of the material resulting in high viscosities at the long-flow-length sections of the mold. The problem is actually exacerbated by the fact that the excess moisture in the molding compound which increased the preheat temperature also lowers the viscosity in the fluid state, thereby further reducing the viscosity. Conversely, low moisture content reduces the preheat temperature and increases melt viscosity independent of the temperature effect.

The remedy to this problem has been infrared temperature sensors built into the preheaters. An infrared source produces a radiation beam incident upon the molding compound charge during preheating, which senses its temperature. The control mechanism is then set to stop the dielectric heating once the temperature reaches a certain value rather than at a fixed timer setting. In this way, the preheat temperature is far more consistent despite differences in moisture content that will occur in most instances. The only difficulty with this control mechanism is that the calibration on the infrared sensor has to be checked regularly. Improved analog amplifiers have reduced drifting significantly over earlier models so that recalibration is required far less often.

## 5.4 TRANSFER MOLDING MACHINERY

The transfer molding machine can have a significant effect on packaging productivity and yield. The effects of machinery can also extend to reliability since accommodations made to compensate for faulty or inadequate machinery, such as low transfer pressure or low clamp pressure, can cause insufficient molding compound density resulting in higher moisture ingress and uptake. For this reason, molding machinery has to be selected, installed, and maintained with as much care as the equipment used in wafer fabrication.

Transfer molding machinery predates the microelectronics industry by decades. The process, one of the early plastic forming methods, has been used to manufacture sturdy electrical and structural parts for more than 40 years. Insert molding of metal conductors and contacts is facilitated in the transfer molding process, hence its attraction in molding parts such as distributor caps and rotors for automobiles. The machinery used for modern microelectronics packaging is not noticeably different than 20 to 30 year old machines used for these much simpler electrical parts, yet the refinements to the machinery can and do make a significant difference in process productivity and yield. The rudimentary com-

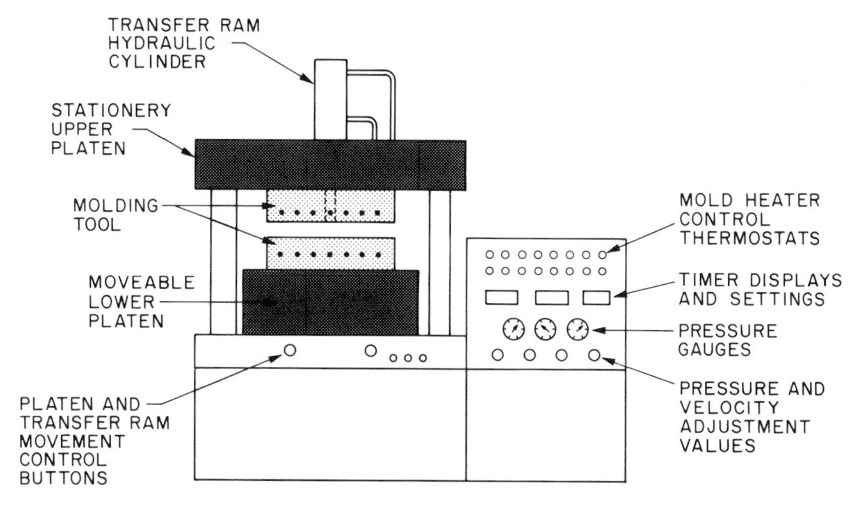


**FIGURE 5-2** Schematic diagram of a transfer molding machine of the type used for plastic packaging of microelectronic devices.

ponents of a modern transfer molding machine used for IC packaging are illustrated in Figure 5-2 and a photograph of a full size press is provided in Figure 5-3.

A transfer molding machine is essentially a large mechanical clamp, usually driven hydraulically, which holds the mold closed during material injection and

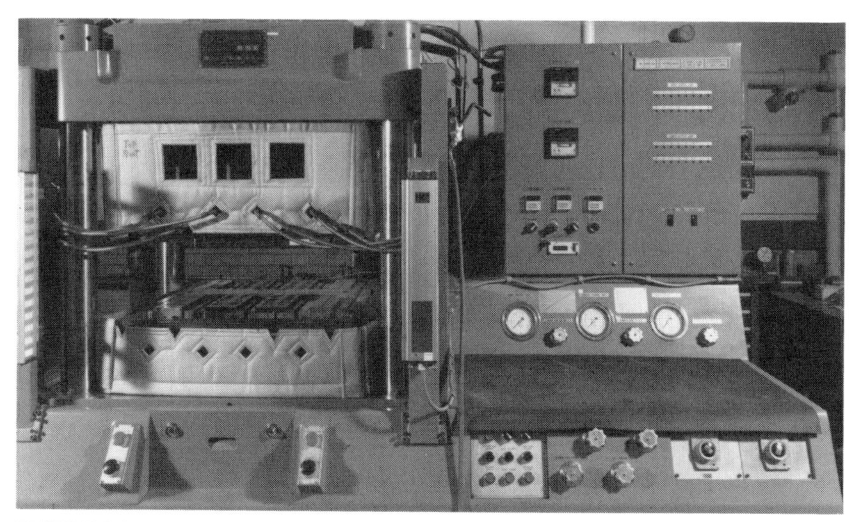

**FIGURE 5-3** Production-size molding press for molded packaging of integrated circuit devices. A large transfer molding tool is shown mounted in the press.

curing, and a hydraulic transfer ram or piston which pushes the molding compound into the mold. The process sequence steps of the machine are described in detail in Chapter 6. The design and features of the machine relevant to machine selection, configuration, purchase and set-up are discussed in the present section. The following is a list of these features and a short discussion of each.

*Machine Size* Transfer molding presses are sized according to the clamping force of the press. In general, a clamp force of 55 $kg_f$ is the minimum required for each square centimeter (1200 psi) of wetted surface area which includes top and bottom surfaces of the molded packages and the runner system. This is not intended as a generalization because the specific size will depend on specific designs and process conditions, but it is a useful scaling parameter to estimate press size needs. Exceptions are for molding tools that require substantially more or less filling pressures. Standard filling and packing pressures are approximately 5.5 MPa (800 psi), but can be as low as 1.7 MPa (250 psi) or as high as 7.6 MPa (1100 psi) depending on tool design and gate size. As an example of a normal case, a large molding tool that molds one hundred 68-pin PLCC packages with 13 $cm^2$ (2.0 $in.^2$) of surface area each, and which has another 260 $cm^2$ (40 $in.^2$) of runner and transfer pot area will require a clamp force of at least 130 metric tons (144 tons). For this reason, the most popular size molding presses for microelectronics packaging are in the vicinity of 200 metric tons.

*Platen Size* The platen area is usually scaled to the clamping force of the press. Platen areas of 0.5–0.8 $m^3$ (5–8 $ft^2$) are standard on large 200 ton presses, whereas platen areas as small as 0.1 $m^3$ (1 $ft^2$) are found on 10–15 ton presses.

*Setup Facilities* Some presses have special facilities for installation and setup of the molding tool. A full size package molding tool can weigh 1400 kg (3000 lbs) making it difficult to move and align on the platens of the press. This alignment is critical because the transfer ram has to pass through a precision bore cylindrical opening in the molding tool with very little clearance. Some large presses are equipped with an air flotation system to facilitate tool movement and alignment. High-pressure air piped to the molding machine is distributed over the surface of the lower platen and forced through a large number of small holes. This high-pressure air partially levitates the molding tool, significantly reducing the effort needed to move the mold to the precise alignment position.

*Mold Temperature Control* The temperature controllers for the molding tool are incorporated on the control panel of the molding machine as was shown in Figure 5-2 or in a separate control box as is shown in Figure 5-4. Precise

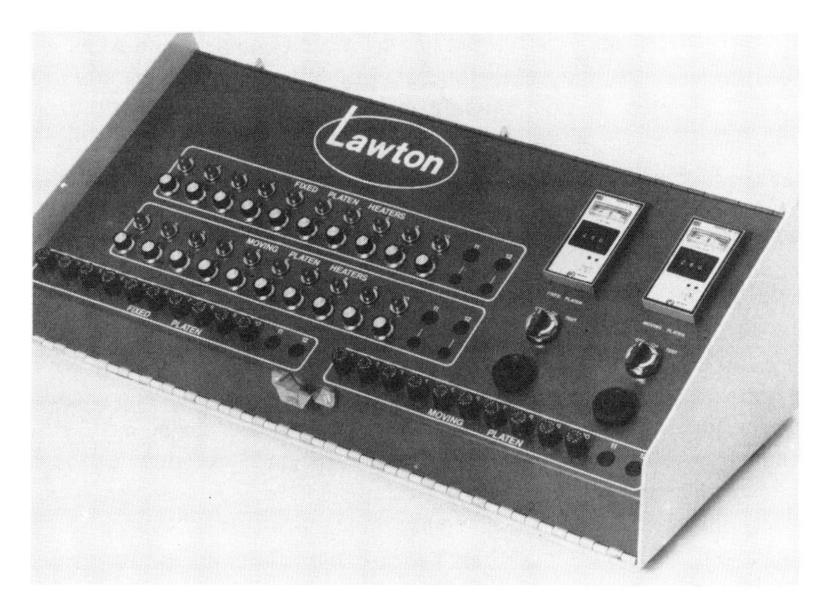


**FIGURE 5-4** Heater wattage controllers for achieving desired temperature profile over the mold surface. In most cases, a uniform temperature profile is required, but because of the internal construction of the molding tool and the different thermal loads on different sections of the tool surface, the heater wattage settings are not uniform. (Courtesy of the C. A. Lawton Co.)

and uniform molding tool temperature is a prerequisite of high-yield molding. For a production size mold, this optimum number is in the range of 8–12 per mold half; the higher number being preferred for more uniform temperature control. The molding machine should have enough controllers to accommodate this number of heater rods.

*Soft Clamp Timer* Soft clamping is a feature of the movement of the lower platen that causes the lower mold half to stop as it approaches and just touches the upper mold half. A delay timer is then activated and maintains this soft clamp configuration until the prescribed time interval expires after which the full clamping pressure is applied and the mold shuts tightly. The purpose of this soft close cycle is to allow time for the leadframes to expand fully on the mold when both the top and bottom of the leadframes are in close contact with the heated mold steel. The locations of the cavities in the molding tool have to be designed appropriately to account for this expansion. Full expansion provides more precise molded bodies and less buckling of the leadframe. Both effects improve the yield of the trim and form operations, hence this soft close feature is a desirable one if the molding tools are dimensioned to take advantage of it.

*Clamp Obstruction Sensor* Molding tools are often damaged when molding compound debris or clean-up tools are inadvertently left on the mold when the clamp is closed. Some molding machines have the feature that the clamp movement is closely monitored. These sensors are highly sensitive to movement resistance prior to reaching the expected full clamp position. On meeting any resistance prior to contacting the other mold half, the clamp movement stops and damage to molding tools is minimized. There is also an important human safety component of this type of sensor.

*Parallelism of the Platens* It is of great importance that the platens of the molding machine be parallel to a very close tolerance. Platens that are not parallel will cause the mold halves to be skewed slightly in the clamp closed position, resulting in excessive flash. Although this problem can be rectified with spacers, a process called *shimming* the mold, it is much preferred that the platens are parallel and remain so during clamping. The number of posts or hydraulic cylinders that close the clamp has some influence on platen parallelism. More posts can minimize the deformation sustained under high-pressure clamping and thereby preserve the parallelism built into the press. Up to five support posts are available on some machines: one in the center and one in each corner of the platen.

Conversely, machines with only one central post can also be engineered to provide very high precision and close tolerances. State-of-the-art equipment can now hold platen planarity to less than one thousandth of an inch.

*Noise Levels* Most transfer molding machines are driven by hydraulics which can be noisy. Newer molding machines have much lower noise levels than older equipment, and even further reductions are sought. Noise from the hydraulic pump has a significant effect on worker comfort and fatigue, therefore machinery with the lowest possible noise level should be sought.

*Working Height* The working height of the press is an important consideration from a worker comfort and safety standpoint, but it also affects productivity. The working height is the distance from the floor to the surface of the lower platen when the press is opened. The surface of the molding tool can be another 30 cm greater than this height since most transfer molding tools are relatively thick. Presses with higher working heights, greater than 125 cm to the lower platen surface, often require a platform built in front of the press to bring the height to a more convenient level. There are productivity, fatigue, and safety implications associated with the operator climbing onto this platform repeatedly. Working heights below 120 cm are preferable for this reason.

*Safety Features* Safety features should be a first-order consideration in selecting molding equipment. The greatest dangers of the transfer molding operation are limbs getting caught in the platens when the press closes, and the dangers associated with bursting of high-pressure hydraulic lines. The press should be thoroughly safeguarded against the possibility of anyone getting an appendage caught within the press. Some of the possible safeguards against this include a light curtain for the front opening of the press, visible in the photograph of the molding press provided in Figure 5-3, complete enclosure of the other three sides of the press, and a two-button activation switch which requires that the operator has one hand on each button to start the press movement, also shown in Figure 5-3. A large red shutoff switch should also be easily accessible. The best safeguards are worthless, however, if they are easily deactivated or if the press operation requires deactivation on a regular basis. There are some instances when the safeguards have to be disabled. The side enclosures may have to be removed when a new molding tool is installed and aligned and the press may have to be operated in this configuration to conduct the alignment effectively. In these instances, there should be a second person present in the event of an accident. With regard to the hazards presented by the failure of a hydraulic line, safety glasses should be worn at all times when the hydraulics are on. Molding areas should be clearly marked as safety glass areas and there should be a supply of visitor glasses on hand.

## 5.5 MOLDING TOOLS

The molding tool probably has more influence on the productivity and yield of plastic packaging than any other piece of equipment. It is also one of the most expensive pieces of capital equipment, so mistakes are not easily obliterated by simply replacing an inferior tool. A production size molding tool can easily cost more than a production molding machine because of the high labor content of the custom designed and machined tool. It is only since the 1980s that different types of molding tools are available. Prior to that time, essentially all transfer molding tools for plastic packaging were the conventional cavity molds that have been in use since the 1940s. The introduction of aperture plate molds and multiplunger molds offers new options in tool technology. This section reviews the types of tools for plastic packaging and the features that could be relevant to productivity and yield. It does not address optimizing the design according to process characteristics which will be addressed in Chapter 7.

### 5.5.1 Cavity Chase Molds

This mold design is the type used for nearly all transfer molds outside the microelectronics industry and it is probably the most popular type of tool for plastic packaging as well. A cavity chase mold is based on conventional split mold design where the cavities are machined out of steel blocks and plates and then assembled to produce a large multicavity mold. The fabrication techniques can include precision grinding, electrode discharge machining (EDM), or milling. The machined components are then assembled to form the cavities and runner system. The mold produced in either case is a split cavity design where one half of the cavity is in one side of the mold and the other half is formed in the other side. The runner system typically resides only in one half. Ejector pins are also required to push the molded part out of the cavity. A cross section of a cavity and runner for a cavity chase mold is provided in Figure 5-5.

The ejector pins are controlled by the movement of the platens. It is usually preferable to have the molded parts remain in the lower half of the mold so that

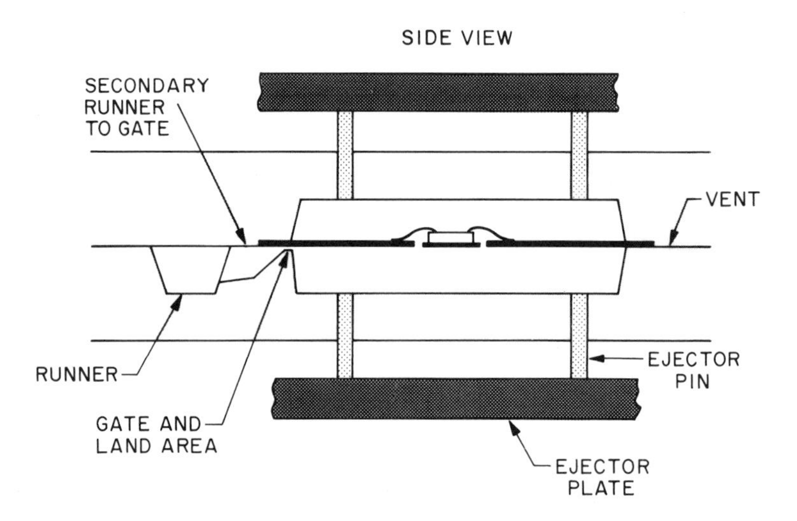


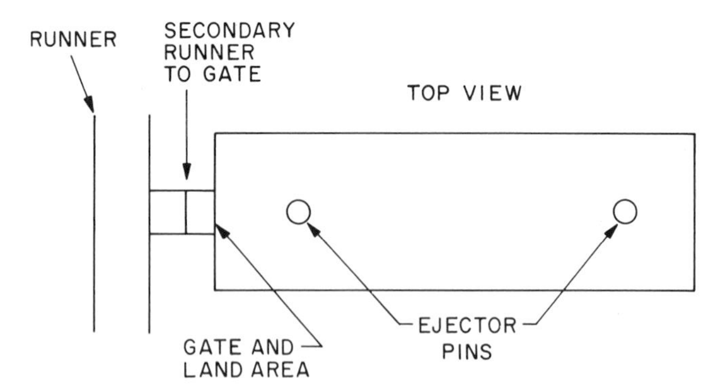


**FIGURE 5-5** Side and top cross-sectional views of the cavity and runner for a cavity chase mold. The side view cuts through the long axis of the cavity and perpendicular to the runner and shows the configuration of the ejector pins and ejector plate. The top view shows the approximate size of the subrunner and gate and the positioning of the ejector pins on centerline of the package.

they are easily accessible to the operator for removal. Upon opening of the press, the ejector pins in the top half of the mold are activated first and they free the parts from the top cavity after which the lower ejector pins activate and release the parts from the cavities. The lower pins retract to the original flush position and the molded parts fall back into the lower cavities where they are removed without difficulty. The upper pins retract when the clamp is closed.

The cavities in the mold are grouped according to the number of sites on the leadframe strip. There can be either one or two leadframe strips per runner arm. The grouping of two leadframe strips fed by a single runner that passes between them is called a *chase*. Large plastic packages such as the 84 PLCC consume too much molding compound to feed two leadframe strips from a single runner, hence these designs usually have only one strip per runner arm.

The ejector system significantly increases the weight and bulk of a cavity chase mold. A typical plastic package design such as a 32-pin DIP will require four ejector pins per package: two on top and two on the bottom. A mold for 32-pin DIPs would typically accommodate 84 cavities providing nearly 336 separate ejector pins. A 68 lead PLCC package could require one ejector pin in each corner of each side of the package for a total of eight pins per package and over 500 for a production mold. Large ejector plates are contained in the upper and lower halves of the mold to push the ejector pins. The large open areas within the tool for the ejector pins would reduce its structural integrity if it were not for the substantial steel posts that reinforce the construction, stiffening the tool and preventing it from flexing under the compressive loading in the clamp closed position. The cumulative weight of all of these components pushes the total weight of a production packaging tool to several thousand pounds. An illustration of a cutaway view of a cavity chase mold is provided in Figure 5-6.

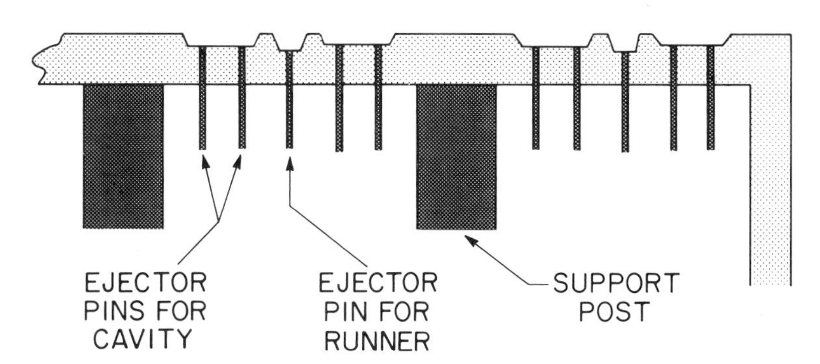


**FIGURE 5-6** Cutaway view of the lower half of a cavity chase mold, showing the support posts and ejector system. The drawing is simplified for illustrative purposes, all structural details are not shown.

*Universal Mold Base* A cavity chase mold is usually dedicated to molding one type of plastic package. The machined cavities, ejector system, and support posts are not easily rearranged or modified to mold a different package design, and even similar designs cannot be accommodated. In some instances of development work, prototyping, or short production runs, a more flexible tooling arrangement may be advantageous. This can be realized in a cavity chase mold through the use of a universal mold base (UMB). A UMB is a special design of a cavity mold where the chase locations are designated and fixed, but the cavity blocks and ejector system are removable and replaceable. With this design, one or more chases can be removed and replaced with a different package design. Smaller UMBs of two to four chases are useful for prototype work and generating sample lots of new package designs. A large UMB is not cost effective when most or all of the chases have to be replaced. In that case, the cost of all six or eight chases will approach the cost of a new dedicated molding tool. Replacement of only one or two chases in a large eight chase UMB with the new design is feasible and it is an effective way of providing the process characteristics of a full production mold. A photograph of a UMB with eight chases of different package designs is shown in Figure 5-7.

The principal disadvantage in using a UMB is that smaller molding tools do not emulate the process characteristics of the full size production tool, hence

**FIGURE 5-7** Photograph of a universal mold base (UMB) which has several different package designs mounted. In a UMB, the chase blocks sit on the mold body instead of being machined into the body. In this way, the chases are easily replaceable.

processing issues are not confidently resolved at this prototype stage. Another problem is that even on full-size UMB molds, the heat transfer and temperature profile over the tool surface will probably be different than a dedicated production tool because of the differences in construction. A UMB has more open volume within the body of the tool and the chases themselves are not as well thermally connected. This can result in hot and cold spots over the surface of the tool. Nonetheless, this tooling option is both technically and cost effective for prototype packaging work.

### 5.5.2 Aperture Plate Molds

Aperture plate molds are a patented transfer molding technology [1] developed exclusively for plastic packaging of microelectronic devices. The plate mold addresses some of the shortcomings of cavity molds that are particular to package molding. Its construction is different from a cavity mold in that it is assembled from a series of stacked plates. The sides of the package body are formed with cutouts in the two aperture plates. The leadframes are loaded between these two plates where they are registered onto pins that also serve to align the plates. The top and bottom of the body are formed by separate plates. The runner system is not in the plane of the molded body as in a cavity mold, but is instead in the plate above the aperture plates; the same plate that forms the bottom surface. The general construction of a plate mold is illustrated in Figure 5-8.

The surface finish plate forms the top surface of the molded body and therefore imparts the intended surface texture, either polished for laser code marking or a matte finish for ink marking. These four plates: the two aperture plates, the runner plate, and the surface finish plate, are the device specific part of the molding tool. The remainder of the tool can be used for many different package designs. These additional parts consist of the heater plates where the heater rods are located, the thermal insulation plates that thermally isolate the heated portion of the tool, and the structural support section of the mold. This structural section also serves as a spacer element since a plate mold can be much thinner than an equivalent cavity chase mold. During the molding cycle, the two aperture plates are removed from the tool and loaded and unloaded at a separate work station. All other plates remain in the press.

There are several significant differences between an aperture plate mold and a cavity mold. A plate mold does not have a built-in ejector system. None is required because the molded bodies are not formed in closed-end cavities. Both the top and bottom surfaces of the molded body are exposed when the aperture plates are removed from the press after the cure cycle. Ejection is accomplished through the use of an ejection fixture. The plates are manually pressed against a ''waffle iron'' device which positions a raised area against each of the cutouts in the aperture plates. The first application of this tool separates the two plates

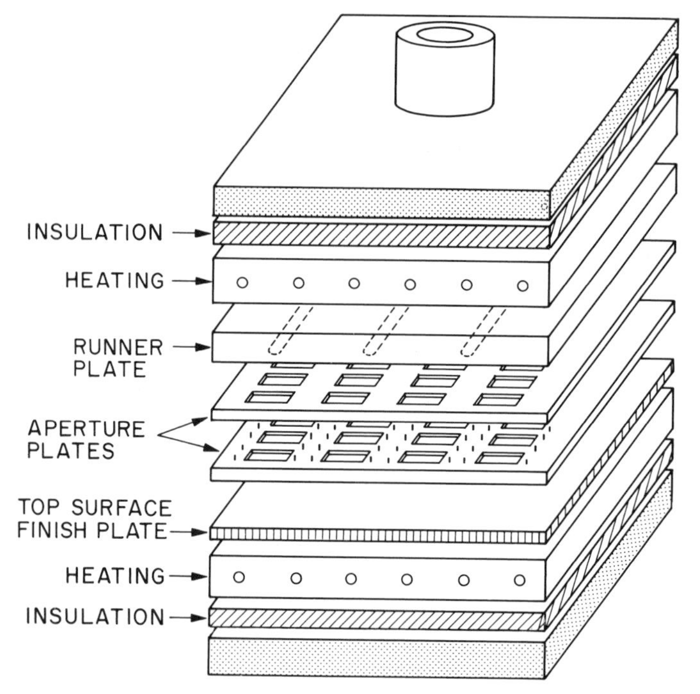


**FIGURE 5-8** A schematic diagram of the construction of a plate mold.

with the molded body remaining on the top plate. The second application ejects the molded parts from this plate. Without an ejector system, an aperture plate mold is considerably simpler and lighter than a cavity chase mold.

A significant difference between a cavity mold and an aperture plate mold is the shuttling of the aperture plates in and out of the molding press during every cycle. The plates are thin, essentially the half thickness of the molded package, their surface area to volume ratio is high which allows them to cool to near room temperature between cycles. Their low thermal mass also allows them to heat rapidly so the conduction time in the press to bring them back up to the mold temperature is not long; in many cases no extra time interval is allotted for this heating, it being assumed that the plates reach an acceptable temperature during the inherent delays in the cycle such as preheating of the molding compound. The implications of this temperature cycling will be discussed in the optimization discussions in Section 7.4.1.

There is a major difference in the gating of an aperture plate mold compared to a cavity mold. In most cases, the runner in a plate mold passes parallel along the bottom edge of the molded body. The gate between the runner and the cavity can be formed anywhere along this intersection as is shown in Figure 5-9. The gate width can be any fraction of the length of this edge of the package, and

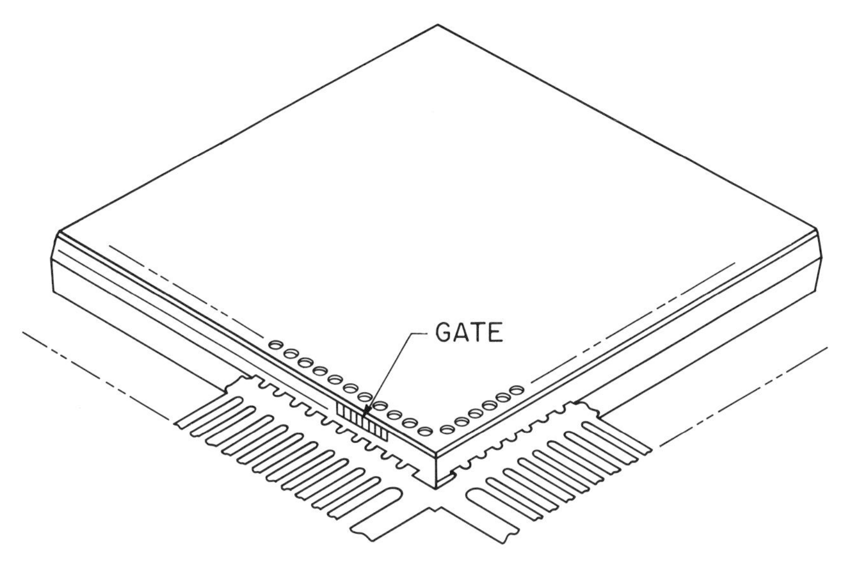


**FIGURE 5-9** The gating of a 68 PLCC package with an aperture plate mold. The gate is along an edge of the package. Top or bottom surface edges are possible depending on whether the package is molded rightside up or upside down.

the gate depth is controlled by the depth of the recessed area in the aperture plate. The gate itself is actually an orifice that connects the runner to the cavity with little if any intervening distance as is shown in Figure 5-9. Compare this gate geometry with that of the cavity mold shown in Figure 5-5. A cavity mold has a short runner segment with a convergent section that feeds the orifice to the cavity. The pressure drop through a cavity mold gate is usually greater than the pressure drop through an aperture plate mold gate because of these differences, resulting in a higher transfer pressure to fill a cavity mold.

One of the major advantages of an aperture plate mold is its adaptability to mold many different types of packages with only small capital expenses for retooling. Packages within the same generic design, such as 0.600-inch-wide DIPs, can be molded in the same mold base using the same runner plate and cover plate. Only the aperture plates and the ejection fixture would have to be changed to mold 24-, 32-, or 40-pin DIPs in the same mold base. Packages with significantly different designs such as 68 PLCC could also be molded in the same mold base with a change of runner plate as well as the aperture plates and ejection tool. Similarly, the top surface finish of the package can be changed easily by removing the lower surface finish plate and replacing it with a different one. In this way, a switch between laser code marking and ink marking can be made in a short time and with minimal capital expense. The operation of the plate mold requires much more operator involvement, cited by some as a disadvantage. Recent automation of plate molds may alleviate this concern. A more complete comparison between the features of plate molds and cavity molds is provided in Table 5-1.

### 5.5.3 Multiplunger Molds

Multiplunger molds are smaller, automated versions of cavity molds. They are also known as gang pot molds. The distinguishing feature of a multiplunger molding machine is that there are a large number of transfer plungers as opposed to a single plunger for a conventional molding machine. One to four cavities are fed from each transfer pot as is shown in Figure 5-10. The machine still molds the packages on leadframes so it is not radically different from conventional plastic package molding. Simultaneous delivery of the large number of preforms to the transfer pots cannot be accomplished manually, so extensive

**Table 5-1. A Comparison of the Features of Cavity and Plate Molds**

| *Feature* | *Cavity Molds* | *Plate Molds* |
|---|---|---|
| Number of Cavities | Several hundred for small DIPs, less than 100 for PLCCs | More than cavity molds because of closer packing of leadframes |
| Molding compound waste | 20–40% | 20–40% |
| Susceptibility to flash | Average, requires careful shimming | Minimal due to self-alignment of flexible plates |
| Ejection | Ejector pins in each cavity | External ejector tool |
| Cavity gating | End slit gate for DIPs, corner gate for QUADs | Orifice on edge of molded body can be length of one side of package |
| Package features | Radii on all edges | Edges are sharp since they are formed by the intersection of the plates |
| Pressure to fill | 600–1000 psi | 200–400 psi |
| Temperature profile and stability | Good | Shuttling in and out of press causes temperature drop which may require extra preheat steps |
| Labor cost | Medium, can be reduced through automation | High, greater number of operator steps |
| Capital cost | High, mostly custom machining | Lower than cavity mold |
| Cost of changeover to different package | High, requires new molding tool, or new UMB chases | Low, mold base is interchangeable, aperture plates are inexpensive. |
| Maintenance costs | Medium, gate, and cavity wear after tens of thousands of cycles | Comparable costs to cavity molds, plates are replaceable |

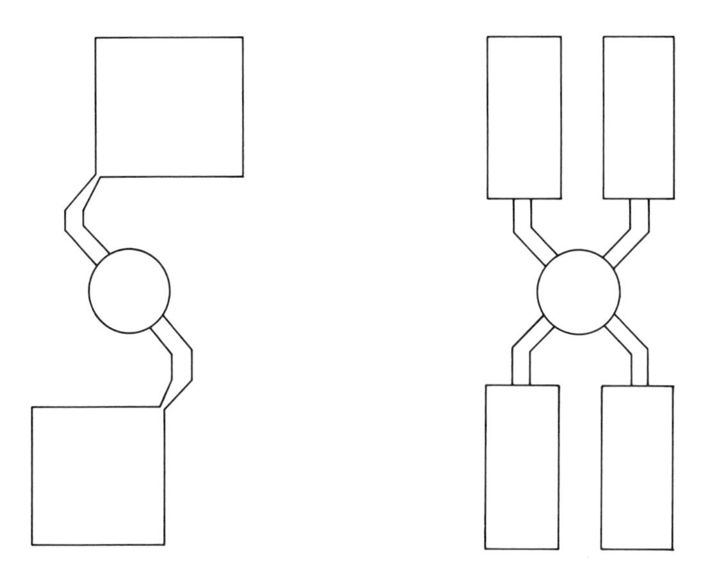

**FIGURE 5-10** A schematic diagram of several configurations of multiplunger molding. There can be one, two or four cavities fed from a single transfer pot.

automation is usually incorporated in a multiplunger machine. Most of these machines require little if any operator attention during normal operation.

There are several economic and technical advantges to using multiplunger molds. The technical advantages include the exceptionally short and uniform flow length inherent to the design. This assures that all of the molding compound arrives at the devices in the same state of cure and at the same temperature. This is a technical advantage that makes optimization much easier; it cannot be achieved on conventional transfer molding tools where the molding compound flows along a runner to reach the cavity locations. The short flow length and attendant short flow time also allow the use of faster curing molding compounds which are able to flow for less time, but can be ejected from the mold sooner. Typical cycle times for multiplunger machines are less than two minutes compared to three to four minutes in a conventional machine. Another advantage is the ability to use long cavity fill times to reduce the flow velocity over the devices and thereby lower the flow-induced stresses. If the allowed flow time of a rapid cure molding compound is 12 seconds, most of this time can be used for cavity filling since only a small fraction of the flow time is used in the flow of the molding compound to the cavity. In a conventional transfer mold almost one half of the 18-second flow time can be used in reaching the far cavities.

The disadvantages of multiplunger molding machines are also important, and have generally prevented the technology from capturing a large fraction of the market. The preforms cannot be handled easily if they are preheated to the normal preheat temperature of 90–95°C. Therefore, most machines use a lower preheat temperature or no preheating at all. The temperature of the molding compound as it reaches the cavities can be too low and the corresponding viscosity too high for delicate devices. The short flow length limits the amount of conductive heating that can be achieved, hence high flow-induced stresses can be a consideration. Conversely, unheated preforms are excessively abrasive on the molding tool and plunger because the glass filler is not thoroughly wetted.

Other considerations in using multiplunger machines include the extensive use of automation to preheat and deliver the preforms, load the leadframes, remove the molded frames, break off the gates, and clean the mold. Dozens of sensors are located throughout the machine to assure that there are no jammed parts or mechanisms. Although most of this automation is simple and very reliable, upkeep and maintenance of such a complicated piece of production equipment is a consideration. The level of complexity of the process usually requires that the total number of cavities is much lower than can be achieved in a conventional transfer mold. Most multiplunger machines mold only two to four leadframe strips per cycle compared to 12 to 24 for large conventional molds. The twofold reduction in cycle time cannot recover this significant drop in throughput, hence overall hourly output is generally lower. Overall economic productivity may be approximately the same however, if the lower labor content of the automated system is included. Table 5-2 is a more concise summary of the advantages and drawbacks of multiplunger molding in comparison to conventional transfer molds, grouping both aperture plate and cavity molds together for this comparison.

### 5.5.4 Design Considerations for Molding Tools

This section reviews the design and construction features of the molding tools used for plastic package molding. These are largely practical considerations that concern tool performance and upkeep rather than technical features of the mold design that influence productivity of the molding operation. These optimization considerations will be discussed thoroughly in Chapters 6 and 7.

**5.5.4.1 Materials of Construction** Transfer molding tools for plastic packaging are made from high-grade tool steels. The cavity blocks and mold body are usually of an air hardened steel denoted by some manufacturers as A2. Air hardening affects the bulk properties of the steel rather than the surface alone, increasing the hardness to a value of 58–60 Rockwell (Scale C). At this value, the steel has good hardness without being brittle. It is fairly resistant to gate erosion, but special surface treatments improve abrasion resistance considerably. The ejector pins, being moving parts, are often made from case hardened

**Table 5-2. Advantages and Disadvantages of Multiplunger Machines**

| *Advantages* | *Disadvantages* |
| --- | --- |
| Highly automated, low labor | Inadequate preheating of the preforms may have deleterious effect on processing and maintenance |
| Greater consistency without operator involvement | Repair and upkeep of the extensive automation |
| Improves packing because it assures gates remain open until packing pressure is applied. | Hourly output is lower than a full-size single-pot mold |
| Uniform flow time and material properties in all cavities | Tooling is relatively inflexible because it is closely tied to the automation and number of plungers |
| Amenable to full in-line automation of assembly and packaging operations | High initial capital cost |
| Lower cost for small molding tools. | |

carbon steel, which gives them a very hard surface and a softer core. Ejector pins which are hardened throughout are also common.

The materials of construction of an aperture plate mold can be different from those of a cavity mold. The aperture plates and runner plate are usually made from an air hardened tool steel denoted by some manufacturers as H13. The thin plates need to be more ductile and so they are hardened to only a 45 Rockwell hardness. The frames or rails of the aperture plates can be either steel or aluminum. Aluminum rails save a pound or two of weight, an important consideration given that the plates are shuttled in and out of the press on each molding cycle.

**5.5.4.2 Surface Treatments for Molding Tools** Heat and chemical treatments of the surfaces of the molding tool can improve abrasion resistance and facilitate release of the molded article from the tool. In many instances the cavities and gates of a cavity chase mold are thermally and chemically treated to increase their hardness. One such process is called *nitriding*. This helps to minimize the erosion of the gates due to the abrasive glass filler in the molding compound. There is some embrittlement of the steel caused by this treatment, but it is insignificant because it is only a surface effect, the bulk of the steel being unaffected. One drawback of hardening is that machining of the hardened surfaces will be difficult if it should be required. Carbide machining tools, which are expensive and difficult to work with, will be required. The need to rework gates or runners in the event of poor yield or change of molding compound is one of the few cases where additional machining work may be required. This scenario is unlikely in production molds, so hardening is a standard feature of most mold manufacturing. The situation for prototype molds is different, however. In this case, changes in gate, runner or package design are far more common. Not all design changes are amenable to reworking as some will require new machined parts, but machinable changes are common enough and the cumulative effects of erosion so small that hardening is often not useful for prototype tools.

Another important type of surface treatment for transfer molding tools is nitro-carborizing. This is a combined heat and chemical surface treatment of the tool to improve its toughness and reduce its susceptibility to erosion. The treatment is quite effective in most applications, but it is also relatively expensive. Treatment for the wetted areas of a production mold can easily cost several thousand dollars. Similar to nitro-carborizing is the titanium-nitriding surface treatment, often known simply as nitriding. This treatment leaves a characteristic bright yellow finish on the surface. Data on the longevity improvement realized with nitriding and nitro-carborizing are not available as the application to plastic package molding is recent. There is some indication that these surface hardening treatments improve the release of the plastic from the mold surface, but no specific citations can be provided. Chrome plating has also been used in transfer molding tools, but the microporous nature of this surface can hamper release of the molded part.

**5.5.4.3 Air Venting** Venting is one of the more important design aspects of a transfer molding tool. The mold filling step is more properly viewed as one fluid displacing another fluid in the cavities; the movement of the original fluid, air, is critical to satisfactory flow and elimination of defects. Vents are provided in the design of the molding tool to facilitate the elimination of air. These vents have to be in the proper location and of the appropriate size. In general, the location of the cavity vents should be on the parting line and at the point the fluid reaches as it completely fills the cavity. In this way, all the air is pushed out ahead of the advancing molding compound. This location is obvious in an end-gated cavity chase mold as is shown in Figure 5-11. For corner-gated PLCC packages, the venting should be provided at the corner opposite from where the molding compound is introduced. Venting at the ends of runner dead-ends should also be provided as necessary. The size of the vent is also an important consideration because it regulates the degree to which air can escape freely but molding compound cannot. Vents should be wider than they are deep because flash is more dependent on vent depth. The typical depth of vents in cavity chase molds are 0.00125 cm (0.0005 inch), with a wide variance depending on the molding compound in use and the mold filling behavior of the particular tool. Vent widths are of the order of 0.40 cm (0.25 inch), with wide variation around this value. Excessive mold filling pressures or sporadic unfilled cavities are signs of inadequate venting. The problem can be corrected by enlarging

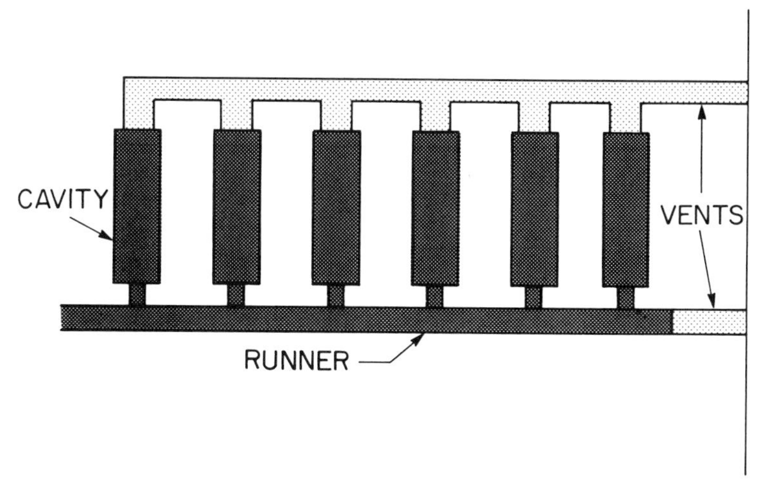


**FIGURE 5-11** Diagram showing the location of the vents for 40-pin DIP packages molded in a cavity chase mold.

both the width and depth of the vent. Increase the gate depth by only several ten-thousandths of an inch at a time as there is often a fine line between adequate venting and a flash problem. Void problems and incomplete fill can also be caused by improper positioning of the vent. Increasing the vent width or changing the vent location may resolve the problem. Some toolmakers incorporate an air vent on the plunger, essentially a small flat section on the cylinder, to allow air venting.

The venting for plate molds is slightly different than that for a cavity chase mold. In general, the venting options are greater and wider vents at more locations can be incorporated. The lower filling pressure of plate molds also allows deeper vents to be used without the threat of excessive flash.

**5.5.4.4 Draft Angles** A draft angle is the angle past perpendicular that the cavity and runner sidewalls make with the bottom surfaces of the molding tool as is illustrated in Figure 5-12. Draft angles are essential in allowing the molded part to be easily removed from the cavity in any type of molding operation. The draft angles common on plastic packages molded in a cavity mold are of the order of 10°. Draft angles on the molded body of 7° are more typical for plate molds because of the more effective action of the manual ejection tool. Smaller draft angles on either plate or cavity molds could cause parts to hang up in the cavities or apertures. There is not unlimited range of draft angles on package bodies since the package dimensions must fall within certain industry standards to be sold and installed. These design criteria as usually wide enough to accommodate small changes needed to solve draft angle problems. Draft angles are

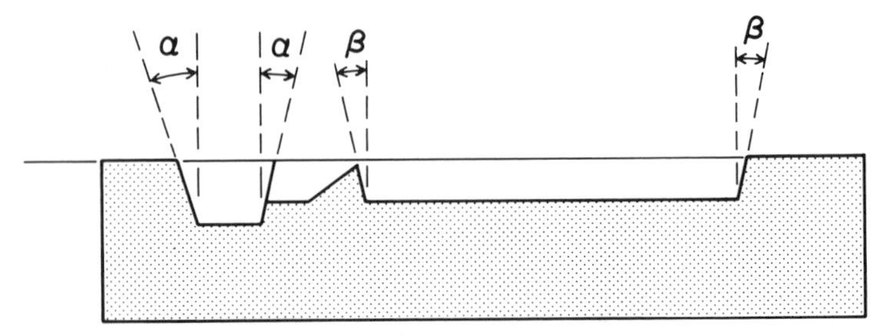


**FIGURE 5-12** Draft angles on the package and runner: $\alpha$ = 15–25°, $\beta$ = 5–10°.

typically greater in the runners, ranging up to 20° in cavity chase molds. Runners in plate molds often have a semicircular or rounded trapezoidal cross section to facilitate the runner dropping out of the runner plate without the assistance of any ejector pins.

### 5.5.5 Mold Maintenance and Cleaning

Residues from the molding compound and the waxes used to prevent sticking can accumulate on the molding tool and ultimately become thick enough to affect surface finish of the part and distort part dimensions, particularly by accumulating in corners and at edges. These stains and residues have to be cleaned on a regular basis. For a molding machine in full production of approximately 200 cycles per day, this cleaning will be required about once per week. This value can vary widely depending on the type of molding compound used, the amount of internal mold release, the specific process conditions, and the amount of external wax used. In general, high mold temperatures and high filling pressures promote greater accumulation of stain and residue. The cleaning process requires scrubbing out the wetted areas with a special cleaning material such as methyl pyrolidone (M-Pyrol), and/or molding with a special melamine molding compound specifically formulated to remove residues and stains. These cleaning materials are available from the molding compound supplier in most instances. The materials themselves are often strong chemicals that should be used only according to the suppliers' instructions and with all appropriate safety precautions. The operators should be thoroughly aware of the hazards of these materials and familiar with the recommended safe handling procedures.

## 5.6 POSTCURE

Postcure is required for essentially all epoxy molding compounds used in molded plastic packaging. The need for postcure is derived from the chemical kinetics of the polymerization reaction. The reaction rate drops off rapidly with concentration, linearly for a first order reaction and with larger exponents on the concentration for higher order kinetic expressions. It requires a much longer time to increase the conversion from 70% to 90% than it takes to increase the conversion from 0% to 20%. There would be a severe penalty in productivity if the reaction was carried to near 100% conversion in the mold cavity, tying up the capital equipment and labor for hours while the reaction approaches the near complete conversion needed for full material properties. Postcure has to be conducted at temperatures above the ultimate glass transition temperature of the epoxy molding compound, typically 140–160°C. Postcure at higher temperature reduces the required time to reach full conversion. Usual postcure treatments are 170–175°C for 4–5 hours. Postcure is conducted batchwise in large convective ovens that are maintained at the postcure temperature. These are fairly conventional ovens that can be purchased in a variety of sizes. A medium size production facility will have to postcure thousands of devices every four hours, requiring several large ovens. The ovens themselves do not require precise temperature control or rapid temperature ramp-up since they are maintained at the same temperature indefinitely, and the postcure temperature can be $\pm 3$°C without causing any noticeable problem. The ovens should be calibrated every several months to ensure that they are not far off the intended set point. Negative temperature deviations of more than 10°C could result in incomplete cure with attendant deficiencies in physical properties, whereas thermal degradation could result from positive deviation of more than 10°C.

## 5.7 DEFLASHING

Deflashing is the removal of the molding compound that has inadvertently flowed onto the leadframe due to design, equipment or process inadequacies. The common location of flash is shown in Figure 5-13. Although common, it is preventable and should be viewed as a productivity problem rather than as a normal aspect of package molding. Nonetheless, essentially all plastic packaging shops have deflashing equipment. The typical thickness of leadframe flash is less than 0.001 cm (0.0004 in.), but if left on the leads it would cause problems in the downstream operations of trim and form and solder dipping. Removal equipment takes advantage of two characteristics of the flash; its brittleness and its thinness. The different types of deflash equipment are reviewed below.

*Media Deflashers* Media deflashers are a common type of deflashing equipment that are used throughout the plastics processing industry as well as in microelectronics packaging. In this equipment, the leadframes pass under and over high pressure pneumatic streams that contain inertial media such as plastic beads, cracked walnut shells, or ground apricot pits. These materials may sound

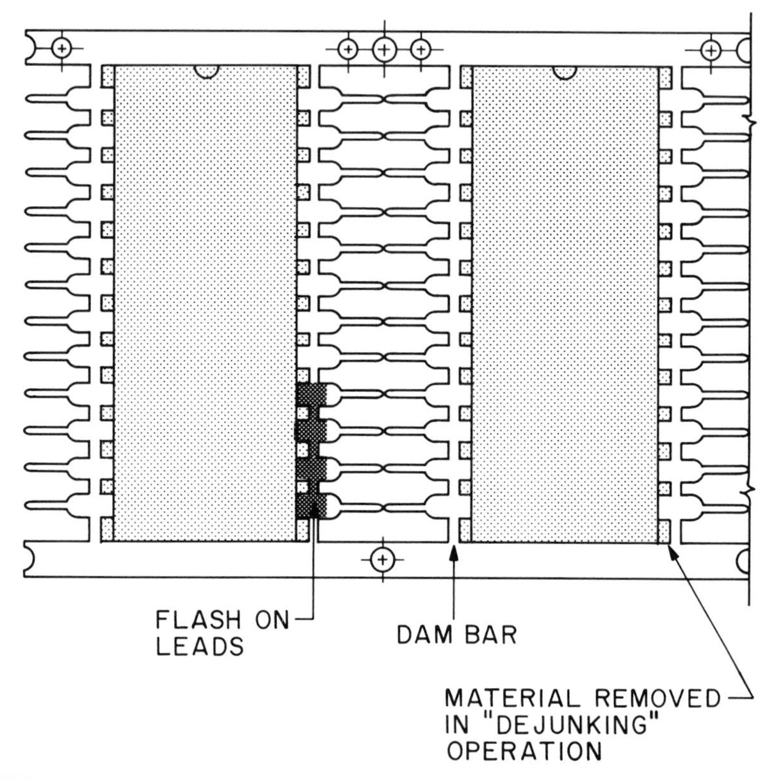


**FIGURE 5-13** Drawing showing the leadframe and package after molding but before deflashing, dejunking and trim and form. The drawing shows where flash appears on the leadframe.

arcane to the uninitiated, but these woodlike materials actually have good hardness and toughness and do not leave impact residues that other synthetic plastics might; they have been used in this application for decades. Air is the usual fluid carrier, although liquid carriers such as water with higher density glass beads as the inertial agent are also used. These machines require careful set-up and adjustment. The streams have to be positioned properly for the particular leadframes, and the stream velocity is also a key parameter. The machines themselves tend to be housekeeping issues since despite the best efforts on the part of machine designers and operators, the media tends to get outside of the enclosure on a regular basis. In addition, there are cases where the properties of the flash such as its thickness, its tenacity in sticking to the leadframe, or its state of undercure or overcure make it difficult to remove despite success with the same molding compound on the same package design in the past. In these cases, careful adjustment of deflashing parameters such as pressures and angles-must be undertaken until the removal is satisfactory, if not complete.

*Solvent Deflashers* Another type of deflash equipment is a solvent deflasher. This operation involves exposing the entire leadframe strip to hot solvent that reduces the adhesion of the flash to the leadframe, and may swell and soften it as well. This type of deflashing is more popular in transfer molding and microelectronics than in thermoplastic injection molding because thermoplastic materials are more susceptible to solvents, and because the gloss or matte finish of appearance parts is usually ruined on solvent exposure. The appreciable solvent resistance of thermoset molding compounds causes only a very thin surface skin of the plastic to be affected by the solvent. In some cases, there is little or no detectable effect on the molded body. Although this short penetration depth does no harm to the molded body, it has a profound effect on the exceptionally thin flash, reducing its structural integrity and adhesion to the leadframe and making it removable by subsequent high pressure streams, usually air jets. The setup and adjustment parameters of solvent deflashers are slightly more extensive than media deflashers because more process steps are involved and solvent temperature and reconditioning must also be considered. Water uptake by the solvents, typically strong hydrophobic agents such as N-methyl pyrilodine (NMP) or dimethyl furane (DMF), is an important consideration. Most systems have distillation columns to remove absorbed water. The advantages of solvent deflashers are the lower deformation it causes to the leadframe, especially important with the thinner leadframes used in fine pitch packages. Other advantages include its high effectiveness in removing even stubborn flash and its better housekeeping character. An important drawback is the use of the solvents themselves which are becoming more undesirable from environmental and safety points of view.

*Water Deflashers* A variation of the media deflasher is water deflashing. In this operation, very high-pressure water streams are directed at the leadframe flash, blowing it apart through a combination of inertial and abrasive forces. Recent innovations in high pressure water hydrodynamics and nozzle design makes this technology possible. It has the advantage of being solventless and probably cleaner than media deflashers.

*Dejunking* Media deflashing removes the small amount of molding compound that is outside the molded body but behind the dam bar, also visible in Figure 5-13. The dam bar has to be displaced from the edge of the molded body to allow it to be cut out cleanly without chipping the package at the parting line, hence molding compound will flow out until it encounters the dam bar. In the case where media deflashing is not used, removal of this material behind the dam bar is still necessary because it could damage the trim and form tool.

This separate operation is sometimes called dejunking. The options for dejunking include removal in a dedicated apparatus, or as an additional step in the trim and form operation. Some improved mold and leadframe designs can minimize the amount of material behind the dam bars to the point where a separate operation may be unwarranted.

## 5.8 TRIM AND FORM

Trim and form has a major impact on packaging productivity and yield. The evolution of the technology to higher lead counts, thinner leadframes, and closer lead spacing places greater emphasis on the engineering time devoted to trim and form design, implementation, and problem solving. Trim and form involves considerable amounts of investment in presses, tooling, floor space, and labor. Defects and problems are as often related to the molding process as to the trim and form process. A thorough understanding of both molding and trim and form is needed to properly design and implement tools as well as solve problems. Optimization of parameters for trim and form will be discussed in Section 7.4; equipment and tooling are discussed below.

### 5.8.1 Trim and Form Machinery

The machinery for trim and form is essentially a reciprocating punch press. Typical punch rates are approximately 80 cycles per minute. The use of progressive dies with one station for each forming step enables one tool to complete a forming operation on each press stroke. A press with a trim and form tool for a specific package design is not dedicated to that design since changeover of the trim tool can usually be accomplished in less than one hour. In a typical operation, the leadframes are loaded into one side of the punch press either manually, through cassettes, or automatically from a larger cassette carriage. The leadframe strips move through the trim and form tool on successive punch cycles ultimately being completely formed, singulated, and separated from the frame. They exit from the opposite end of the press where they may be loaded in tubes or trays. Extensive automation of the leadframe and package handling operations are more recent features of these machines.

### 5.8.2 Trim and Form Tooling

Trim and form tools are custom-made machine tools manufactured in the same way and often by the same vendors as molding tools. Their materials of construction are usually air hardened steel such as D2 tool steel or carbide steel. The action of a trim and form tool for plastic packaging is fairly complicated, and this complexity is reflected in the design and construction of the tool. Most dies are progressive in that the tool consists of a series of stations or sites where one specific action is performed. In many tools, the first action is cutting out the dam bars, a process also called dewebbing. After subsequent removal of the shorting bars, which connect the lead tips together, the leads are electrically isolated and can be tested. Therefore, the package does not have to complete the entire trim and form process to assess reliability. This option can save money because the forming stations add additional cost. In some cases, the packages may be postcured, deflashed, or solder plated after the deweb operation. The sequence of process steps depends on the particular requirements of the molding compound and the postmolding operations.

The forming steps are different for various lead and package designs. The simplest lead forming is for through-hole and butt packages (see lead configurations in Figure 1-6) where only a single action is needed to form the lead after dewebbing in most cases. The tool action for this lead bend, common to all lead configurations, is shown in Figure 5-14. In instances of very tight bends, however, a two-station, two-stroke bend may be employed to reduce the deformation sustained on each press stroke. In all cases, the lead is not bent against the plastic molded body since this would crack the plastic under the lead. An anvil has to be inserted under the leads to provide a bending fulcrum. The need for this anvil and the minimal thickness it must have to withstand the bending forces mandate a minimum distance that the lead must be positioned away from the molded body. In most cases, the thinnest anvil is 0.060 cm (0.025 inch),

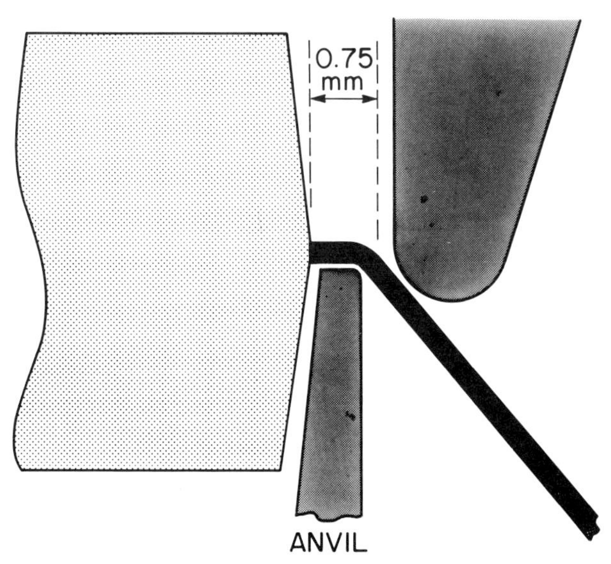


**FIGURE 5-14** The tool and lead configuration for a simple lead bend typical of through-hole packages, but also a forming step in both J and gull wing leads.

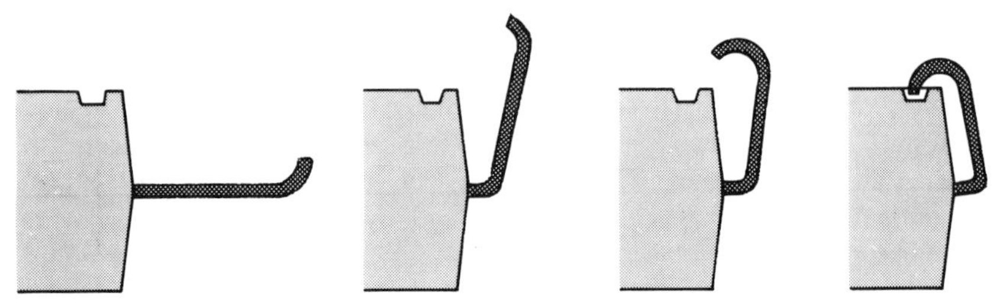

FIGURE 5-15 Schematic drawing of the forming steps for a J lead.

which translates to a minimum lead standoff of approximately 0.075 cm (0.030 inch). The forming action for J and gull wing leads are fairly more complicated than for the straight leads. As an example, the forming actions for a J lead are shown in Figure 5-15 and a photograph of a leadframe undergoing the trim and form progression is shown in Figure 5-16. Another type of forming tool is one based on cam actions. These cam tools are more capable of forming tight bends such as those found on J leads.

## 5.9 SOLDER DIPPING, SOLDER PLATING

Most integrated circuit devices are ultimately soldered to a circuit board. Solder coating of the leads is required to facilitate this solder attachment. It can be achieved in two ways. The leads can be solder plated after molding, or the leads can be dipped in molten solder to achieve coverage. In either case an extra step is added to the operation for this purpose. In the special case where there is no need to wire bond to the leadframe, such as in TAB or other thermocompression attachment, it may be possible to solder plate the leads prior to molding, thereby avoiding the solder plating or dipping operations that must be conducted at the assembly and packaging plant. Special, dedicated equipment is available for solder dipping or solder plating. The machinery also includes the fluxing and

FIGURE 5-16 Photograph of a leadframe in the trim and form progression for a J lead configuration.

washing operations, hence the equipment is fairly large, one of the largest on the assembly floor.

## 5.10 CODE MARKING

Code marking adds printed information to the top surface of the plastic package. This information would include the manufacturer, the country of origin, and the device code. There are two primary means of printing this information: ink marking or laser marking. Ink marking is accomplished through a rubber stamp that is dipped in a polymer based ink and then stamped onto the molded body. Screen and pad printing are other feasible options with pad printing garnering more attention. Some inks, particularly the common types based on epoxy or phenolic, may require a heat cure or UV radiation cure that would be conducted soon after application. If code marking is done before post-cure, the ink curing and post cure can be conducted simultaneously. Most marking operations are at least semi-automated in that the packages move on a conveyor that delivers them under the stamper. In a single line system, the marking frequency is greater than one per second. The most common problem associated with ink code marking is the smearing or removal of the ink upon handling or other operations conducted after code marking. In some cases, this problem can be alleviated by improved drying or curing of the ink, closer quality control of the ink, or closer attention to the surface of the molded body. Inadvertent soiling of the surface will preclude proper adhesion of the ink. A matte finish on the upper surface of the molded body is preferred for ink marking, and a more textured finish may improve ink adhesion.

Laser code marking has been growing in importance, but it has not displaced ink marking or even been able to garner the majority of the market at the time of this writing. Laser marking consists of writing the information onto the package by burning with a laser. The most common lasers used for this application are $CO_2$ or YAG lasers. The marking frequency is also about one package per second. An important advantage of laser marking is that it is indelible. Also, quality control problems of the ink, its application, drying, or curing are eliminated. The major problem associated with laser marking is the poorer contrast between the exposed area and the background. The exposed area has a light brown or tan color which does not contrast well with the black, actually very dark brown, molding compound. Special epoxy molding compounds have been developed to enhance the contrast and these are quite effective, although they still do not achieve the contrast achieved with white ink.

## 5.11 INSPECTION

Inspection is usually conducted after trim and form and code marking. This is also a semi-automated operation in that the packaged devices are moving on a conveyor, possibly the same line that brought them through the code marking station. The criteria for this inspection are primarily the position and forming of the leads, although it could also include defects in the molded body or code mark. Leads can often be formed improperly or bent in the trim and form die or in the subsequent handling operations. The inspection equipment will vary based on the specific criteria being applied and the type of package, but it often involves gauges for sensing the lead positions. Vision systems for detecting bent leads or checking for lead planarity are also used. Most inspection equipment is commercially available, although many production facilities have custom and semi-custom inspection lines.

## 5.12 BURN-IN

Burn-in is a reliability test of the packaged device to identify defective devices that would fail in the short term, a type of failure known as infant mortality. The principal target is silicon defects that manifest themselves during this initial heavy duty cycle. Section 8.1 covers the short term failures more thoroughly. During burn-in, the device is connected to a circuit board under bias and placed in an oven at elevated temperature. The temperature, time duration and the voltage loading vary according to device type and manufacturer, but typical conditions call for temperatures of 125–150°C for approximately 24–48 hours at a voltage of 6.2–7.0 volts, well above the standard 5 volt operating bias of most ICs. Burn-in is usually conducted on some fraction of the population, with some companies using burn-in on 100% of the devices. Burn-in ovens are available from a number of vendors. Older equipment requires special wiring that passes through the wall of the oven to access the circuit packs. Newer burn-in ovens are computer controlled with intrinsic power and logic. You can program certain voltages and voltage cycles for different circuit packs. The burn-in capacity must be several times that of the postcure oven capacity, since burn-in times are as much as 10 times greater than postcure. A medium-sized assembly and packaging operation will need 10 to 20 full size burn-in ovens, each measuring approximately 6 $m^3$. Burn-in consumes an appreciable fraction of the production floor space of an assembly and packaging facility. The equipment produces an inordinate amount of heat, so extra air conditioning capacity will be required to make the space habitable. The expense of burn-in is considerable, and there is strong motivation to eliminate or modify it as a strict requirement. The options are to replace 100% population burn-in with statistical sampling, and reduce the fraction sampled in cases where less than 100% are subjected to burn-in.

## 5.13 TESTING

A full function test on 100% of the packaged devices is standard in the microelectronics industry. The human and capital resources required for this testing contribute a significant fraction of the cost of the assembly and packaging. Testing is conducted in computer controlled machines that evaluate all functional parameters of the device according to prepared testing algorithms. The devices are fed to these machines either in tubes, cassettes or manually. The test boards, essentially printed wiring boards, adapt different devices to the same test equipment. They are custom to the type of device, its function, and its testing procedure, and have to be made specially for a particular device code. Integrated circuit testing equipment is commercially available from a number of vendors. An excessive dropout rate after burn-in could indicate a serious reliability problem that may warrant scrapping the entire lot.

## 5.14 FAILURE MODE ANALYSIS

Defective devices are segregated into separate collection bins for subsequent failure mode analysis (FMA). Some typical FMA patterns include "open and shorts," current leakage, surface charge accumulation, dielectric breakdown, and contact degradation. Failure mode analysis is an invaluable tool in understanding and improving the plastic packaging process. When conducted thoroughly and regularly, and when the proper remedial action is implemented, FMA can lead to improved productivity, yield and quality. There are several instruments that are important FMA tools. Not all of these are needed in all assembly and packaging operations, and some of them are very expensive, but most provide important insights in manufacturing optimization.

*X-Ray Analysis.* X-ray analysis is used to assess the degree of wire bond deformation and/or paddle shift caused by flow-induced stress. Broken wires appear as opens in the device testing, whereas crossed wires or wires pressed onto the edge of the paddle or conductors of the chip are discovered as shorts. X-ray analysis can also be used to assess the occurrence and location of air voids. Larger voids, those greater than 1 mm in diameter, are detectable in full size X-ray photographs. Most smaller voids will require photomicrographs or examination of the X-ray photograph with a medium power lens. The best X-ray equipment has a movable stage that is large enough to accommodate one or more leadframe strips. In this way, X-ray analysis of the entire strip can be conducted in real time without frequent repositioning. The progression of wire sweep in the direction toward or away from the transfer pot often indicates the nature of the material, design, or process deficiency that is responsible for the deformation, hence this examination of the entire leadframe strip is an important part of FMA. Paddle shift is more difficult to detect with X-ray analysis because the package must be viewed edge on, the X-ray beam traversing a much greater thickness of molding compound. Also, the paddle is often skewed rather than displaced parallel to its initial position, further obscuring the image of the paddle displacement. Sectioning of the package is a preferable way to assess paddle shift, albeit much more labor intensive.

*De-Encapsulation.* In this type of failure mode analysis, the thermoset molding compound is removed from the device to facilitate visual inspection of the wire bonds, bond pads, and device features. This technique is particularly useful for conductor line shifts and deformations introduced by shrinkage of the molding compound, and also to detect passivation layer cracking. A simple form of de-encapsulation, also called decapsulation, is to swell the package with chemical treatments and then tear the plastic carefully from the leadframe and chip. This will destroy the wire bonds but the chip surface will be preserved. Other forms involve a complete dissolution of the molded body in suitable acids such as sulfuric or nitric, or impinging a hot acid stream at the package to erode and dissolve the molding compound. The equipment for this type of decapsulation consists of standard wet chemical facilities and a fume hood. The decapsulation process requires strong chemicals that have to be handled carefully, observing all relevant safety precautions. Specific decapsulation procedures for a molding compound can be recommended by the molding compound supplier. It is unwise to concoct special formulations or to deviate from the suppliers recommendations. Another method of decapsulation for small outline packages or memory devices is easier and more direct, but it is much more likely to damage the fragile parts of the device. In this procedure, the package is immersed in molten solder until the molding compound softens. The plastic is then quickly torn off the leadframe and chip with pliers. The more robust defects such as passivation layer cracking and conductor path shifting will not be disturbed, whereas the wire bonds and ball bonds are likely to be affected.

*Microscopy* Microscopy is an invaluable FMA tool in microelectronics. Many of the features and defects of the microcircuits are by definition microscopic and undetectable by the human eye. Several different types of microscopes will be needed. A low- to medium-power optical microscope is essential to see ball bond liftoffs and passivation layer cracking. One that has both photographic capability and a video monitor is preferable. A scanning electron microscope (SEM) is another essential FMA instrument. SEM can enlarge the defect sites that the optical microscope can only detect. With it, the probable causes of a failure are more easily deduced. An SEM instrument that also has X-ray fluorescence mapping capability is an important asset. With this instrument, you can detect and map the presence of several important elements on the surface being examined; an important tool in delamination problems to determine where the adhesive failure occurred.

*Scanning Laser Acoustic Microscopy* This technique, often known by the acronym SLAM, is a relatively new but exceptionally powerful technique for detecting voids, cracks, and delaminated interfaces. The technique is a nondestructive test using ultrasonic waves with a frequency of 400 kHz to 100 MHz, with frequencies between 1 and 5 MHz being particularly important [2]. The waves are generated using piezoelectric crystals and transmitted to the specimen by water. The reflected waves are received by piezoelectric crystals and converted back to electrical signals from which an image is constructed. When an ultrasonic wave propagating through one medium encounters another medium, some of the wave is reflected and some is propagated into the new medium, the ratio of reflection to propagation being controlled by the acoustic impedances of the media. The acoustic impedance of a medium is the product of the density and the speed of sound in the medium. Therefore, the reflectance/transmission ratio on encountering the leadframe is different than the ratio on encountering a void, hence the two can be distinguished. SLAM works primarily with transmitted acoustic energy, whereas a different type of scanning acoustic microscopy known as SAT or C-SAM works in the transmission/reflection mode. C-SAM provides resolution in the z direction since it detects reflections from planes perpendicular to the z axis. SLAM provides information only in the x-y plane since it sees through the entire thickness of the specimen.

*Dynamic Scanning Calorimetry* This is a thermal analysis technique, known often as DSC, which is essential for investigating material-related package failures. Using just a small chip of the molded body, the FMA engineer can determine degree of undercure, low $T_g$, excessive or insufficient mold release. With other thermal analysis equipment such as thermomechanical analysis (TMA), the coefficient of thermal expansion can be checked against the vendor's specification. A thermogravimetric system (TGS, also known as TGA) can be used to detect thermal degradation that may have resulted from incorrect post-cure.

*Sectioning* A saw to section the molded plastic package is both an FMA and yield improvement tool. Sectioning shows the extent of paddle shift and lead deformation. The epoxy molding compound is filled with silica glass making it very difficult to cut. The best instrument is a carbide saw immersed in liquid cutting oil. Other diamond cutting wheels or band saws are also effective. The surface should then be polished to improve contrast and delineation of the features.

## REFERENCES

1. U.S. Patent No. 4,332,537, "Encapsulation Mold with Removable Cavity Plates" (June 1, 1982).
2. T. Tabata, H. Suzuki, T. Hamada, and M. Yamaguchi, "Internal Defects Observation of IC Package by Scanning Acoustic Tomography," Nitto Technical Reports, September 1987, p. 70, Nitto Electrical Industrial Co., Ltd. 1-2, 1-chome, Shimohozumi, Ibaraki Osaka 567 JAPAN.

# 6 Process Description

Transfer molding is the process method for essentially all plastic packages in the integrated circuit industry. Although it is one of the older plastic processing methods, it has often been associated with low technology and mundane applications. It also accounts for a much smaller production volume, both in pounds of plastic used and dollar value of product, than the major plastic process operations of injection molding, extrusion and blow molding. Additionally, it is a more complicated and difficult process to treat analytically than either thermoplastic extrusion or injection molding because of the time-dependent behavior of the molding compound, the irregular cross sections of the runners, and the presence of inserts in the cavities in most applications. As such, it has garnered very little attention by research and development engineers. The major thermoplastic material suppliers have largely stayed away from developing transfer molding compounds because the volume of the business is only a small fraction of the injection and extrusion plastic volumes. This situation has begun to change as transfer molding has assumed such a prominent role in the high technology areas of packaging of microelectronic devices and the manufacture of optical fiber connectors. Although usage of thermoset molding compounds still represents only a small fraction of the volume of thermoplastic materials, they often require sophisticated formulations that have high value-added content, making them more competitive with the commodity thermoplastics for R&D attention. This chapter reviews the process characteristics of transfer molding for plastic packaging of microelectronic devices. Both practical and fundamental concepts are discussed. In most instances, the fundamental underpinnings of the practice are provided.

## 6.1 MOLDING COMPOUND PREFORM VOLUME AND SIZE

Both the volume and size of the molding compound preforms have to be properly selected for the molding tool capacity. The overall volume can be computed by summing up the volume of the individual cavities and subtracting the volume of the chip and leadframe which can be significant in many instances, particularly small outline packages. The volume required for the runners has to be estimated from the total runner length and the cross section area of the runners. There should be an additional cushion of molding compound left in the transfer pot to prevent the plunger from bottoming out, a condition that would preclude the application of the packing pressure. This cushion height, also known as the cull height, should be a minimum of 0.5 cm. Extra cull height wastes molding compound and extends the time of cure. Once the overall molding compound volume is determined, it has to be converted to a weight by multiplying by the density. Molding compound densities are approximately 1.80 gram/$cm^3$.

The molding compound supplier can usually provide a wide range of preform diameters and weights. The diameter is selected by accounting for the transfer pot size. A relation between air space around the preform in the pot and the occurrence of voids is often speculated, though no specific citations can be made. The diameter of the preform should be as close to the diameter of the transfer pot as can be readily inserted without deformation. This is approximately 90% of the transfer pot diameter. The required size has to be matched against the preform diameters that are available. If no diameters are available in the optimum range, then it is preferable to err on the small side since undersized preforms of only 70% of the transfer pot diameter provide adequate molding yields in most cases, whereas preforms that are too large will cause insertion problems on every single transfer cycle. The preform diameter selected will often be available in more than one weight. For a given molding tool, the preform weight that provides the fewest number of preforms is preferred. It is advantageous, however, to minimize the number of preform sizes and weights that are carried in inventory. Different size production molding tools will require between three to six preforms. From this perspective, the smaller preforms have an advantage in minimizing the waste of molding compound when one preform size is used in a number of molding tools that have different preform charge requirements.

Preform density is an important productivity, yield and quality consideration in plastic packaging. In general, preforms with greater density and more uniform density are preferred. The higher preform density produces a fluid melt with a higher density. Surprisingly, the pressures reached during preforming usually exceed the pressures reached in the transfer and packing stages of the process, so that low density and porous preforms can contribute directly to greater porosity of the molded body. In many cases, the density gradient over the length of the preform is large, often large enough to the detected visually. Preform density and density uniformity are considerations in choosing among molding compound vendors.

## 6.2 THE PROCESS SEQUENCE

The layout of a typical package molding operation is shown in Figure 6-1. The process sequence begins with the loading of the leadframes onto the mold. In both cavity chase and plate molds, this is done away from the molding press at a separate work station or loading apparatus. A cavity chase mold uses a loading

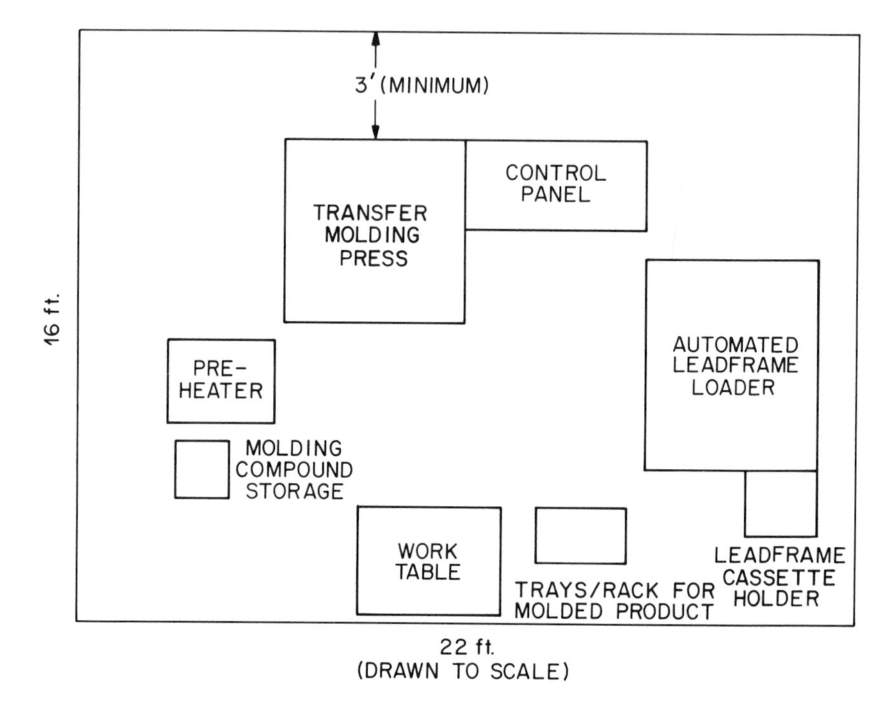


**FIGURE 6-1** Layout of plastic package molding operation. The difference between a cavity molding layout and a plate mold layout is in the frame loading work station. Automated frame loading onto the loading fixture is common in cavity mold operations, whereas the aperture plates themselves are shuttled in and out of the press in plate molding.

fixture that facilitates this loading. Many molding operations have automated leadframe loaders which take the leadframes from cassettes and place them on the loading fixture. The loading fixture is then carried by the operator to the press and placed on the lower mold half where it registers into position, precisely locating the leadframe sites over the cavities of the mold. The loading fixture stays on the mold although the leadframes themselves are now supported by the mold rather than the loading fixture. In an aperture plate mold, the plates themselves are loaded with the leadframe strips as they shuttle in and out of the molding press. The press is clamped closed after the leadframes are loaded. If there is dwell delay time or other soft-close feature to allow the leadframes to heat to the mold temperature, it is activated at this time. The operator then picks the appropriate number of preforms from the nearby bucket, or from the humid conditioning chamber if one is in use, places them on the rollers of the preheater, and starts the heating. The preheater opens automatically when the temperature or time interval has been reached. The operator removes the preforms, transports them to the mold and drops them into the cylindrical transfer pot. He or she then activates the transfer plunger which begins pushing the softened molding compound in the mold. The packing pressure, usually higher than the transfer pressure, is applied once the mold has been filled. In a standard press without microprocessor control, the transfer pressure is controlled by throttling the higher packing pressure through a speed control value during transfer. Once the mold fills, the transfer pressure builds up to the packing pressure since the flow through the throttling valve slows to a stop. For presses with microprocessor control systems, the transfer ram switches from transfer to packing pressure either by setting a transfer time interval, a travel distance, or through a feedback sensor that indicates when the mold is filled. The packing pressure is applied throughout the cure time which may be as much as three minutes, but in a press with a process controller the packing pressure may be reduced during the cure cycle to minimize resin bleed. The press opens after the cure interval timer expires at which point the operator removes the molded leadframes from the press with the loading fixture. In an aperture plate mold, the aperture plates themselves are removed from the press and transported to the work station. The operator then cleans the mold of any molding compound debris or removable residue. The loading fixture or aperture plates are returned to the work station, the leadframes are removed and loaded into cassettes or storage bins where they will be transported to the trim and form operation. A flow chart illustrating this process sequence is provided in Figure 6-2.

## 6.3 PREHEATING

The molding compound is below its glass transition temperature at room temperature. The purpose of the preheating step is to increase the temperature of the molding compound to a temperature at which it can be pushed into the mold under the action of the transfer plunger. Typical glass transition temperatures of the molding compound are 40–50°C, whereas preheat temperatures are in the range of 90–95°C. Typical preheat times are of the order of 20–40 seconds. Longer times can overly advance the polymerization reaction on the rollers of the preheater, reducing the flow length and flow time of the material in the mold and increasing the flow-induced stresses. Also, the molding compound will begin to cool once removed from the preheater, so any delay on the part of the operator in moving the preforms into the transfer pot and starting the transfer will allow the material to cool and the reaction to advance. Operators should have thermal insulating gloves to handle the hot molding compound after preheating. It is not uncommon for the operator to work the preforms manually so they can be inserted in the transfer pot more conveniently. Although this practice is not clearly harmful, it can introduce impurities and fibers from the gloves into the molding compound. The preferred practice is to properly size the diameter and the volume of the preforms so that this forming is unnecessary. The

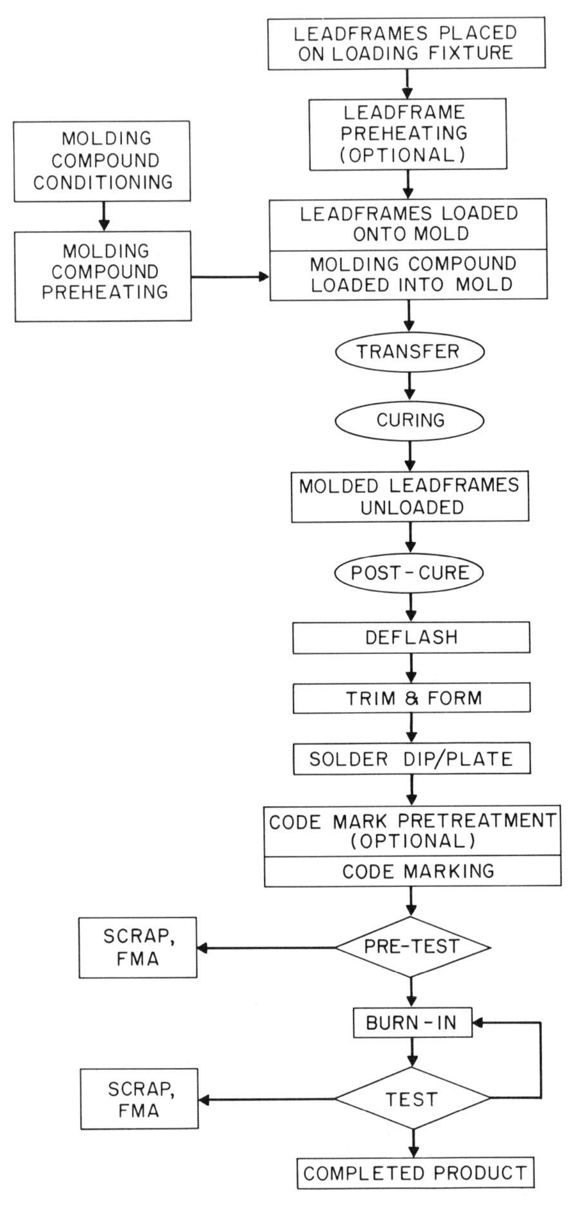


**FIGURE 6-2** Flow chart of the process sequence for plastic package molding, including all of the process steps after assembly. Some manufactures will use different sequences depending on the specific materials and processes they employ. (*Note:* Solder plating would normally be conducted prior to trim and form. For brevity, solder dipping and solder plating have been combined in one block which is shown after trim and form, the sequence used for solder dipping.)

transfer pot diameter should also be properly sized to accommodate the required volume of molding compound.

## 6.4 TRANSFER OF MOLDING COMPOUND IN THE MOLD

The transfer of the molding compound into the mold is the most important process step with regard to molding productivity and yield. It is during this process that the flow-induced stresses that cause wire sweep and paddle shift are generated. High productivity molding requires both skillful mold design and optimum process parameters; ideally the design, process and material should all be integrated in an optimized configuration. The process characteristics of the mold filling step are reviewed in this section. There are several distinct flow regimes during the transfer and each has been described separately, although the relationships among them are reviewed as well.

### 6.4.1 Transfer Pot

The flow and heating in the transfer pot begins the transfer of the molding compound into the mold. In most tool configurations, the pot is a simple cylindrical shape with a diameter between 4 and 7 cm. The runners emanate directly from the transfer pot in most designs, although distribution rings or enlarged runner entrances are also used. A review of the different transfer pot and runner configurations is provided in Figure 6-3. The distribution ring geometry shown in Figure 6-3b provides a more uniform flow field in the transfer pot with no angular coordinate dependence, but it is unknown whether there is any benefit to this flow as compared to the direct runner coupling shown in Figure 6-3a.

The temperature increase that occurs in the transfer pot is significant with regard to downstream processing and overall yield. The molding compound is loaded into the transfer pot at a temperature of 90–95°C, whereas the cylindrical cavity itself is maintained at the mold temperature of 170–175°C. It is important that the transfer ram itself is heated to the mold temperature to achieve consistent and satisfactory results. Typical material residence times in the pot range from a few seconds for the material that is first extruded, up to 25 seconds for the last molding compound that is transferred into the mold. The temperature increase that is incurred during this period is an essential part of the overall time/temperature history of the molding compound. A published study of heat transfer and flow in the transfer pot used a combination of flow visualization, experimental temperature measurements and a mathematical model to determine this history [1]. The geometry analyzed was a four-runner configuration, shown in Figure 6-4. The flow visualization work was performed with a bright gold colored molding compound layered into conventional black molding com-

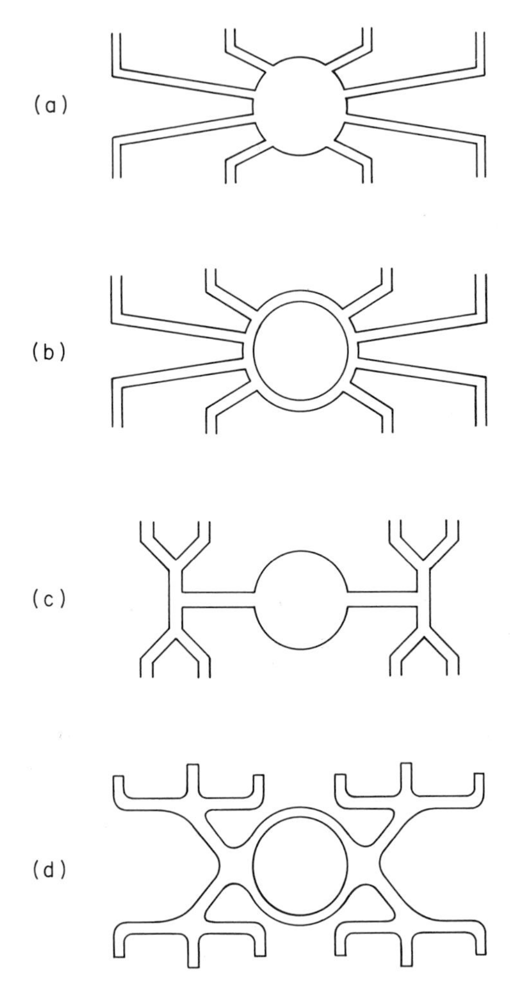


**FIGURE 6-3** A variety of different transfer pot and runner entrance configurations: (a) direct coupling of runners to transfer pot, (b) runner coupling through distribution ring, (c) a manifold runner system where larger feed runners supply a smaller manifold which supplies smaller runner arms, (d) runner coupling through enlarged runner entrances with manifold.

pound preforms, as shown in Figure 6-5. Two preforms were placed in the transfer pot with the first gold layer against the bottom of the pot. The molding compound was then pushed out of the transfer pot and the transfer halted after prescribed time intervals of flow. The cure was completed at each time interval. The remaining cull was removed, sectioned on a diameter that passes through two runner ports, and then polished to enhance the contrast and definition of the tracer bands. The sequence of tracer patterns representing different transfer

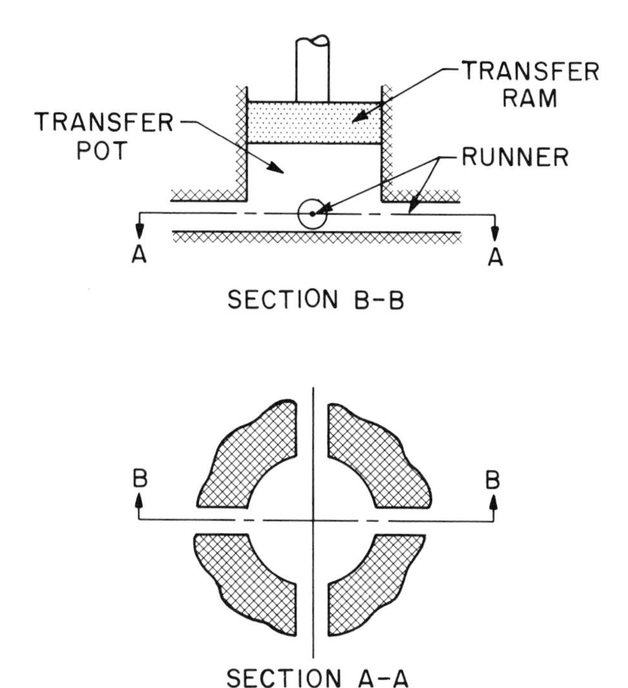


**FIGURE 6-4** Configuration of the transfer pot used in the flow and heat transfer study in Reference 1. This is a four-runner design, a low number for a production mold which would commonly have six to eight runners exiting from the transfer pot [1]. (Reprinted with Permission of the Society of Plastics Engineers.)

flow intervals is shown in Figure 6-6. Interpretation of the flow pattern indicates that the upper bands, those closer to the plunger, show very little distortion besides some expected sticking to the wall of the pot. At later flow times there is significant thinning of the lower bands as they are drawn out transversely toward the exit ports. The flow geometry and the tracer pattern suggest a stagnation point flow, a well known fluid mechanics solution. A stagnation point flow gives rise to a stagnation point where both the axial and transverse velocities go to zero. The flow visualization studies indicated that there was a stagnant zone along the bottom surface of the transfer pot indicated by a relatively undisturbed gold band of molding compound at the bottom of the cull as shown in Figure 6-6. The streamline pattern and the stagnant zone are depicted in Figure 6-7.

The streamline pattern deciphered from the flow visualization studies was used as the basis of a heat transfer model to predict the extrudate temperatures. The model was based on a two-dimensional finite difference solution of the molding compound in the cylindrical cavity. This solution was coupled to a

**FIGURE 6-5** Photograph of the multicolored preforms and the configuration as used in a flow visualization study [1]. (Reprinted with Permission of the Society of Plastics Engineers.)

mathematical solution for the temperature distribution in the cylindrical steel pot to provide a more realistic boundary condition at the interface. The fluid motion was treated in a heuristic fashion in that the nodes of the numerical model, representing the fluid elements, were required to follow the streamline pattern that was discerned from the flow visualization studies. In addition, the fluid and heat transfer solution were decoupled in that the streamline pattern did

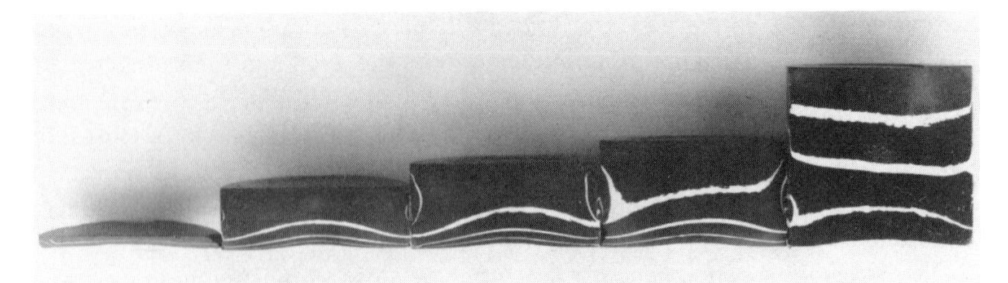

**FIGURE 6-6** Sequence of tracer patterns for transfer flow times of 3, 7, 12, 16, and 24 seconds. The thinner sections correspond to longer transfer flow times [1]. (Reprinted with Permission of the Society of Plastics Engineers.)

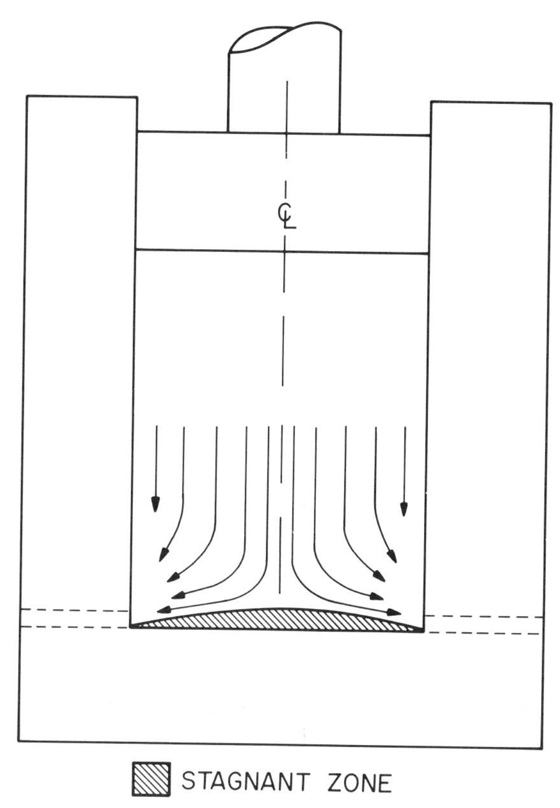


**FIGURE 6-7** Streamline pattern proposed for flow within the transfer pot as deduced from the flow visualization studies shown in the previous figures. (After Reference 1.)

not change during the time of transfer, an assumption that was supported by the flow visualization which indicated no major streamline pattern shifts during the flow. The model was used to predict the temperature, degree of conversion and viscosity of the extrudate streams. Extrudate temperatures were also measured just as the molding compound exited the pot. The results indicated that the extrudate temperature was in the vicinity of 100°C, except for brief startup and termination transients. The degree of chemical conversion determined from the mathematical model was low, not exceeding several percent during the entire transfer, although the specific value depends on the kinetics of the material in use. The viscosity, however, is very high at this low temperature, exceeding several thousand poise, although this value also depends on the rheology of the specific molding compound in use. A plot of smoothed extrudate temperature, conversion and viscosity data for a conventional size transfer pot at standard process conditions is provided in Figure 6-8.

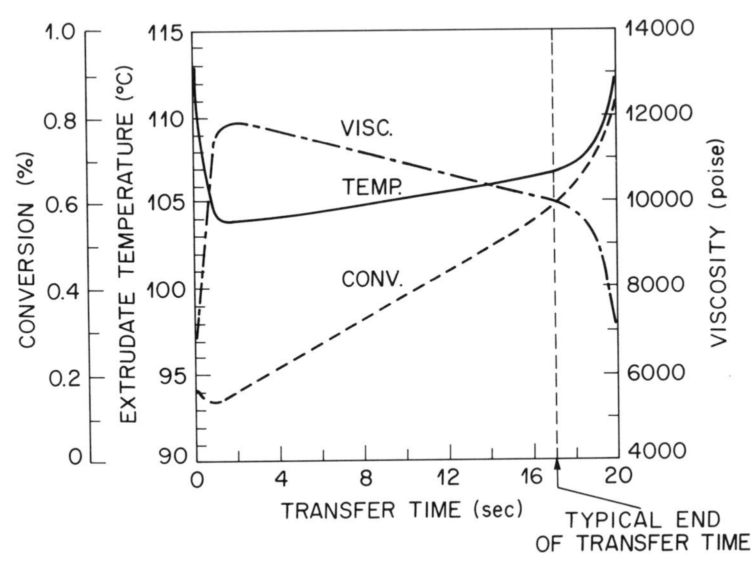


**FIGURE 6-8** Experimental extrudate temperature data from a transfer pot plotted as a function of time for standard process conditions. The chemical conversion and viscosity derived from a mathematical model of the heat transfer and flow in the pot are also presented on the same plot.

The effect of process parameters on extrudate temperature was explored. The extrudate temperature was only slightly influenced by rate of transfer; slow transfer rates providing a temperature increase of a few degrees compared to high transfer rates. The effects of mold and preheat temperature were explored with the mathematical model and the results are presented in Figure 6-9. The preheat temperature, which is the initial temperature of the molding compound when it is placed in the transfer pot, has a direct and essentially linear effect on extrudate temperature. The effect is not as useful a process aid as one would expect, however, because the molding compound residing in the pot at the higher temperature will show a reduced flow time. Techniques where the temperature is elevated later in the process sequence are preferred because they have the least effect on flow time. The influence of pot temperature on extrudate temperature, displayed in Figure 6-9b, is also relatively weak given that a 30°C range of pot temperature is considered. The pot temperature is the same as the mold temperature in most cases, and the flow length in a tool at 185°C is likely to be shorter than at a more typical mold temperature of 170°C, although the specific change on flow length would depend on the effects of reaction rate acceleration and viscosity reduction with increased temperature. The influence of the diameter of the transfer pot on the extrudate temperature was also explored. For these studies the mathematical model was again used in place of the experiment since only one experimental pot diameter was available. As

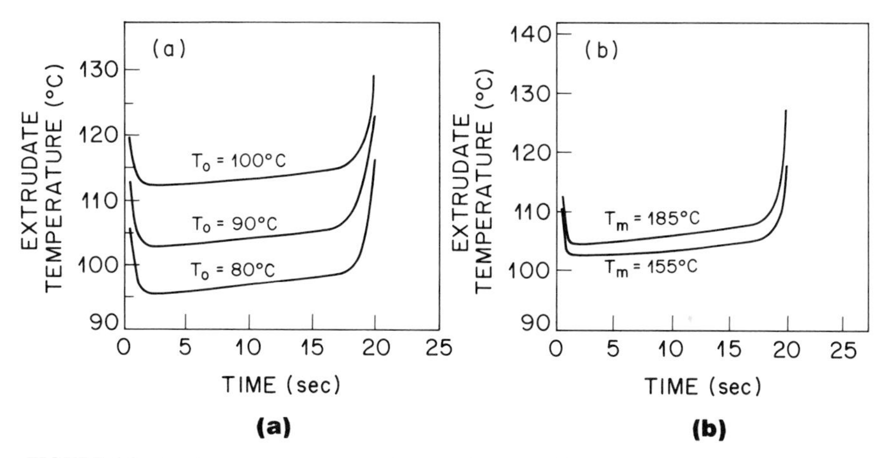


**FIGURE 6-9** Predicted extrudate temperature as a function of time for different process parameter changes: (a) extrudate temperature for three different preheat temperatures, $T_o$, (b) extrudate temperature for two different transfer pot temperatures, $T_m$ (the transfer pot temperature corresponds to the mold temperature in most molding tools) [1]. (Reprinted with Permission of the Society of Plastics Engineers.)

expected, smaller pot diameters do increase extrudate temperature as shown in Figure 6-10, but the effect is again small. The greatest difference is at the later stages of transfer.

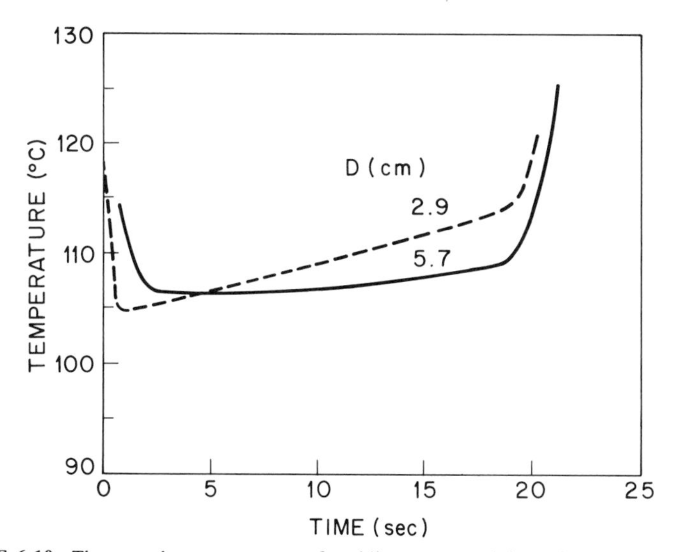


**FIGURE 6-10** The extrudate temperature of molding compound from the transfer pot into the runners as a function of pot diameter as predicted from a mathematical model of heat transfer and flow in the pot. (After Reference 1.)

These results combined with the previously discovered small effect of transfer rate on extrudate temperature indicate that the heating is not dominated by conduction into the large mass of preform charge, but instead the extent of heating depends more on the flow into the constriction itself. It is in this region that the surface-to-volume ratio of the flow increases, exposing more of the molding compound to the high-temperature metal. This is confirmed by a mapping of the temperature profile in the transfer pot at an elapsed time of flow of 12 seconds, shown in Figure 6-11. The bulk of the molding compound located in the core of the cylindrical geometry is still at the initial preheat temperature of 90°C, whereas the high-temperature material is limited to a thin skin of molding compound that is directly in contact with the heated metal. The important effect of the stagnant zone at the bottom of the transfer pot is now apparent. The molding compound is a poor thermal conductor, hence the stagnant zone serves to insulate the molding compound flowing over it from the high temperature metal; low extrudate temperatures are the result of this phenomena. In general,

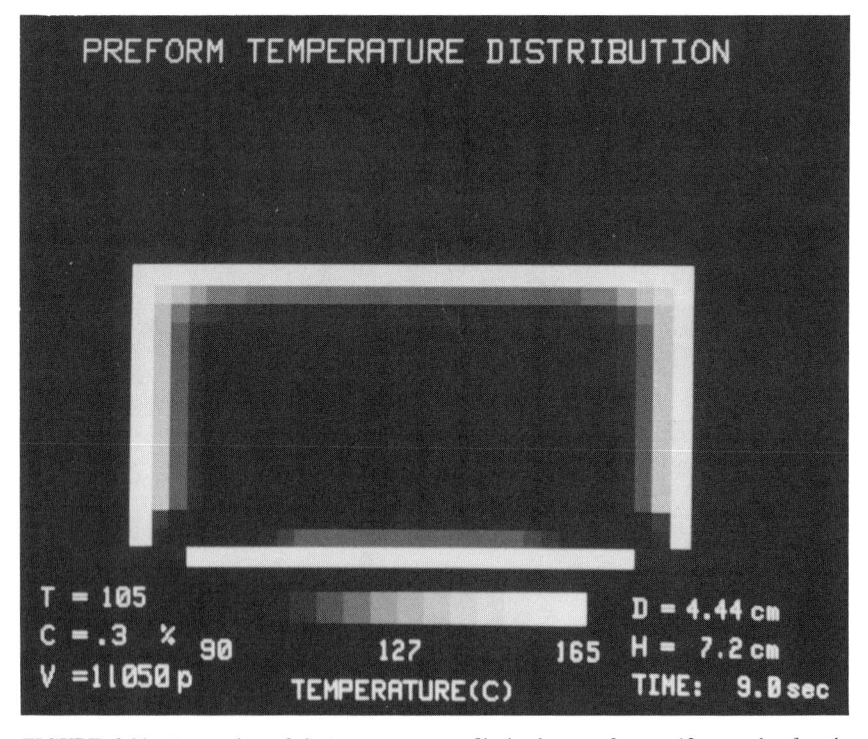


**FIGURE 6-11** A mapping of the temperature profile in the transfer pot 12 seconds after the molding compound began extruding from the pot into the runners [1]. (Reprinted with Permission of the Society of Plastics Engineers.)

however, this discussion of transfer pot heating and flow has established the temperature and degree of conversion of the molding compound as it enters the runner system of the mold. Although the degree of conversion is not a concern for subsequent high-yield molding, the temperature is too low and the viscosity much too high to provide acceptable yield based on these extrudate conditions. Therefore, downstream heating of the molding compound plays a crucial role in increasing the temperature and reducing the viscosity to levels that do provide acceptable yields.

### 6.4.2 Flow in the Runner

The molding compound condition as it exits the transfer pot was described in the previous section. In the general case of a production-size pot 5–6 cm in diameter with nominal transfer times between 12 and 20 seconds, the extrudate temperature of the molding compound is between 100 and 110°C, with a time averaged value of approximately 103°C. This establishes the initial condition for flow into the runners of the mold. The runner cross section will vary among different mold designs, but the general shape in cavity chase molds is that of a trapezoid. Semicircles and rounded trapezoids are also used in some designs. A sampling of runner cross sections is provided in Figure 6-12. The trapezoidal geometry is easily machined and also provides easy ejection, but it is not optimum in terms of heat transfer or for any analytical treatment of heat transfer

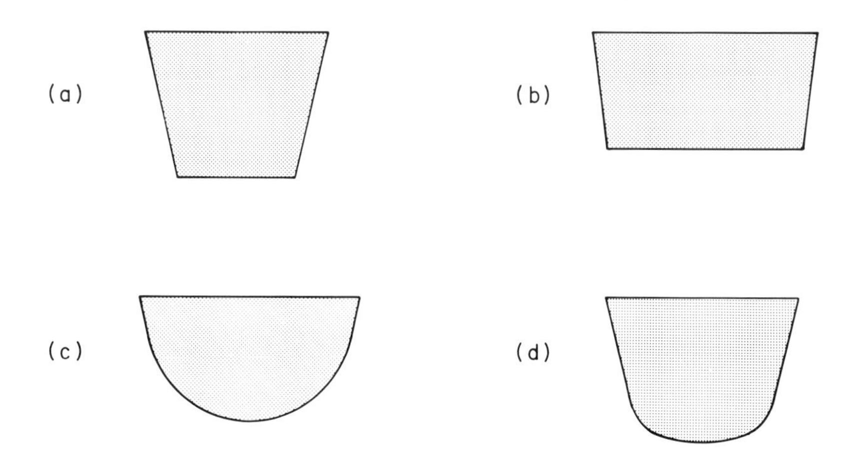


**FIGURE 6-12** Typical runner cross sections of both cavity chase molds and aperture plate molds: (a) trapezoidal, (b) near-rectangular, (c) semicircular, (d) rounded trapezoidal. Typical runner widths are approximately 0.5 cm with variations around this value.

and flow behavior. The approach of most investigators has been to combine both experimental and analytical approaches to determine important process information such as pressure drop and temperature increase during flow.

**6.4.2.1 Pressure Drop in the Runner** An understanding of pressure drops for flow in the runner is needed to design molds that have predictable and reasonable fill pressures and filling profiles. Instrumenting the molding tool with pressure transducers is one way to obtain this information. There are several commercial pressure transducers that can be used for the pressure ranges encountered in transfer molding. Pressures are typically low, 200–1200 psi (1.4–8.2 MPa), compared to the much higher pressures encountered in injection molding, which can approach 20,000 psi (137 MPa) for some materials. Temporary installations can be made by placing the pressure transducer in an ejector pin hole. At least two pressure transducers are needed for accurate pressure drop measurements since it is unsatisfactory to assume that the pressure can be determined from the applied pressure anywhere in the mold. Trying to assign the pressure in the runner entrance from the hydraulic pressure on the plunger is difficult and subject to significant errors. It is, however, safe to assume that the pressure at the end of an open channel is the atmospheric pressure. Pressure drop measurements should be made at different volumetric flow rates so as to change the shear rate through the channel and influence the shear thinning viscosity of the thermoset molding compound. Care must be exercised to assure that the volumetric flow rate remains constant throughout a measurement, and that there are no transients in the flow such as an advancing flow front. Be aware, however, that the temperature profile in the runner will probably change with different volumetric flow rates. Pressure drop data for flow in a trapezoidal runner are presented in Figure 6-13 [2]. It is evident that over a distance of 26 cm, the pressure drops encountered are in the range of 100–500 psi (0.7–3.5 MPa) depending on the volumetric flow rate. Higher flow rates require higher pressure drops to sustain them despite the shear thinning character of the molding compound which tends to reduce the viscosity and the pressure drop.

The volumetric flow rate of molding compound in a runner is usually between 0.2 and 3.0 $cm^3/sec$; the rate changes with both time of flow and position in the mold. The highest volumetric flow rates and corresponding velocities are encountered just outside the transfer pot soon after the molding compound enters the runners. The flow length is short at this time, and the applied pressure can provide a very high velocity because of the minimal flow resistance encountered over the short flow length. The full transfer pressure is not applied at these short flow lengths because of the limiting velocity of the transfer ram. As the flow length increases, the back pressure at the transfer pot increases and reaches the applied transfer pressure, at which point the flow is pressure driven rather than constrained by the maximum velocity of the ram. Lower velocities

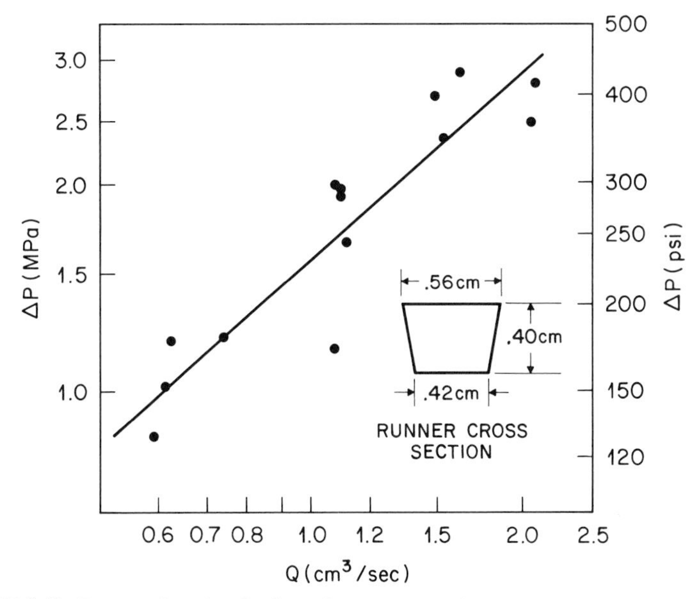


**FIGURE 6-13** Pressure drop data for flow of an epoxy molding compound in a trapezoidal-cross-section runner. The flow length was 26 cm. The flow was nonisothermal and the material was heating as it flowed. The mold temperature was 170°C and the initial molding compound temperature was 110°C. The exit molding compound temperature depended on the volumetric flow rate and ranged between 140 and 165°C [2]. (Reprinted with Permission of the Society of Plastics Engineers.)

are encountered as the flow length increases and molding compound gets diverted into the cavities. A diagram showing approximate velocities in a production size mold is provided in Figure 6-14.

Increasing flow length slows the velocity since the applied pressure has to push a longer column of fluid. Expressions can be written which relate the velocity to the other parameters of the flow. The velocity profile in the axial direction $u(r)$ for fully developed flow of a power law fluid in a circular cross section channel is:

$$u(r) = \frac{nR}{1+n}\left(\frac{R\Delta P}{2K(T)L}\right)^{1/n}\left(1-\left(\frac{r}{R}\right)^{1+1/n}\right) \qquad (6\text{-}1)$$

The volumetric flow rate $Q$ is easily obtained from the above relationship:

$$Q = \frac{n\pi R^3}{1+3n}\left(\frac{R\Delta P}{2K(T)L}\right)^{1/n} \qquad (6\text{-}2)$$

These relationships show how the volumetric flow rate and velocity depend on the applied pressure, $\Delta P$, flow length, $L$, fluid viscosity and the radius of the

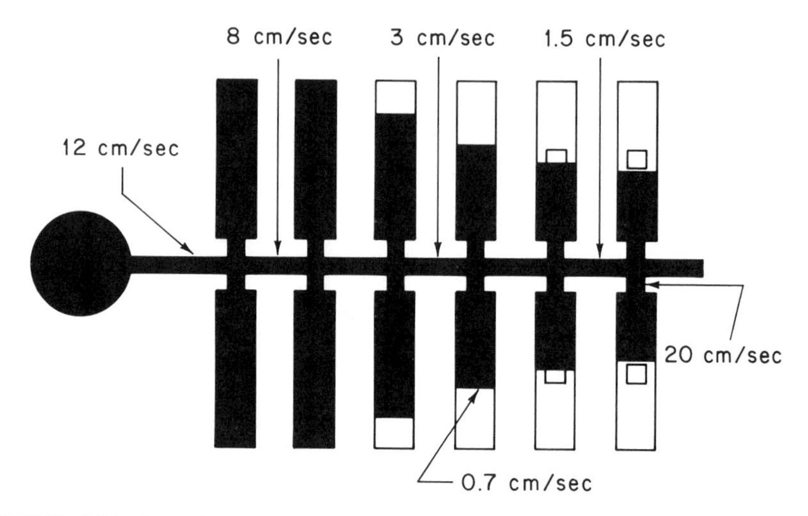


**FIGURE 6-14** Approximate velocities in a production-size molding tool at several points throughout the flow length. The velocities slow due to increasing flow length at early times, and then slow again significantly when the molding compound gets diverted into the cavities, reducing the volumetric flow rate and velocity in the runner.

channel, $R$. The rheology of the non-Newtonian molding compound is incorporated in the power-law index $n$ and the viscosity coefficient $K(T)$. The temperature dependence of the viscosity is found in this coefficient. Different approaches to expressing the temperature dependence of viscosity were discussed in Section 2.4.4. For the usual case of noncircular-cross-section runners, the above relationships can be used by determining a hydraulic radius, which is the radius of the circular channel which provides the same flow behavior as the noncircular-cross-section channel being considered. Use of the hydraulic radius principle is well established and widespread in fluid engineering studies. See Reference 3 for a thorough analysis of the use of hydraulic radius for polymer flow in noncircular cross sections. Numerical treatments of non-Newtonian flow in noncircular channels are available in Reference 4. The hydraulic radius was previously defined in Section 2.4.7, Equation (2-19), and is shown here as Equation (6-3) in the form that was used for results shown in this book. In general, it is preferable to represent semicircular or rounded trapezoid cross sections such as those shown in Figure 6-12c,d by a hydraulic radius and then use Equation (6-1) for circular channels to conduct pressure drop calculations.

$$R_h = \frac{2A_{xs}}{P_w} \tag{6-3}$$

For trapezoidal or near rectangular-cross-section runners such as those shown in Figure 6-12a,b, it is preferable to represent them with flow in a rectangular channel instead of a circular channel. The solution for the rectangular geometry is far more difficult and involves nonlinear partial differential equations which cannot be solved analytically. The result is most conveniently presented using a shape factor that accounts for the aspect ratio of the channel as is shown in the expression for volumetric flow rate provided in Equation (6-4). The shape factor which also depends on the power law index is provided in Figure 6-15 [5].

$$Q = WB^2 \left( \frac{B \Delta P}{2KL} \right)^{1/n} S_p \tag{6-4}$$

In general, these relationships apply only to fully developed, steady flow and not to the transient flow that exists before the runner is completely filled. Fully developed means there is no axial dependence of axial velocity, the velocity profile down the channel is constant. In practice, however, the deviation may not be large enough to completely invalidate the analysis, particularly if the velocity is not assumed to be constant during the filling, but is instead adjusted

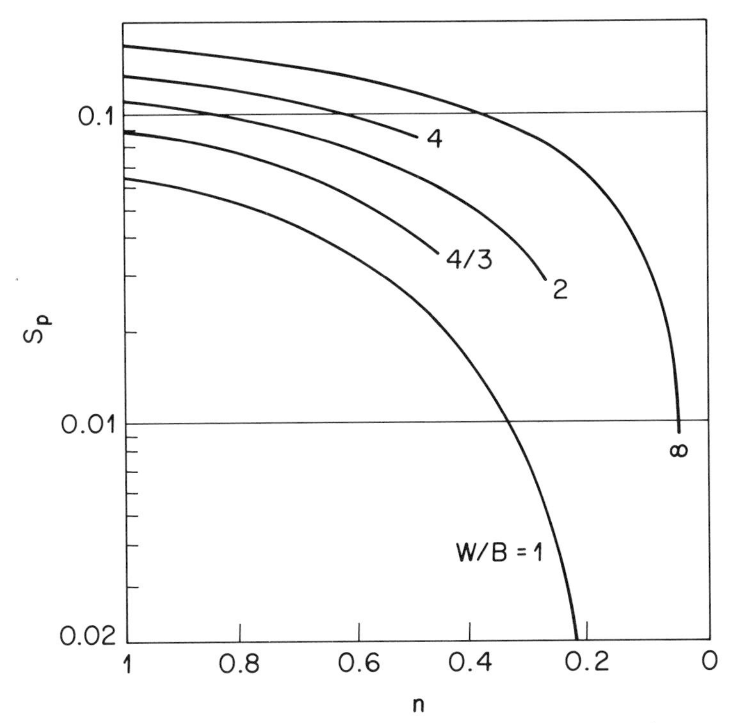


**FIGURE 6-15** The shape factor for flow of a power law fluid in a rectangular cross section channel [5]. (S. Middleman, Fundamentals of Polymer Processing, Copyright © 1977 McGraw-Hill, Inc, Reprinted with Permission.)

in a step-wise manner as the flow front moves down the runner. The analysis of transient and not fully developed flows is largely intractable, and certainly not amenable to the closed form solutions provided in Equations (6-1) to (6-4).

These relations provide a means to predict and analyze flow in the runners of the mold. The pressure drops predicted, however, depend on the temperature profile in the runner which can be significantly different based on process conditions.

**6.4.2.2 Heating During Flow in the Runner** Transfer molding is nonisothermal with the typical situation being a lower-temperature molding compound flowing in a higher-temperature channel. The molding compound viscosity is highly sensitive to temperature, so that the pressure drop required to sustain a given volumetric flow rate can change drastically with a temperature change of 10°C or less. The temperature, and therefore the viscosity, at which the molding compound flows through the gates of the mold and over the devices can vary significantly based on process parameters and mold design, making a major impact on flow-induced stresses and subsequent yield. It is, therefore, essential to understand the heating behavior of the molding compound as it flows through the runner system in order to optimize design and process variables.

*6.4.2.2.1 Analytical Treatments of Flow in the Runner* The heat balance for confined laminar flow in a runner is fairly complicated because it has to include convective as well as conductive heat transfer to the moving fluid. High-viscosity polymers may exhibit shear heating as well, and there is also the exotherm of the polymerization reaction. In addition to these modes of heat transfer and generation there is also the strong dependence of viscosity on shear rate, temperature and time which can significantly alter the velocity profiles and the convective heat transfer. Using cylindrical coordinates appropriate for the hydraulic radius representation of the runner, and assuming constant physical properties, conduction in the axial direction is negligible compared to convection in the axial direction, and only the axial velocity is important, the energy balance expression becomes:

$$\rho C_p \left( \frac{\partial T}{\partial t} + v_z \frac{\partial T}{\partial z} \right) = k_T \left( \frac{1}{r} \frac{\partial}{\partial r} \left( r \frac{\partial T}{\partial r} \right) \right) + 2\eta \left( \frac{\partial v_z}{\partial r} \right)^2 + \Delta H_r \left( \frac{\partial C}{\partial t} \right) \tag{6-5}$$

It is apparent that this relation cannot be solved explicitly for the temperature distribution, but is instead coupled to the momentum balance to determine the velocity profile, and also coupled to the mass balance to determine the spatial concentration of the reactive species so that the generation term can be properly determined. This is a series of three coupled partial differential equations that must be solved simultaneously. An analytical solution is not possible, hence numerical methods must be used. Even with a numerical approach, the solution of these coupled relations which incorporates both non-Newtonian viscosity, time dependent viscosity, viscous heating, and heat generation through chemical reaction, is at the state of the art of computational fluid mechanics. For the purpose of an engineering analysis of the heating and flow in the runner of a transfer mold, it is preferable to make simplifying assumptions that reduce the problem to a manageable endeavor. Some of these effects and related assumptions are analyzed in the following sections.

*Heat Generation from Polymerization* The first factor to assess is the importance of the heat generation term. The resin comprises only 25% of the molding compound by weight (less than 50% by volume) and the epoxy material has already been partially advanced in preparation and compounding so that the concentration of functional groups is much lower than in a common liquid epoxy resins that can show very high exotherms. The adiabatic exotherm can be determined from the energy balance shown in Equation (6-5) by neglecting the convective and conductive terms, and representing the partial differential operators with differences, similar to a finite difference treatment. The relation shown below indicates that the adiabatic exotherm for 100% conversion is approximately 20°C [6]. The molding compound can only flow in the runner while its degree of conversion remains below the conversion to gel, which for most materials is less than 30%. Hence the maximum adiabatic exotherm that can be realized during flow in the runner is approximately 6°C. In the more common case where the conversion during flow does not exceed 15%, the adiabatic exotherm would be in the vicinity of only 3°C.

$$\Delta T_{\text{adb}} = \frac{\Delta H C_0}{\rho C_p} = \sim 20^\circ\text{C} \tag{6-6}$$

The significance of this exotherm in a nonadiabatic system such as the runner where heat is conducted from or to the runner walls can be assessed through a dimensionless group that can be derived by casting the energy balance relation shown in Equation (6-5) into dimensionless form with the convective terms omitted.

$$\frac{\partial \bar{T}}{\partial t} = \left(\frac{k_T}{\rho C_p L^2}\right)\frac{\partial^2 \bar{T}}{\partial x^2} - \left(\frac{\Delta H r_{\text{ref}}}{\rho C_p T_{\text{ref}}}\right)\bar{r} \tag{6-7}$$

The group multiplying the dimensionless reaction rate $\bar{r}$ is the time scale of heat generation, denoted $\lambda_g$. If the time variable is now made dimensionless by dividing it by the quantity of $\rho C_p L^2/k_T$, which is the thermal conduction time scale, $\lambda_c$ then Equation (6-7) is rendered completely dimensionless and a dimensionless group which is the ratio of the heat generation and thermal conduction time scales emerges. This group, known as the Damkohler Number, $Da$, is a metric of the relative importance of heat generation to heat conduction [7]. Values greater than 1 indicate that heat generation is important. Values much greater than 1 indicate a thermal runaway condition where the heat liberated from the reaction can overwhelm the heat removal capacity of the system resulting in an uncontrolled acceleration of the temperature-controlled reaction rate.

$$\lambda_c = \frac{\rho C_p L^2}{k_T} \qquad \lambda_g = \frac{-\Delta H r_{\text{ref}}}{\rho C_p T_{\text{ref}}} \tag{6-8}$$

$$\frac{\partial \overline{T}}{\partial t} = \frac{\partial^2 \overline{T}}{\partial x^2} + \frac{\lambda_c}{\lambda_g} \overline{r} \tag{6-9}$$

$$Da = \frac{\lambda_c}{\lambda_g} = \frac{-\Delta H L^2 r_{\text{ref}}}{k_T T_{\text{ref}}} = \frac{\text{heat generation}}{\text{heat conduction}} \tag{6-10}$$

For a fluid with inert filler such as the molding compound, the Damkohler number has to be modified to account for the volume fraction of filler, $f_v$ [7]:

$$Da = \frac{-\Delta H L^2 r_{\text{ref}} (1 - f_v)}{k_T T_{\text{ref}}} \tag{6-11}$$

Substituting in values for the reaction exotherm and the physical parameters of the molding compound, the value of the Damkohler Number is determined to be 0.003, indicating that the time scale of heat conduction is much shorter than for generation, hence heat generation plays a small role in the overall energy balance and subsequent temperature profile in the runner. The generation term can therefore be conveniently neglected in all but the most rigorous analyses.

*Shear Heating During Flow in the Runner* Viscous heat dissipation occurs whenever a fluid is sheared because some fraction of the mechanical energy of the motion is converted to heat energy. The degree to which this effect is important, however, will depend on the rheology of the fluid, the deformation rate, and the ability of the system to remove the heat generated. The limiting case for maximum heating is an adiabatic system. The First Law of Thermodynamics provides the temperature increase that can be expected when a pressure drop is experienced. Given that the overall pressure drop in transfer molding is not great and significantly lower than other polymer process operations where shear heating is important such as injection molding, the shear heating in an adiabatic system does not provide a large temperature increase.

$$\text{Shear Heating: } \Delta T_{\max} = \frac{\Delta P}{\rho C_p} \tag{6-12}$$

The overall importance of this term in a nonadiabatic system can be assessed through the Brinkman Number, $Br$, a scaling parameter which gauges the relative importance of viscous heating to conductive heat transfer. Values far below 1.0 indicate that shear heating will be overwhelmed by conductive heat transfer, hence no noticeable temperature increase will be found. Conversely, Brinkman numbers much greater than one indicate that shear heating is important and significant temperature increases can be expected. For a Newtonian liquid in a slit geometry, the Brinkman number is expressed as [8]:

$$Br = \frac{\eta V^2}{k_T(T_w - T_0)} = \frac{\text{shear heating}}{\text{conduction}} \tag{6-13}$$

For the case of a power-law fluid, only the circular-cross-section geometry result is available [9]:

$$Br = \left(\frac{1 + 3n}{n}\right)^{1+n} \frac{KU^{1+n}R^{1-n}}{k_T(T_w - T_0)} \tag{6-14}$$

Using Equation (6-14) to assess the importance of viscous heating in the runners, a Brinkman number of 0.05 is computed for typical process conditions of a 10 cm/sec velocity, a power-law index of 0.5, a viscosity coefficient of 500, and a hydraulic radius of 0.20 cm. The viscosity parameters provide a viscosity value of 400 poise at the 300 $sec^{-1}$ shear rate typical of a runner. This low Brinkman number indicates that shear heating is not an important effect for this material at these process conditions. The viscosity of the molding compound and the shear rates encountered in the runners are both too low to cause significant shear heating.

*Temperature Dependence of Viscosity* The other important effect is the viscosity dependence on temperature. For the case of flow in a channel maintained at a higher temperature, the usual case is that there are lower viscosities at the wall compared to the centerline. The exception is when the cure has advanced substantially during the filling stage, an abnormal condition that often results in an incomplete fill. The effect of lower viscosity at the wall is similar to the shear thinning behavior of the molding compound in that the higher shear rates at the wall also tend to reduce the viscosity there compared to the centerline. This has important implications for the velocity profile. Lower viscosities at the wall cause higher velocities than would otherwise be observed there. The flow profile becomes more pluglike and moves away from the parabolic profile that occurs with uniform viscosity. Therefore, the temperature dependence of viscosity will affect the temperature profile in the runner in the same way that the shear thinning behavior affects the temperature profile.

*Time Dependence of Viscosity* This effect becomes much more important as the degree of conversion approaches to within several percent of the conver-

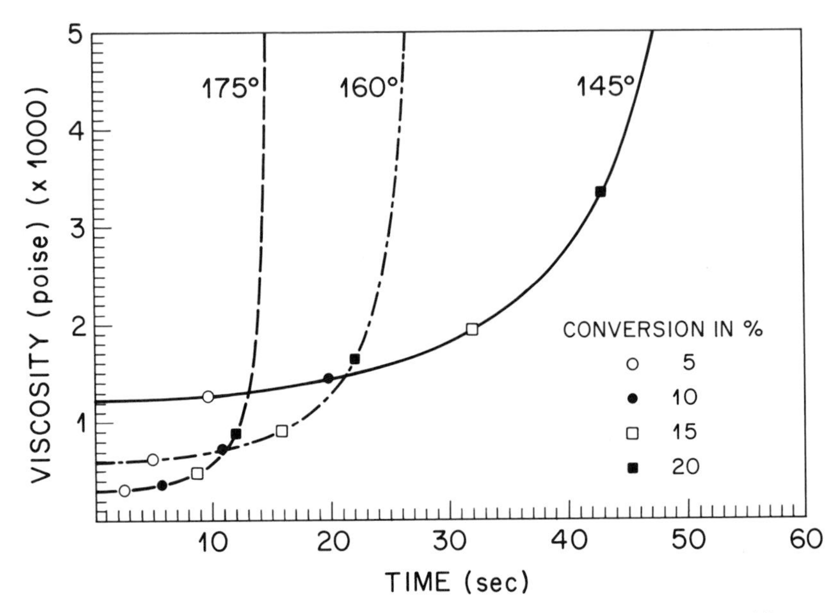


**FIGURE 6-16** Plot of viscosity versus time and degree of conversion for an epoxy molding compound at isothermal conditions. Regular increments of degree of conversion are shown on the curves. The conversion to gel is 25%.

sion to gel. Figure 6-16 provides a plot of viscosity versus conversion and time for a typical epoxy molding compound. It is evident that the viscosity remains nearly constant during the initial stages of the polymerization where only small molecules are being coupled together. As the conversion approaches the gel point, larger molecules are now being joined and the viscosity increase is much greater for the same increment of conversion increase that provided so little viscosity increase at low conversion. Therefore, the time dependence of the viscosity is most important when flow behavior up to the gel point is required, such as the assessment of ability to completely fill a large molding tool with a fast cure molding compound. For most other types of analyses, and especially where premature gelation is not a concern, the effect of time dependent viscosity can be neglected with minimal error. More rigorous approaches and those that seek to compute viscosities and forces at the devices themselves may require a time dependent viscosity treatment.

*Analytical Solutions for Heating in the Runner* Flow of a fluid initially at one temperature in a channel maintained at some different temperature has been one of the most studied engineering problems, long associated with the names of Nusselt and Graetz. In the case of a Newtonian fluid without viscous heating and a temperature independent viscosity, the most well known solution for the average, or mixing cup, temperature is that by Graetz, which can be written in the form of an infinite series [10, 11, 12]:

$$\langle \tilde{T} \rangle = \sum_{i=1}^{\infty} A_i e^{-a_i \zeta} \tag{6-15}$$

where the mixing cup temperature and the axial variable $\zeta$ are defined as:

$$\langle \tilde{T} \rangle = \frac{T_w - T}{T_w - T_0} \tag{6-16}$$

$$\zeta = \frac{k_T z}{\rho C_p R^2 U} \left( \frac{1 + n}{1 + 3n} \right) \tag{6-17}$$

The coefficients for flow of a power-law fluid were determined by Lyche and Bird [12], and are provided along with the Newtonian case in Table 6-1 [10]. No solutions for noncircular cross sections can be cited.

This solution can be used to predict the mixing cup temperature in the runner during fully developed, steady flow. With proper attention to assumptions and adjustments for deviations, however, this solution can be extended to some transient problems and not fully developed flow fields. Predicted results for the temperature in a circular cross section channel that is a representation of a trapezoidal runner under constant velocity conditions are presented in Figure 6-17. Temperature measurements in the runners of actual molding tools [2] confirm these predictions. Results for power-law indices of 0.5 and 0.33 are presented in Figure 6-17, and it is apparent that the effect of this degree of shear thinning behavior on the axial temperature profile is minimal.

In Figure 6-17, the rate of temperature rise is steep at the beginning of the runner where there is a large difference between the inlet fluid temperature and the runner wall temperature. The rate of temperature increase falls off as the molding compound temperature approaches the runner wall temperature. It is important to note the distances over which the heating occurs for these typical process and design parameters. Approximately 10 cm of flow length is required to heat the molding compound to 140°C, and almost 25 cm before the molding compound temperature comes within 10° of the mold temperature.

**Table 6-1. Coefficients for the Graetz Solution of Flow in a Circular Cross Section Channel for Newtonian and Non-Newtonian Fluids [10, 12]**

| | $a_i$ | | | | $A_i$ | | | |
|---|---|---|---|---|---|---|---|---|
| $i$ | $n = 1$ | 0.5 | 0.33 | 0 | 1 | 0.5 | 0.33 | 0 |
| 1 | 7.31 | 6.58 | 6.26 | 5.78 | 0.82 | 0.81 | 0.81 | 0.69 |
| 2 | 44.6 | 39.1 | 36.4 | 30.5 | 0.10 | 0.11 | 0.11 | 0.13 |
| 3 | 114 | 99.5 | 92.3 | 74.9 | .032 | 0.039 | 0.046 | 0.048 |

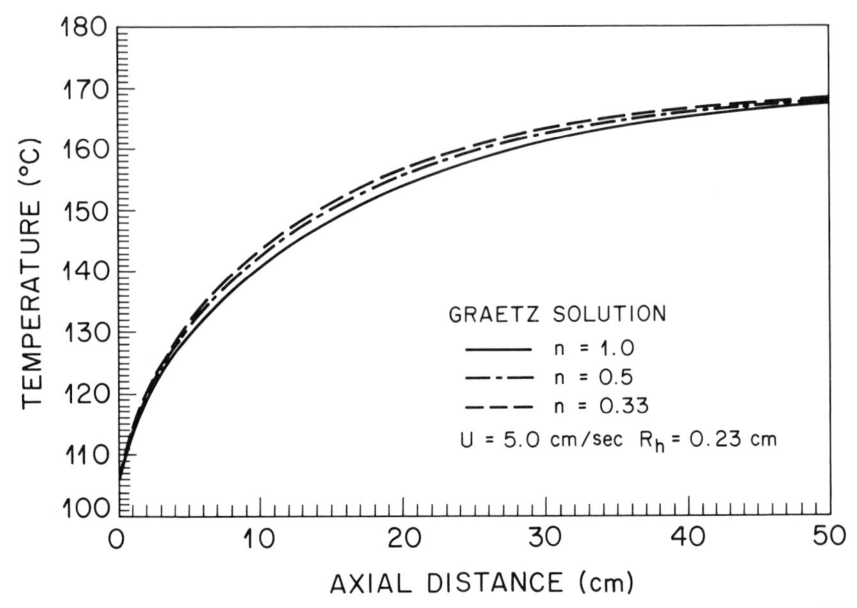


**FIGURE 6-17** Predicted temperature for flow of an epoxy molding compound in a trapezoidal runner represented with a circular cross section using a hydraulic radius value of 0.23 cm, computed using an analytical solution for flow of non-Newtonian fluids. Results for a power-law index of 0.33, 0.5, and 1.0 are presented. The average molding compound velocity was constant at 5 cm/sec and the mold temperature was maintained at 170°C.

*6.4.2.2.2 Heat Transfer Coefficients and Empirical Treatments.* Another approach to computing the temperature profile along the runner is through the use of heat transfer coefficients. These coefficients are well established in the chemical industries for predicting the performance of heat exchangers and other complicated flows that do not lend themselves to analytical treatments. The general form of the heat transfer coefficient relation is similar to, and based upon, Newton's Law of Cooling. It states that the rate of heat flow across an interface is proportional to the product of the area of the interface, the characteristic temperature difference, and a proportionality factor known as the heat transfer coefficient. These treatments are one-dimensional in that they provide no information on the radial or transverse temperature distribution, but instead can be used to determine bulk averaged, or mixing cup, temperatures:

$$Q = hA(T_w - T_0) \tag{6-18}$$

Be aware that $h$ is not a characteristic property of a fluid or a heat transfer system, but is instead a complicated function of many variables such as material properties, geometry, temperature difference, temperature distribution, and flow behavior. Correlations for heat transfer coefficients abound in the technical literature with a good representation of them found in the standard engineering handbooks [13]. The simplicity of the relation, however, makes experimental determination relatively straightforward, especially if the temperature gradient and the thermal conductivity are known. In these cases, the heat transfer coefficient is often expressed as a Nusselt Number, $Nu$, a dimensionless group which is defined as $hD/k_T$. As a scaling parameter, the Nusselt number is an indicator of the relative importance of interfacial heat transfer to conduction in the bulk fluid. For example, a high heat transfer coefficient coupled with a low thermal conductivity provides for an accumulation of heat with resulting high temperatures at the interface, thereby sustaining a higher temperature gradient. In this way, the Nusselt number is often thought of and expressed as a dimensionless temperature gradient. For flow in a tube, the relation for the heat transfer coefficient expressed as a Nusselt number is provided below; the starred variables are dimensionless, where $r^* = r/D$, $z^* = z/D$, and $T^* = (T - T_0)/(T_{b1} - T_0)$:

$$N = \frac{hD}{k_T} = \frac{1}{2\pi L/D} \int_0^{L/D} \int_0^{2\pi} \left( - \frac{\partial T^*}{\partial r^*} \right) \bigg|_{r^*=1/2} d\theta \, dz^* \qquad (6\text{-}19)$$

The above relation is useful to define the Nusselt number and understand its physical significance, but it is relatively useless in determining the heat transfer coefficient because of the difficulties of measuring the temperature gradient along the runner surface. There are a number of correlations for heat transfer coefficient and Nusselt number in the technical literature [13]. A widely accepted correlation for laminar flow in tubes based on an empirical modification of the Graetz solution discussed above is [13, 14]:

$$\frac{hD}{k_T} = 1.86(Re\ Pr\ D/L)^{0.33} \left( \frac{\eta_b}{\eta_0} \right)^{0.14} \qquad (6\text{-}20)$$

The $b$ subscript indicates parameters evaluated at the average bulk conditions, whereas the subscripts 0 and 1 indicate parameters evaluated at the average initial and exit conditions. $Re$ and $Pr$ are dimensionless groups which characterize the momentum and thermal transport characteristics of the flow field and are defined below:

$$Re = \frac{\rho D U}{\eta} \qquad (6\text{-}21)$$

$$Pr = \frac{C_p \eta}{k_T} \qquad (6\text{-}22)$$

The Reynolds number $Re$ represents the ratio of inertial forces to viscous forces. Flows with Reynolds numbers greater than 2100 are turbulent or chaotic in nature, those below are laminar. Flows with Reynolds number of order one or lower are called *creeping flows*, where inertial effects may be neglected. The Reynolds number for flow of epoxy molding compound in a runner is order 0.01, placing it well within the creeping flow regime. The Prandtl number $Pr$ is the ratio of momentum transfer to heat transfer. The numerical value of the Prandtl number for flow in the runner is of the order 50,000. Equation (6-20) is applicable for flows where the triple product of $Re\ Pr\ D/L > 10$: the numerical value of this product for flow in the runner is approximately 25, close to the inapplicable region, indicating that the relation should be used cautiously. In general, correlations such as Equation (6-20) are more suited to water and other low-viscosity media that do not show shear thinning behavior. The use of the bulk properties at entrance and exit conditions provides a way to account for shear thinning and temperature effects, but applying these relations to transfer mold flow is tenuous. The presentation of the heat transfer correlations shown in Equation (6-20) is intended more to illustrate the dependence of the coefficient on material and process variables so that adjustments to known coefficients can be made, rather than as a means to computing the coefficients from physical properties.

Experimental determination of the heat transfer coefficient is relatively easy, and where possible is much preferred over the use of the correlations. The experimental apparatus to determine heat transfer coefficient includes a runner segment that will experience nonisothermal flow instrumented with temperature sensors, and a means of estimating the volumetric flow rate. The temperature sensors should be spaced at such a distance so that there is a significant temperature increase between them, but not so far that the molding compound approaches the mold temperature. The means of monitoring the volumetric flow rate can be collection of the extrudate over a time period, monitoring the filling rate of a closed cavity of known volume, or through the use of a ram-follower device that was discussed in Section 5.4.1.11. A diagram of a molding tool specially designed to make runner temperature measurements is shown in Figure 6-18.

The temperature sensors can be purchased commercially, or they can be custom fabricated from conventional thermocouples. In fabricating sensors, it is preferable to use bare thermocouple junctions that will be in intimate contact with the molding compound. An ejector pin can be hollowed out to create a small cavity for the thermocouple. It is important that the thermocouple be thermally isolated from the mold. Embedding the thermocouple in an epoxy potting compound within the hollowed ejector pin is one way to achieve this electrical and thermal isolation. A glass-filled epoxy will stand up to the abrasive molding compound better than an unfilled material. One design for this type of temperature sensor is shown in Figure 6-19. The positioning of the thermocouple in the runner cross section is also very important. A flush-mounted

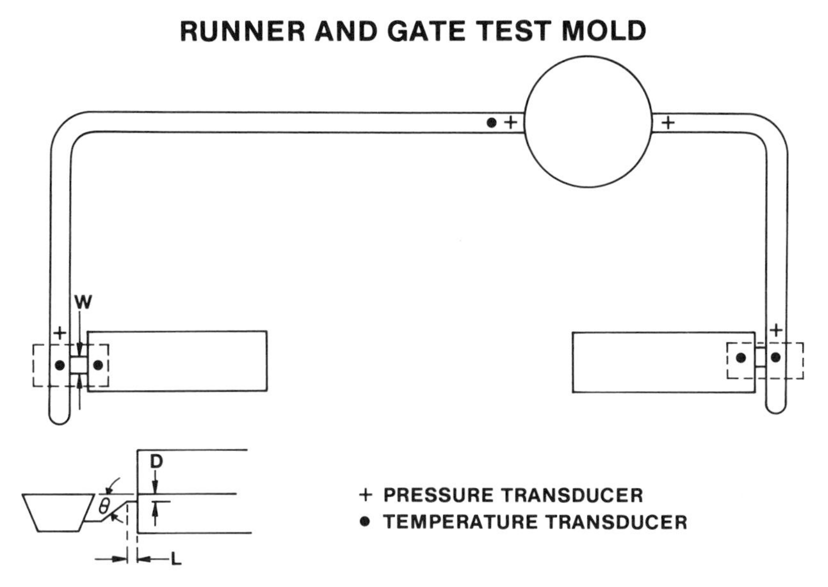


**FIGURE 6-18** Transfer molding tool designed to make temperature and pressure drop measurements in the runner and through the gates. In this configuration the volumetric flow rate is determined by a ram follower device attached to the plunger movement. A process controller was used to provide a constant residence time in the runner approaching the gate, and then to switch to a prescribed volumetric flow rate through the gate.

sensor would provide a reading that would be higher than the bulk mixing temperature since the material there has been in intimate contact with the high temperature walls. A measurement on the centerline of the flow would provide a low reading since the material flowing there is the furthest from the heated wall. The optimum position is probably at the one-quarter line of the flow channel, also shown in Figure 6-19. The temperature of the molding compound here most closely approaches the mixing cup temperature, although the sensor probably causes significant redistribution of the velocity profile. Nonetheless, temperature measurements at this location can be taken to be the mixing cup temperature with only minimal error.

The temperature during flow in the runner can also be determined through measurements made in production molding tools that have been instrumented for temperature measurements. In most cases, the installation of temperature sensors can be temporary unless they are specifically process hardened for production use. The sensors should be positioned in the mold cavity according to the precepts mentioned above: far enough apart to provide a significant temperature difference, yet close enough together to prevent the temperature at the far point from reaching the mold temperature. Temperature data acquired in a

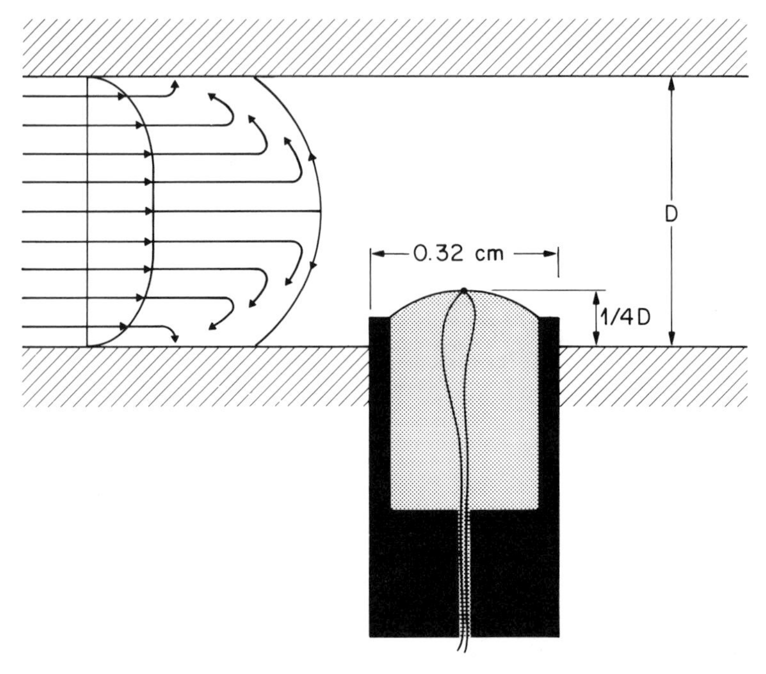


**FIGURE 6-19** Custom sensor for measuring temperature in the runner of a transfer mold.

single chase of a 40-pin DIP mold are presented in Figure 6-20 [2]. These data show a significant temperature increase over the 15.5 cm distance between the first cavity on the leadframe and the last cavity. This indicates that there is fairly rapid heat transfer in the small cross section of the runner. Also, there is a significant velocity decrease in passing from the first cavity to the sixth cavity because of the loss of molding compound diverted into the intervening cavities. These data, combined with temperature data from the molding tool described in Figure 6-16, provide enough information to compute an approximate heat transfer coefficient for flow in a conventional size runner within a range of flow velocities encountered in package mold filling. A value of $h = 0.031$ cal/cm$^2$-°C was derived empirically and is appropriate only for the given molding compound, runner geometry (hydraulic radius of 0.23 cm), and flow velocity range (approximately 3–5 cm/sec). It may, however, be used with other molding compounds with similar rheology and in similar flow configurations, particularly if the numerical value is modified for deviations using the heat transfer correlation provided in Equation (6-20).

*Numerical Solution of Heating Based on the Heat Transfer Coefficient* An approximate solution can be derived to determine the temperature in the runner

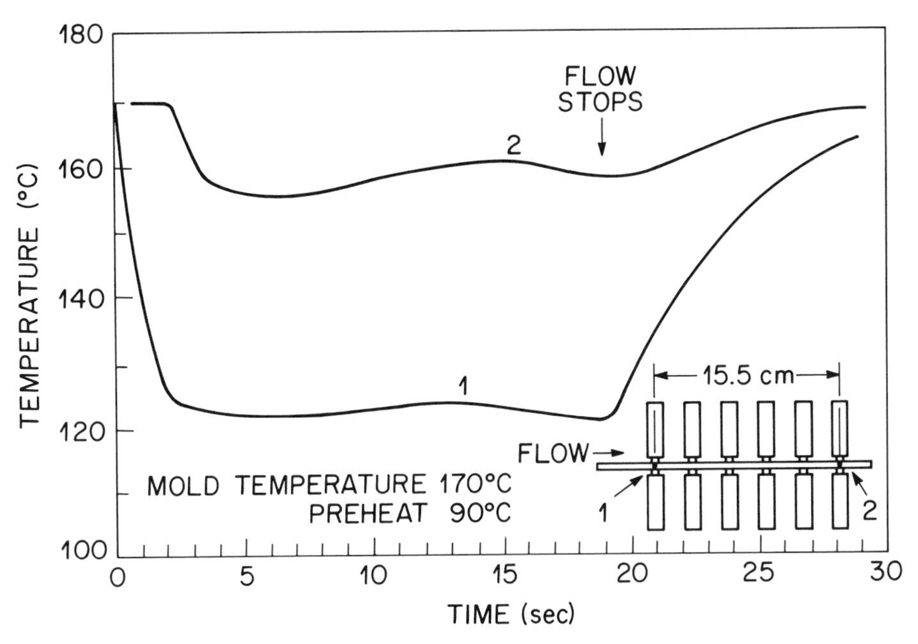


**FIGURE 6-20** Temperature data for flow in the runner of a 40-pin DIP mold operating under normal process conditions. The distance from the transfer pot to the first thermocouple is 5 cm, whereas the distance between the first and second thermocouples is 15.5 cm. [2] (Reprinted with Permission of the Society of Plastics Engineers.)

during transfer molding. The simplest approach is to consider a one dimensional solution that would provide the mixing cup temperature as a function of axial position in the runner. This solution is derived from the one dimensional relation that was used to define the heat transfer coefficient in Equation (6-18). The solution is based on equating the heat transferred across the liquid/solid interface to the rise of temperature in the molding compound:

$$Q = mC_p \frac{dT}{dt} = hA_{xs}(T_w - T_i) \tag{6-23}$$

In macroscopic balances, the above relation would be applied over the length of the tube, where $T_i$ is the inlet fluid temperature. The situation with molding compound flowing in the runner is more complicated, however, since the temperature rise is significant and the fluid temperature will approach the mold temperature after only a modest flow length. Also, the fluid velocity changes down the length of the runner. An iterative, or stepwise, application of Equation (6-23) is preferred in this case. The flow channel is divided into a large number of axial sections or cells. The heat transfer relation is then applied progressively to each individual cell, computing an outlet temperature from the inlet temperature which is the computed outlet of the previous cell. Starting at the runner entrance where the temperature is known from the previous discussion of the transfer pot, the solution works its way forward until it reaches the advancing flow front or the end of the runner. The velocity of the fluid is accounted for in the mass of material heated. The velocity can be varied down the flow length to accommodate diversion of molding compound into the cavities since Equation (6-23) is applied at each length increment. Lower material velocities decrease the mass of fluid heated in the given time increment, and thereby increase the temperature rise. An important assumption of this solution is the plug flow velocity profile that it requires; the entire contents of one cell move into the next cell. There is actually some velocity profile over the cross section of the flow channel, with a maximum velocity in the center and zero velocity at the wall if the fluid sticks at the wall. For an approximate solution, however, the plug flow velocity assumption is not a serious misrepresentation. The molding compound is a shear thinning material which will tend to provide a more plug-like velocity profile. Even more significant is that the heated mold wall will further lower the viscosity at the wall, in some cases overwhelming the shear thinning reduction. In addition, rheometer studies indicate that these highly filled materials show slippage against machined metal surfaces at shear rates greater than 50 $sec^{-1}$, providing strong evidence that the stick at the wall hypothesis does not extend to these highly filled materials. The combined effects of the shear thinning viscosity, temperature decrease of the viscosity, and wall slippage at the shear rates in the runner provide a velocity profile that could approach plug flow depending on specific process and material characteristics. These effects are summarized in Figure 6-21.

Another important assumption of this model is that the viscosity is not affected by the chemical conversion. This is again not a serious misrepresentation of the actual phenomena since the viscosity does not increase significantly due to conversion until the conversion is within a few percent of the conversion to gel as was seen in Figure 6-16. Hence for most of the fill time, it can be assumed that viscosity is independent of conversion without serious error. The temperature effect on the viscosity and the velocity profile has already been accommodated by assuming that the velocity profile remains plug flow throughout the transfer time. Using this approach, the temperature profile in the runner can be computed and the result compared to the analytical solution in Figure 6-22. One advantage over the analytical solution is that this heat transfer coefficient solution can accommodate velocity changes along the runner length.

*6.4.2.2.3 Effect of Design and Process Parameters on Heating in the Runner* The heating during flow in the runner is strongly influenced by the runner design and the process parameters, less dependent on the selection of molding compound since most IC packaging materials have similar thermal

| PROCESS/MATERIAL PARAMETER | VELOCITY PROFILE |
|---|---|
| NEWTONIAN FLUID, ISOTHERMAL FLOW, WALL STICKING | 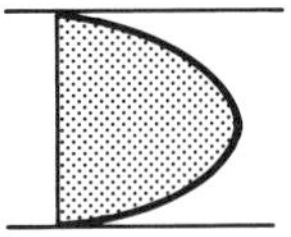 |
| NON-NEWTONIAN, SHEAR THINNING VISCOSITY (n = 0.5); ISOTHERMAL FLOW; WALL STICKING | 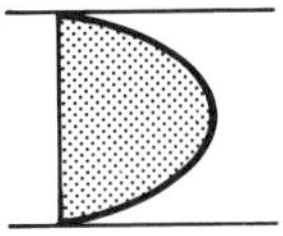 |
| NON-NEWTONIAN, SHEAR THINNING (n = .5); HEATED WALL; WALL-STICKING | 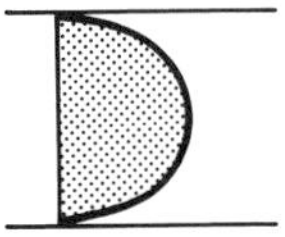 |
| NON-NEWTONIAN, SHEAR THINNING (n = .5); HEATED WALL; WALL SLIPPAGE | 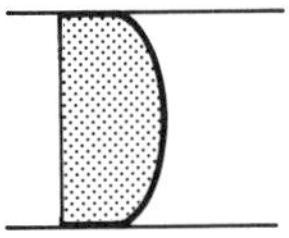 |

**FIGURE 6-21** A summary of the material and process phenomena that justify a pluglike velocity profile in approximate solutions of temperature in the runners of a transfer mold.

properties. The differences in molding compound rheology, particularly the shear thinning behavior, will not have a major impact on heat transfer because these differences will cause only modest changes in the velocity profile, which has a minimal effect on heat transfer as was illustrated in Figure 6-17.

The runner dimensions have an important effect on the heating of the molding compound. This dependence on runner size can be assessed by applying the approximate solution discussed above, with the dimensions of the noncircular runner cross section represented by a hydraulic radius. Both the analytical solution and the iterative approach offer the capability to change the tube radius. Figure 6-23 shows the predicted temperature profile in the runner for three different hydraulic radii derived from the heat transfer coefficient solution described in Section 6.4.2.2.2.

The effect of velocity in the runner is assessed using the heat transfer coef-

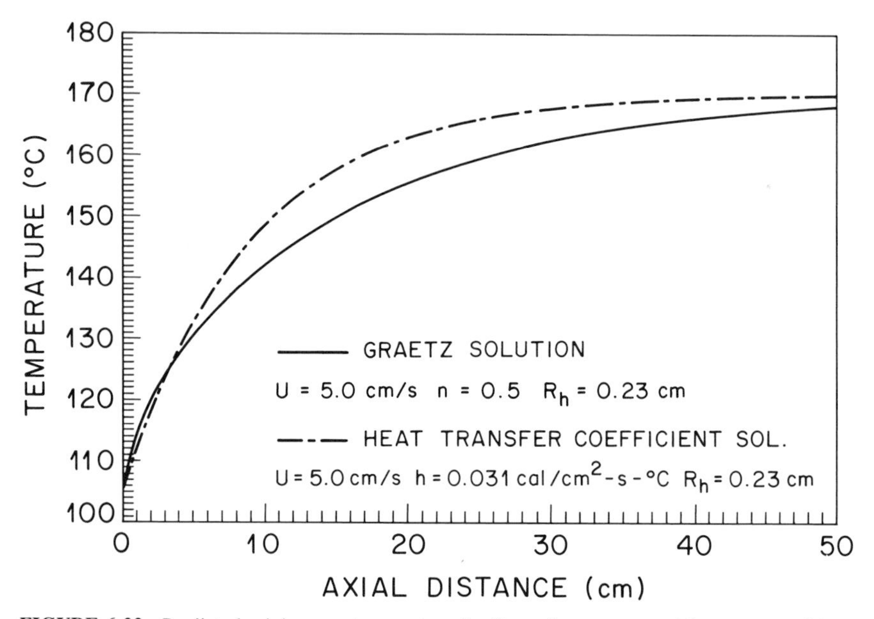


**FIGURE 6-22** Predicted mixing cup temperature for flow of an epoxy molding compound in a trapezoidal runner represented with a hydraulic radius value of 0.23 cm computed with a numerical solution using an experimentally determined heat transfer coefficient. The results for the analytical solution described previously are shown on the same plot. The difference between the predictions narrows as the power law index for the analytical solution approaches the plug flow limit of zero.

ficient model. Figure 6-24 shows the temperature profile in the runner for three different velocities that span the range of what is found in a package transfer mold. The highest velocity is typical of flow in the runner near the transfer pot and before the molding compound has reached any of the cavities. The lowest velocity is typical for the far end of the runner when the molding compound is flowing into all of the cavities upstream. There is a very significant effect of velocity in determining the flow lengths that must be met to achieve certain temperatures. This result indicates that even minor changes in transfer rate can cause significant changes in the temperature profile in the runner. Both high- and low-temperature excursions are serious. Low temperature increases the viscosity and the flow-induced stress, potentially increasing device damage and lowering yields. Premature high temperatures prolong the time of rapid polymerization rate and can prematurely gel the molding compound, which is to drive its viscosity toward infinity, as shown in Figure 6-16.

Another process parameter that has an important effect on heating in the runner is the mold temperature. Figure 6-25 shows a plot of mixing cup temperature plotted against flow length for three different mold temperatures that span the feasible range. These results show the importance of maintaining the mold

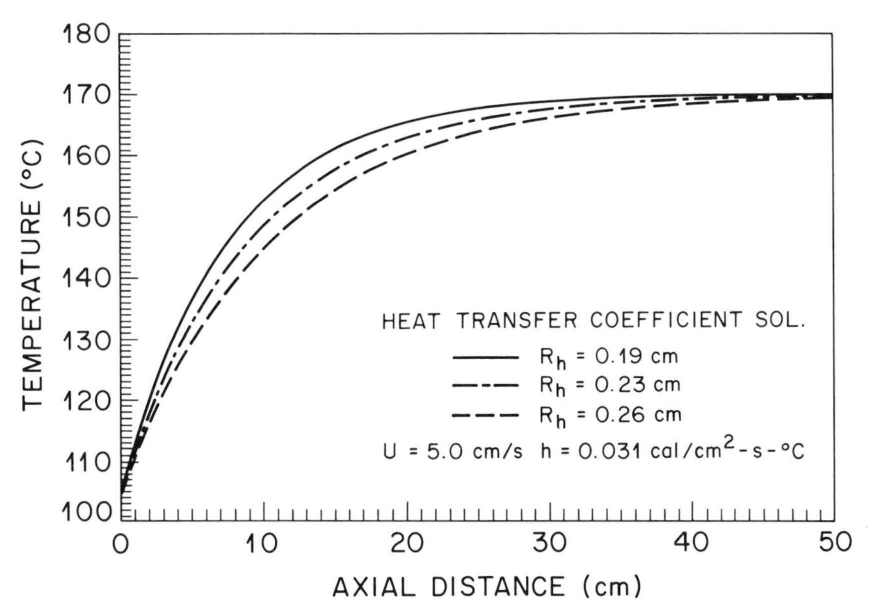


**FIGURE 6-23** Plot of the mixing cup average temperature in the runner for three different hydraulic radii of 0.19, 0.23, and 0.26 cm. The velocity is maintained at 5 cm/sec and the mold temperature is 170°C.

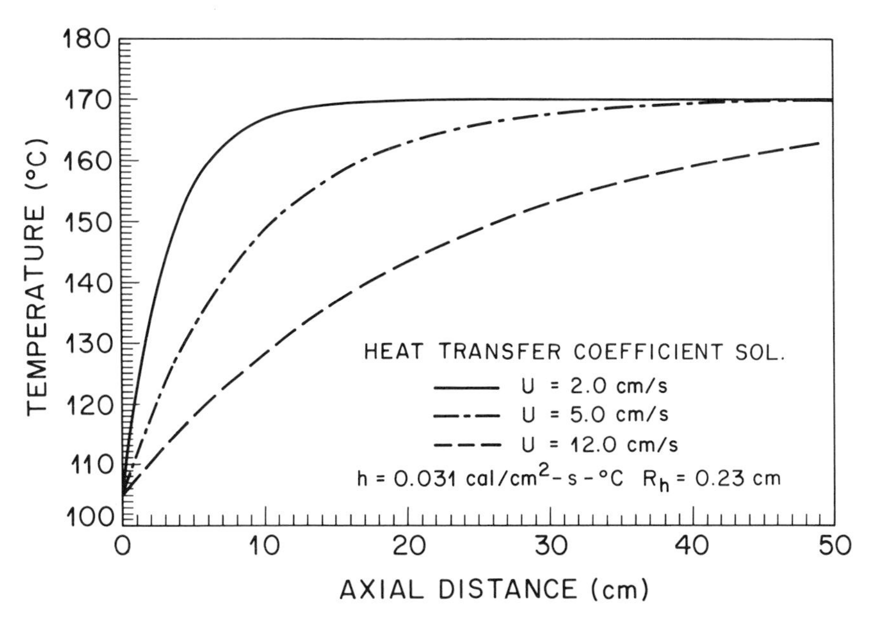


**FIGURE 6-24** Plot of the mixing cup temperature plotted against axial flow length for three different velocities in the runner. The mold temperature is maintained at 170°C. The three velocities shown span the range found in a packaging transfer mold.

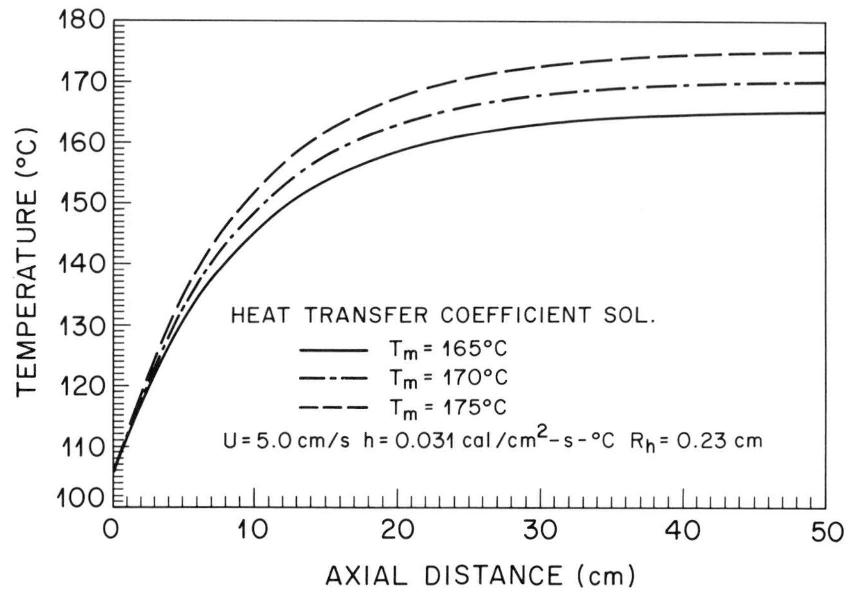


**FIGURE 6-25** Plot of the mixing cup temperature for three different mold temperatures. The runner temperature is equal to the mold temperature in transfer molding. The velocity in the runner is 5 cm/sec.

temperature at the specified value, both temporally and spatially. Large excursions in mold temperature often occur because the mold is much hotter than the ambient air, so that even drafts on the tool or an extended clamp open time can lower the temperature by as much as 10°C. The number and placement of the mold heaters can also cause significant temperature variations over the surface of the tool as will be discussed in Section 7.3.2.

**6.4.2.3 Viscosity During Flow in the Runner** The viscosity of the molding compound as it flows in the runner determines the flow-induced stresses over the devices to a large extent. There is a temperature increase and viscosity decrease experienced upon flow through the gate, but this effect is not readily influenced by design and material variables, leaving the viscosity in the runner as one of the more important metrics of the subsequent flow-induced stress on the device. There is no feasible method to measure the viscosity of the molding compound as it flows in the runner of a transfer mold, hence simulation techniques must be employed. If the rheology of the molding compound is determined, as was described in Section 4.2.1.1, then the temperature and time history of flow in the runner can be used to compute the viscosity profile. In this way, the viscosity profile is a characteristic of the material properties, the design variables of the runner, and the process parameters. It is a unique fingerprint of a molding system that can be used to evaluate or select different materials, runner geometries and process conditions.

One approach to characterizing the viscosity in the runner is to follow a single fluid element as it moves out of the transfer pot, through the runner, and up to the gate. Figure 6-14 shows approximate velocities of the molding compound in the runner. Coupled with the solution of temperature in the runner based on the heat transfer coefficient analysis, the time-temperature history of the fluid element in the runner can be derived. By factoring in the rheology of the specific molding compound, the viscosity history of the fluid element can be determined. This treatment is approximate because the rheology should not be decoupled from the temperature profile in the runner, but as we saw in Figure 6-17, the shear thinning behavior has only a minor effect on the temperature profile over the range of power law indices encountered in typical epoxy molding compounds. Figure 6-26 shows the viscosity of the single fluid element during flow in the runner as computed with this treatment. The viscosities shown are for a specific molding compound rheology, which is provided below in Equation (6-24):

$$\eta = 1.82 \times 10^{-5} \exp\left(\frac{17369}{RT}\right)\left(\frac{0.30}{0.30 - \dfrac{X^2}{0.30}}\right)\dot{\gamma}^{-0.63} \qquad (6\text{-}24)$$

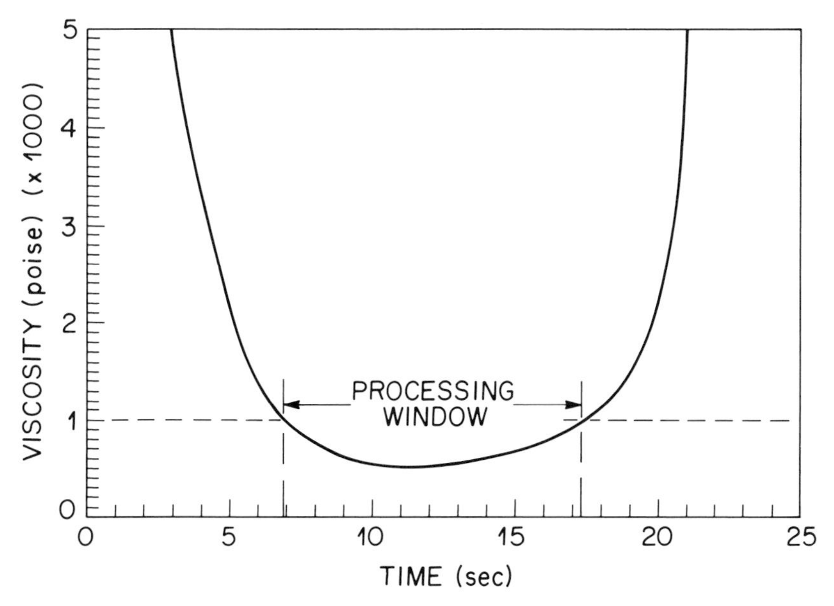


**FIGURE 6-26** Plot of the predicted viscosity of a single fluid element during flow in the runner. The results are for a runner with a hydraulic radius of 0.23 cm and a shear rate of 100 $sec^{-1}$.

These rheological parameters are typical of IC-grade molding compounds, although most molding compounds appear to have conversions to gel closer to 25%. This rheology should not be used for specific design work or feasibility estimates since different grades could show enough difference to invalidate the findings.

The shape of the curve shown in Figure 6-26 provides important insight in optimization of the transfer molding process for IC packaging. The molding compound enters the runner at relatively low temperature (approximately 105°C) and high viscosity, as was determined from the transfer pot study discussed previously. The molding compound heats rapidly once it begins flowing in the narrower runner, and its viscosity drops significantly over the first few seconds of flow. The molding compound temperature then approaches the mold temperature and the viscosity reaches a low plateau value, but the polymerization reaction rate is near its maximum because of the high temperature and the high concentration of unreacted epoxy groups. The viscosity does not increase rapidly, however, because only small molecules are being coupled together at this early stage of the reaction. The viscosity remains at a relatively low level for a period of time until the polymerization has proceeded far enough that larger molecules are now being coupled, resulting in more rapid increases in molecular weight and viscosity as was detailed in Section 2.4.3. The viscosity increase is steep as the chemical conversion approaches the conversion to gel where a three dimensional molecular network is formed. The material is now a crosslinked rubber that can no longer flow as a true fluid. It is apparent that the flow-induced stresses on the device would be minimized if the molding compound contacted the leadframes only when it was within the low-viscosity process window. Similarly, it is inappropriate to flow over the devices when the viscosity is far out of the low viscosity window, either before it has flowed and heated appreciably or when it is approaching gelation. Both of these scenarios will result in excessive flow-induced stress and possible device damage.

### 6.4.3 Flow Through the Gates

The flow of the molding compound through the gates determines the mold-filling profile to a large extent. Flow through the gates also has an influence on the temperature and viscosity of the molding compound as it contacts the leadframe and device.

**6.4.3.1 Pressure Drop Through the Gates** Understanding and quantifying the pressure drop through the gates is necessary for controlling the mold-filling profile, minimizing or eliminating short shots, and gaining control of overall filling and packing pressures. This pressure drop is strongly dependent on molding compound, process, and gate design parameters. The shear thinning behavior of the molding compound means that even when the pressure drops are known or the filling behavior understood at a set of process conditions, it can not be easily scaled to predict different process conditions. For example, doubling the rate of transfer through the gate does not double the pressure drop; in many cases the pressure drop is much less than expected. Compounding the issue is the difficulty in conducting rheological experiments at the same conditions of shear and elongational deformation that are encountered in the gate. Experimental measurement of the pressure drops experienced on flow through the gate is one of the more effective ways to characterize this important behavior. An experimental configuration developed to conduct this measurement was shown in Figure 6-18 [2]. The pressure drops over the range of volumetric flow rates encountered in actual molding processes were measured. In addition, the effect of gate dimensions on the pressure drops was studied, varying four key gate parameters that are illustrated in Figure 6-27 [2]. These data shown in Figures 6-28, 6-29, 6-30 and 6-31 [2] provide a useful compendium of design options to control the pressure drops through the gates.

The pressure drop data indicate that gate width and depth are the most effective way to control gate pressure drop. The angle of the convergent section is fairly ineffective, reinforcing the hypothesis that the pressure drop and filling profile are largely controlled by the constriction size rather than by any of the upstream features such as the runner, the sub-runner, or the convergent section. Between the gate width and the gate depth, the gate width is the preferred means to control gate pressure drop because it has little or no effect on subsequent process steps. The gate depth has a significant effect on gate breakoff; deeper gates do not break as easily and gouge out a bigger section of the molded body when they do break. Land length also effects the mechanics and aesthetics of the break. In some instances of very narrow packages such as small outline designs there is simply not enough range of width to significantly alter the pres-

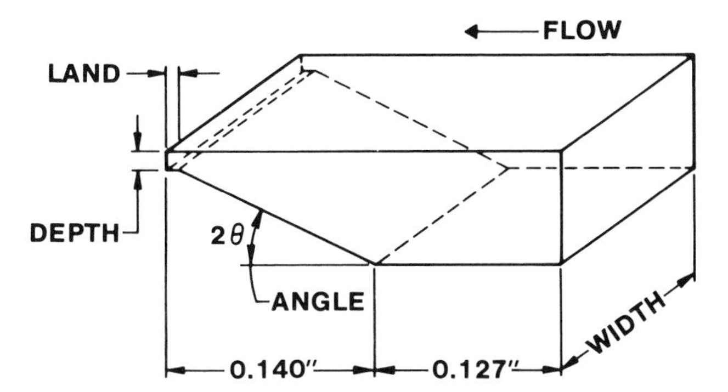


**FIGURE 6-27** Gate geometry for the experimental work on temperature increase in the gate described in Section 6.4.3.2

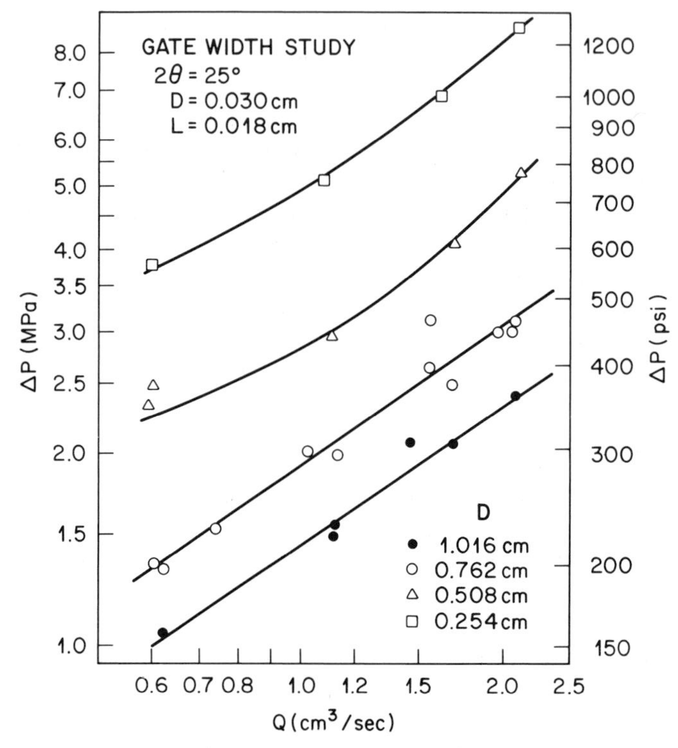


**FIGURE 6-28** Plot of the pressure drop versus volumetric flow rate through the gate for four different gate widths that span the feasible range for a cavity chase mold. [2] (Reprinted with Permission of the Society of Plastics Engineers.)

sure drop through the gate. In these cases, the angle of the convergent section must be used in tandem with gate width, with alteration of the gate depth and the land length used as a last resort.

The pressure drop through the gate can be predicted by applying the analysis of flow of a non-Newtonian fluid through a convergent channel that was developed to characterize molding compounds in Appendix 4A and Section 4.2.3.2.

**6.4.3.2 Heating on Flow Through the Gate** Significant heating occurs when the molding compound passes through the gate. This heating is largely conductive heat transfer, facilitated by flow in the constricted gate geometry, which greatly increases the surface area of hot metal exposed to the colder molding compound. The contribution of viscous heating can be assessed by applying the Brinkman number scaling parameter, as was described for flow in the runner in Section 6.4.2.2.1. Repeating the Brinkman number analysis described in Equation (6-14) for flow of the non-Newtonian molding compound through the gate

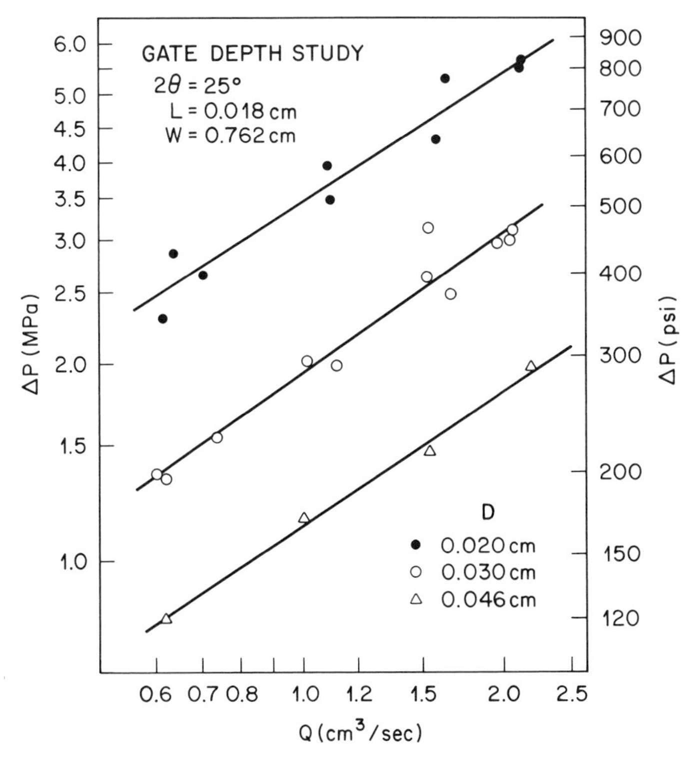


**FIGURE 6-29** Plot of the pressure drop versus volumetric flow rate through the gate for four different gate depths that span the feasible range for a cavity chase mold. [2] (Reprinted with Permission of the Society of Plastics Engineers.)

represented with a hydraulic radius of 0.015 cm provides a Brinkman number of 0.10. The low Brinkman number indicates that shear heating through the gate is insignificant, despite the widely held belief that it is an integral part of the transfer molding process. The misunderstanding is reasonable, given that there is a temperature increase on passage through the gate, accompanied by a viscosity decrease. The viscosity goes down due to two unrelated effects; the shear thinning behavior of the molding compound in the high shear gate geometry, and the conductive heating of the material on passing through the narrow flow channel.

Experimental determination of the temperature increase on flow through the gate is not difficult if in-mold thermocouples are in place. One experiment found the molding compound temperature increases from 125° to 140–145°C for a conventional convergent gate using a mold temperature of 170°C. No significant transient behavior or temperature gradient is likely with this slit flow geometry. Therefore, the heat transfer can be described by a heat transfer coeffi-

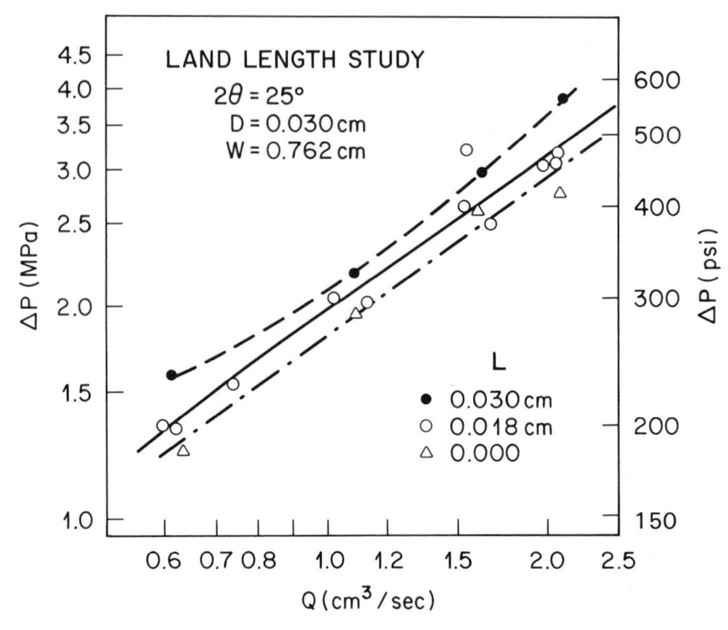


**FIGURE 6-30** Plot of the pressure drop versus volumetric flow rate through the gate for four different gate land lengths that span the feasible range for a cavity chase mold. [2] (Reprinted with Permission of the Society of Plastics Engineers.)

cient, similar to the treatment for flow in the runner, that would provide a simple step change in temperature:

$$\dot{m}C_p\Delta T = h_g A_g(T_{\text{mold}} - T_{\text{runner}}) \qquad (6\text{-}25)$$

The parameters include $\dot{m} = 0.5$ gm/sec, $C_p = 0.288$ cal/gm-°C, $A_g = 0.5$ cm$^2$. It is preferable to lump the mass flow rate, specific heat and area together since for most transfer molds this cluster remains approximately constant. This provides a modified expression with a lumped dimensionless heat transfer coefficient:

$$\Delta T = \frac{h_g A_g}{\dot{m}C_p}(T_{\text{mold}} - T_{\text{runner}}) \qquad (6\text{-}26)$$

$$h_g^* = \frac{h_g A_g}{\dot{m}C_p} = 0.30°\text{C}^{-1} \qquad (6\text{-}27)$$

Simple expressions like Equation (6-26) approximate well the step change increase in temperature experienced on flow through the gate. The temperature increase falls off as the molding compound temperature in the runner ap-

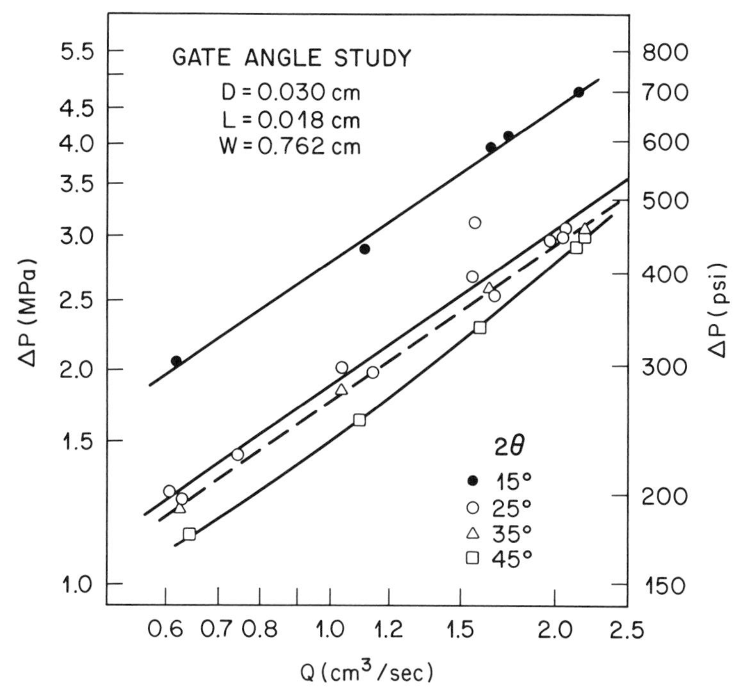


**FIGURE 6-31** Plot of the pressure drop versus volumetric flow rate through the gate for four different angles of the convergent section that span the feasible range for a cavity chase mold. [2] (Reprinted with Permission of the Society of Plastics Engineers.)

proaches the mold temperature as it does in the later cavities of a large mold. The temperature in the runner can be estimated from the treatments of heat transfer in the runner described in Section 6.4.2.2. The numerical constant in Equation (6-27) is presented to illustrate the magnitude of the temperature increase and its drop off as the temperature in the runner converges on the mold temperature. It is appropriate only for the process parameters and gate geometry of the experiment (geometry shown in Figure 6-27). It may be extended to other design and process parameters only for approximate analysis, and then with proper caution and possible modification.

In most cases, the same gate geometry is used throughout the mold. The pressure in the runner, however, is not uniform with the pressure being greater closer to the transfer pot. The filling profile is controlled by the volumetric flow rate through the gate which depends on the pressure drop across the gate and the viscosity of the material in the gate. Thus, cavities that are closer to the transfer pot usually fill sooner than those far from the pot because there is a higher pressure drop across the gate and because the molding compound simply arrives there sooner. This typical filling profile is shown in the short shot profile provided in Figure 6-32. This nonuniform filling profile is not absolute, however, in that the filling profile is also strongly influenced by the temperature profile in the runner, the transfer rate, and the molding compound rheology. The dependencies of mold filling profile are thoroughly explored in Section 7.1.4.

### 6.4.4 Flow in the Cavities

The flow in the cavities determines the extent of the flow-induced stresses that are imposed on the leadframe and wire bonds, thereby affecting the molding yield. The mold designer and molding engineer have very little influence on the flow of molding compound in the cavities once decisions on upstream variables, such as runner geometry, filling profile, and molding compound selection are made. The geometry of the cavity is usually specified by standards committees, and although these standards usually provide some tolerances, particularly on package thickness, the dimensions of the flow channel are often determined by considerations other than molding. It is for this reason that so much emphasis was placed on the upstream phenomena; the flow behavior in the cavity cannot be influenced significantly once all of the upstream parameters have been specified. Nonetheless, it is worthwhile to examine the flow in the cavity to learn how to adjust the upstream variables so as to improve package molding. The leadframe can split the flow into two separate flow fronts moving across the cavity above and below the leadframe as is shown in Figure 6-33. The existence of these two flow fronts and the distance between them depends on many factors which include the flow porosity of the frame, the sizes of the cavities above and below the leadframe, the material rheology, and the position the molding compound is introduced into the cavity. A photograph of the molding compound flowing through the leadframe and wire bonds in response to the lead-lag in the flow fronts is shown in Figure 6-34.

In some cases, it is preferable for the cavity which contains the wire bonds to lag the other flow front so that the molding compound flows through the frame and through the wire bonds from below, billowing them out like wind in a sail. If the flow in the cavity with the wire bonds were to lead, then the flow front would contact the wire bonds from the top. In some cases, this flow field could crush the wires onto the edge of the chip and the paddle support, as shown in Figure 6-33b. For this reason, it can be important to predict which cavity will have the leading front so as to assure that the device is oriented so that the molding compound flows up through the wire bonds instead of down onto them. This is not easily determined or even influenced by mold design parameters. The position of the gate does not control which is the leading front; there are

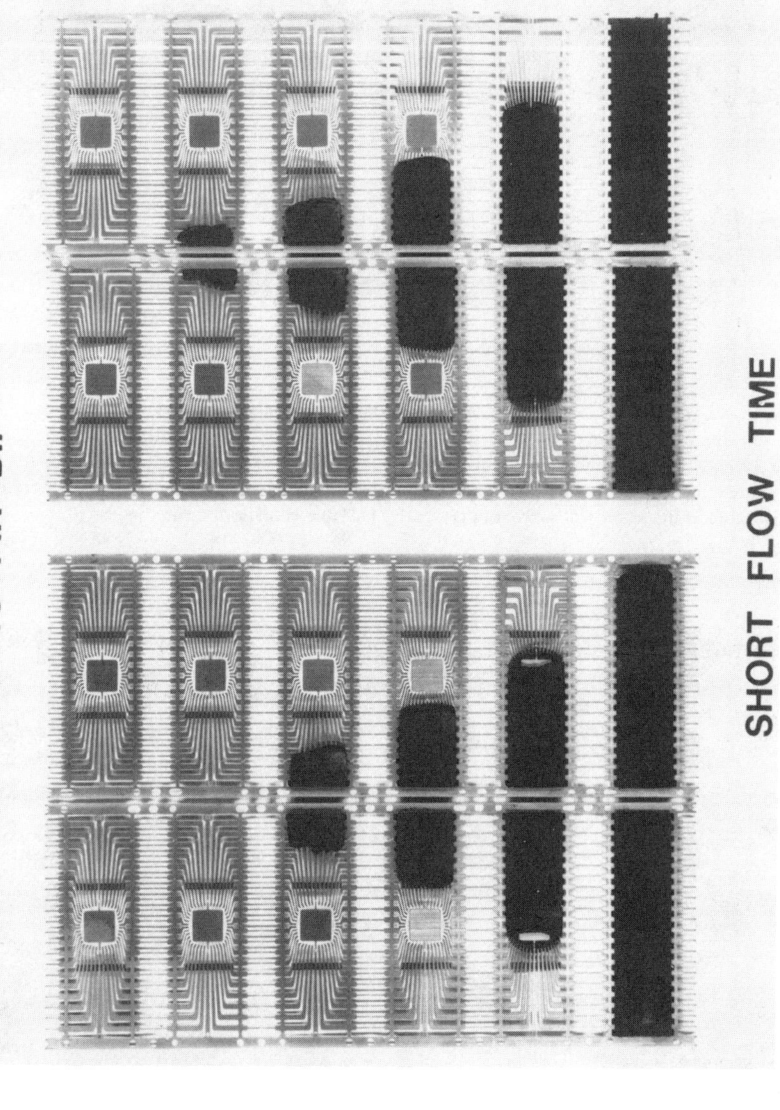


**FIGURE 6-32** Short shot filling profile of a production size molding tool for 24-pin DIPs in which all of the gates are uniform in size.

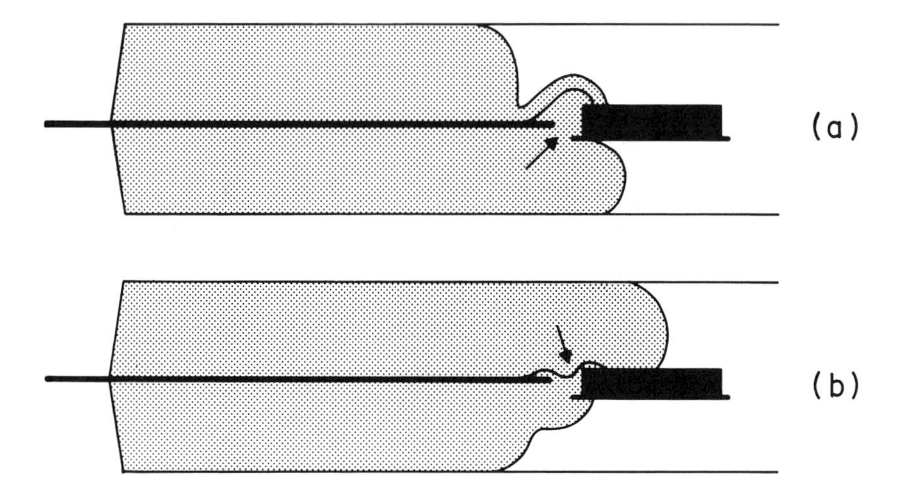


**FIGURE 6-33** Cross section view of flow of the molding compound in the cavity of the packaging mold showing the advancing flow front split into two separate fronts as it divides around the lead-frame: (a) the preferred configuration for molding wire bonded devices is for the flow in the lower cavity, the one without the wire bonds, to lead the flow in the upper cavity, (b) the less desirable configuration of the leading flow front in the cavity half that contains the wire bonds.

**FIGURE 6-34** A photograph showing the molding compound flowing up through the leadframe and wire bonds of a PLCC package in response to the lead-lag of the separate flow fronts.

many instances where the cavity half without the gate contains the leading flow front.

The flow of a non-Newtonian fluid in a rectangular cavity can be analyzed with the solution described previously in Equation 6-4 which uses the shape factors provided in Figure 6-15. The difficulty in using this expression directly is that the flow through the leadframe is essentially intractable, not amenable to mathematical analysis. Therefore, it is impossible to predict the lead-lag of the fronts confidently from first principles. In general, though, the flow in the upper and lower cavities is largely determined by the flow resistance encountered and the volume to be filled in the separate cavities, with the flow through the frame simply responding to pressure imbalances that develop behind the advancing flow fronts. In this way, the flow through the frame influences the extent of the lead-lag, but it is unlikely to change which front leads or lags. One approach to analyzing this lead-lag phenomenon would be to apply Equation (6-4), reproduced here as Equation (6-28) in a form to provide average velocity, to each cavity half separately using the same temperature or temperature profile:

$$\langle V \rangle = B \left( \frac{B \Delta P}{2KL} \right)^{1/n} S_p \tag{6-28}$$

Estimate this temperature in the cavity from the temperature of the material as it exits the gate as determined by Equation (6-26). If this gate exit temperature is more than 15°C below the mold temperature, then the molding compound will experience a significant increase in temperature as it flows in the cavity. This temperature increase and the ensuing temperature profile it causes should be accounted for in the application of Equation (6-28). A typical temperature gradient down the length of the cavity would be approximately 10–20°C depending on the rate of filling, the cavity thickness and the difference between the gate exit temperature and the mold temperature.

Equation (6-28) has significant nonlinear effects so it is important to know the molding compound rheology and the expected cavity fill time. Fill times can vary widely. For nonuniform filling profile of a 24-pin DIP with a mold fill time of 18 seconds as is shown in Figure 6-32, the average cavity fill time is 5 seconds. More uniform filling profiles provide longer cavity fill times for the same overall mold filling time. Smaller cavities such as those for small outline packages can fill in as little as 2 seconds, whereas very large package volumes such as an 84 PLCC can take an average of 8–10 seconds to fill. The cavity fill time can be estimated by conducting a series of short shots and estimating the fraction of the total mold fill time required to fill the cavities in different positions in the mold. In a standard unbalanced mold, individual cavities will fill for between one fifth to one third of the mold fill time. In a balanced mold (discussed in Section 7.1.4) with uniform cavity filling, the individual cavity fill times approach the mold fill time. Equation (6-28) can be used to develop a simple cavity filling algorithm that provides the positions of the flow fronts in time above and below the leadframe. An example of such an algorithm is provided in Figure 6-35, where Equation (6-28) was used in conjunction with the temperature profile to predict the lagging of the flow fronts as the flow encounters a colder chip and paddle assembly, accurately simulating the flow behavior in the molding tool.

Equation (6-28) can also be used to predict the difference in the velocities in the two halves which is an indication of the degree of the lead and lag. The actual lead and lag will be less because the molding compound flows through the frame in response to the pressure gradient that develops across the frame. Changes in process parameters or molding compound rheology suggested by manipulation of Equation 6-28 can resolve some lead-lag problems. As an example, a different molding compound with different rheology may provide different relative velocities in the upper and lower cavities when the cavities are of different thicknesses. In cases where it is important that the leading front should be on the other side of the wire bonds, there are some measures that can be considered to achieve this flow configuration. Inverting the chip and having it face in the opposite direction is one possibility, but various device codes may preclude this rather simple solution. Moving the gate over a section of leadframe that is either more open or more closed is also fairly effective. Moving the gate to the other side of the leadframe is often not as effective, especially if the hydraulic resistances of the cavity halves are significantly different.

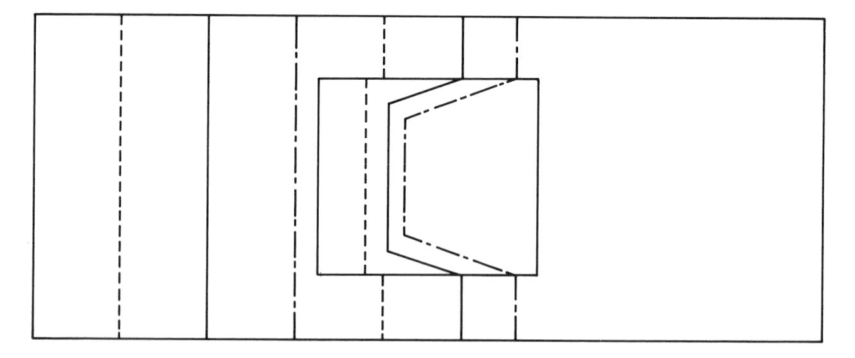

**FIGURE 6-35** The results of a simple cavity flow simulation based on Equation (6-28) to predict the flow front location above the leadframe are shown. The flow front movement is retarded as it encounters the thinner flow channel above the chip. The lower temperature of the chip and pad assembly has also been taken into account, significantly increasing the flow resistance over the chip and allowing the flow in the side channels to move ahead.

### 6.4.5 Packing in the Cavity

It is widely accepted that the packing step is one of the most important with regard to molding compound integrity, permeability and ultimate reliability of the package. With the growing emphasis on the elimination of voids, it also has a major influence on packaging yield and vendor quality standards. There is, however, very little known and understood about packing, as it involves the compressible flow of a fairly porous and nonuniform material. Compressible flows are difficult to treat mathematically, and the complicated structure, rheology, and time dependence of the molding compound make this analysis even more daunting. Much of what is understood is derived from qualitative study of short shot sequences with subsequent cross sectioning and examination of the molding compound consistency.

The molding compound flows in the cavity in a much more porous state than its final density and uniformity would indicate. This porosity can be detected by sectioning the package short-shots at several locations at increasing distance from the gate. The higher pressures are closer to the gate, so higher density material will be found there. The molding compound does not compress uniformly during packing, but instead the high density material moves out away from the gate toward the end of the cavity with continued application of the packing pressure. This behavior is illustrated in Figure 6-36. The higher pressures closer to the gate and the transient density gradient during the application of the packing pressure will cause most voids and low density material to be found away from the gate since continued application of the packing pressure will reduce the size and number of voids through compression, adsorption and absorption as described in Section 7.2.2. These problems are especially relevant when the molding compound in the gate gels before the packing pressure has been applied as is discussed in Section 6.5.3. The transition from the filling pressure to the packing pressure also has important implications for device damage. A rapid step increase in pressure could cause the material to rapidly compress and create a sudden velocity surge as more molding compound rushes into the cavity. Wire sweep and paddle shift can result from this surge, even though the cavity has ostensibly been filled. Conversely, if the application of the packing pressure is too slow, then there is a greater probability of gelling the molding compound in the gates and isolating the material from any additional packing pressure. These concepts are addressed in greater detail in Section 7.2.3.

## 6.5 POLYMERIZATION IN THE MOLD

Polymerization in the mold has important productivity and reliability implications. The cure in the mold requires the longest time interval of any of the process steps in molded packaging, and is therefore the part of the process

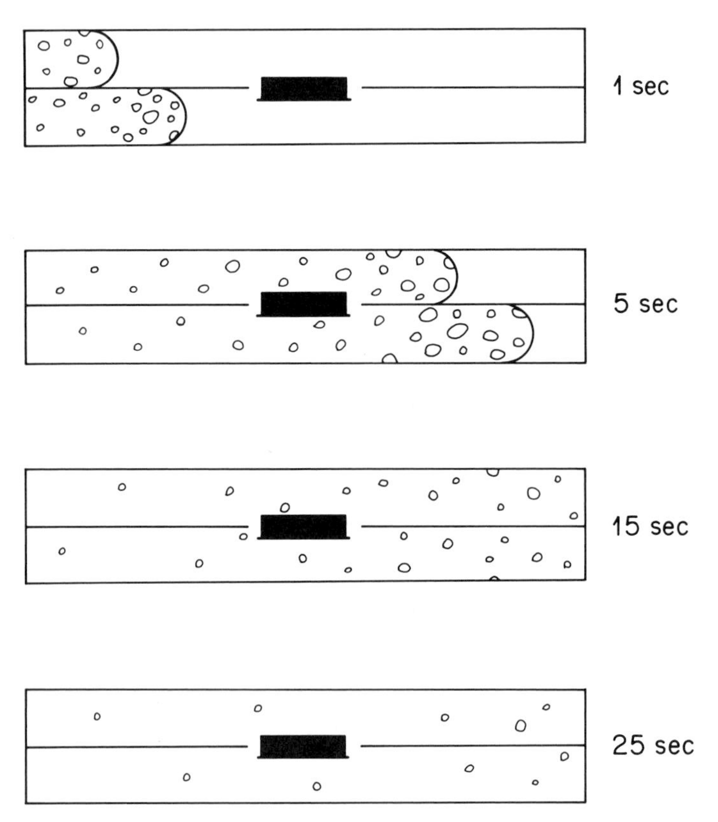


**FIGURE 6-36** An illustration of the porosity gradients that would be found in the cavity of a packaging mold during filling and with continued application of the packing pressure up to the time of gelation in the gates.

where productivity improvement efforts are most effectively targeted. During the course of the polymerization, the molding compound is transformed from a low-viscosity liquid to a rubbery gel to a stiff rubber. If the ultimate glass transition temperature of the molding compound is near or above the mold temperature, then the molding compound can vitrify, or become a hard glassy material, in the mold. This may or may not be desirable depending on how the parts are ejected from the tool. Treatments of polymerization in the cavity can be more rigorous than the approximate treatments of heating and flow in the runner and cavities that were discussed previously because the complications of flow are absent. Detailed analysis will require knowledge of the polymerization kinetics, but there are many other important analyses that can be carried out without a full kinetic study. The following sections describe both rigorous and qualitative treatments.

### 6.5.1 Polymerization Kinetics

Techniques for determining the kinetic rate constants were described in Section 4.2.2.1, and are also documented in the technical literature [15, 16]. Thermal analysis methods which assume that the fraction of the total heat of reaction liberated is proportional to the fraction of complete chemical conversion are preferred for these types of highly filled opaque systems. The rate expressions for epoxy molding compounds that have been reported are generally empirical in that they do not reflect the molecular dynamics of the reaction, but are instead phenomenological in that they predict the engineering behavior of the reaction without a theoretical basis for the reaction mechanism or reaction order. Hale et al. report a rate expression of the form shown in Equation (6-29) for a model system they developed to emulate an actual packaging molding compound [17] (shown previously in Equations (4-11) and (4-12)):

$$\frac{dX}{dt} = \{\exp(12.672 - 7560/T) + \exp(21.835 - 8659/T)X^{3.33}\}(1 - X)^{7.88} \qquad (6\text{-}29)$$

Manzione et al. have reported a rate expression for an actual epoxy molding compound based on an $n$th-order reaction rate [1]:

$$C_0 \frac{dX}{dt} = -\frac{dC}{dt} = 9.40 \times 10^5 \exp\left(\frac{-15957}{RT}\right) C^{1.0}$$

$$C_0 = 0.0020 \text{ equiv/cm}^3 \qquad H_r = -6000 \text{ cal/equiv}$$

$$R = 1.987 \text{ cal/°C} \qquad (6\text{-}30)$$

Many other alternative forms are also appropriate and are likely to appear as more studies are conducted. Either of the expressions provided in Equations (6-29) and (6-30) are suitable for mathematical treatments of cure in the molding operation. Bear in mind, however, that these expressions are for specific molding compounds and should not be used to predict the behavior of other materials. They are provided here because their general behavior will be similar to a large number of molding compounds in use at the time of the citations. Future molding compounds or materials for specific applications, such as multiplunger molding, will have significantly different kinetics.

### 6.5.2 Analyzing the Cure in the Cavities

The cure in the cavities of the mold tool does not advance to 100%. The reaction rate drops off sharply as the degree of conversion advances so it is counterproductive to continue the cure in the limited capacity molding tool. Instead, the parts are ejected when they reach an ejectable hardness which is usually at approximately 80% conversion. The cure is then completed batchwise in the postcure operation. It is sometimes apparent when the cure time in the mold is inadequate. The molding compound can be unusually soft and rubbery, and there may be greater tendency to stick to the mold surface. In other cases, the degree of undercure may not be apparent, but may still be a source of trouble. Undercured molding compound has a lower glass transition temperature and poorer mechanical properties than properly polymerized material. There are several approaches to analyzing the degree of cure achieved after a specified time interval. In most cases an isothermal cure analysis at the mold temperature is adequate for engineering purposes. For more rigorous or research level study, the nonisothermal course of the reaction should be taken into account.

**6.5.2.1 Nonisothermal Cure Analysis** There will be both spatial and temporal gradients in the temperature of the molding compound at the time when it completely fills and packs the cavity. In general, the temperature will be below that of the mold and higher at the far end of the cavity away from the mold gate. For one of the middle cavities of a leadframe strip, Figure 6-25 indicates that the temperature in the runner outside of the gate will be in the vicinity of 140°C. Equations (6-26) and (6-27) can then be applied to estimate a cavity inlet temperature of 149°C. For a moderate flow rate into the cavity, the molding compound could be heated an additional 10°C in transversing the cavity length, as per the discussion in Section 6.4.4. There will also be a temperature gradient in the direction perpendicular to flow because of the molding compound contact with the higher-temperature cavity walls and the leadframe. In most cases, it can be assumed that the leadframe is at the same temperature as the molding tool. From these considerations, an approximate temperature profile in the cavity at the start of cure cycle can be deduced. It can be assumed that the conversion profile is negligible, being below 5% throughout the cavity in most cases. If the process conditions push the degree of conversion close to the conversion to gel, then the degree of conversion in the cavity can be much higher and the gradient in conversion can also be significant.

There are commercial software packages that can provide a mapping of the temperature profile within the cavity during and at the end of the fill. These treatments are often based on important simplifying assumptions that allow the problem to be treated by a general purpose algorithm [18]. Some of these assumptions concern the mathematical description of the rheology and kinetics of the molding compound, whereas others concern the flow geometry. An application of one of these commercial software packages to package molding is shown in Figure 6-37 where the flow front locations during cavity fill in a multicavity molding tool are illustrated. Temperature and pressure profiles can also be obtained from this analysis.

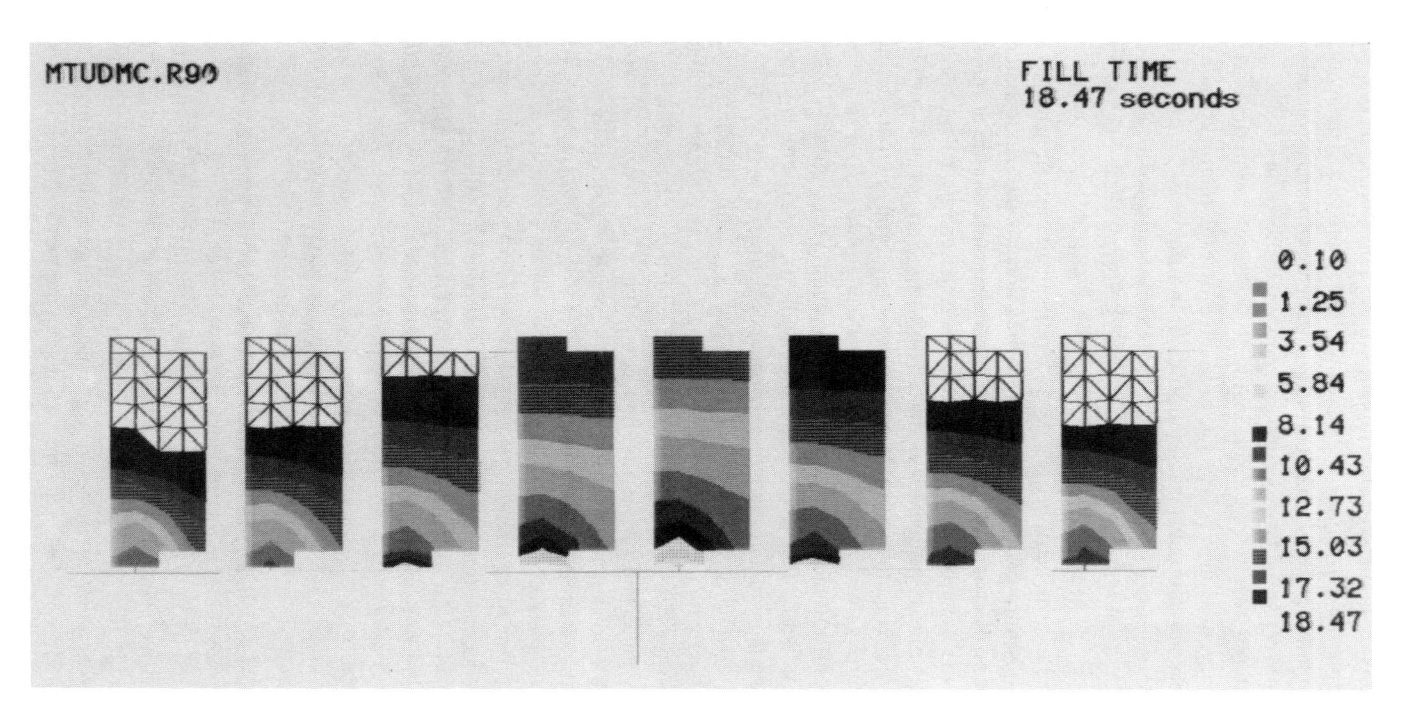


**FIGURE 6-37** The flow front positions in the cavity of a package molding tool during cavity fill as determined by the MOLDFLOW commercial software package. (Analysis by M. A. Zimmerman, AT&T Bell Laboratories)

The temperature and conversion profiles during polymerization can be analyzed by writing the mass and energy balance for the process. The energy balance must include temporal, conductive and generative terms. Viscous dissipation and convective or flow dependent energy transport can now be ignored since the material is quiescent. Rectangular coordinates are preferred for the cavity geometry. It is also acceptable to assume that the thermal conductivity and the density are constant for the purposes of this analysis:

$$\frac{\partial T}{\partial t} = k_T \left( \frac{\partial^2 T}{\partial x^2} + \frac{\partial^2 T}{\partial y^2} + \frac{\partial^2 T}{\partial z^2} \right) + H_r \left( \frac{\partial C}{\partial t} \right) \tag{6-31}$$

The energy balance in the cavity has dependencies in all three dimensions, since there are temperature gradients in the flow direction, in the depth dimension, and even across the width of the cavity. The temperature profile at the end of cavity fill is the appropriate initial condition for the energy balance solution. This can be obtained through a commercial software package as shown in Figure 6-37, or the temperature profile can be assigned through an approximation of the gradients, as was described above. For many purposes, it is acceptable to assume a uniform temperature as the initial condition. Typically, a temperature 10–20°C below the mold temperature will not be in serious error. The energy balance is coupled to the mass balance of the reactive species. The rate of the reaction which consumes reactive species, typically the epoxy groups, can be expressed with an $n$th-order kinetic expression, as is shown in Equation (6-32), or another kinetic expression, either of which would be derived from a kinetic study conducted on the molding compound. The local concentration, however, can also depend on diffusion of the reactive species as well as its consumption by reaction. In the most rigorous sense, the temperature profile in the cavity will give rise to a reaction rate profile, which will cause a concentration gradient. The concentration gradient will then foster a diffusive process to equalize the concentration, actually the chemical potential, throughout the cavity. The expression for the concentration term in Equation (6-31) then becomes as shown below where simple $n$th order kinetics were used to represent the chemical reaction:

$$\frac{\partial C}{\partial t} = D \left( \frac{\partial^2 C}{\partial x^2} + \frac{\partial^2 C}{\partial y^2} + \frac{\partial^2 C}{\partial z^2} \right) + k_o \left( \exp \frac{E_a}{RT} \right) C^n \tag{6-32}$$

In many cases including the present one, the diffusion terms are negligible, leaving the much simpler relation that the concentration of reactive species is simply depleted by the chemical reaction. The initial condition for the concentration of reactive species can also be assumed to be uniform without serious error. A value in the vicinity of 95–98% of the initial concentration is reasonable, with adjustments up or down for unusually long or short flow times, or exceptionally slow or fast reacting molding compounds. There have been a number of reports of numerical solutions to this system of coupled energy and material balances in the cure of thermoset polymers in molding tools [7, 19, 20, 21]. In many cases, one or two of the dimensional dependencies can be omitted to simplify the solution as well as the interpretation of the results. Figure 6-38 provides a mapping of the temperature and conversion of reactive groups during the polymerization of a package molding compound in a short DIP package using the kinetic rate expression provided in Equation 6-30. The solution is two dimensional in that the profiles in the width direction have not been considered. The results presented in Figure 6-38 show that the nonisothermal character of the cure washes out after only a small fraction of the overall cure time has been expended. This suggests that an isothermal cure at the mold temperature will provide information that is as useful and effectively as accurate as the much more involved nonisothermal analysis that requires solution of the energy balance relation.

**6.5.2.2 Isothermal Cure Analysis** Isothermal analysis of cure in the mold cavity is far simpler than the nonisothermal mapping described above and shown in Figure 6-38. For this treatment, we neglect the temperature gradients that are found in the cavity, properly assuming that the effect of the temperature profile and its transient character are insignificant to the overall cure which continues over several minutes. As a consequence, we may also neglect the concentration gradients that are the result of the temperature profile and the residence time distribution of the molding compound in the cavity. These assumptions allow us to apply a simple kinetic expression, examples of which are provided in Equations 6-29 and 6-30, to assess the extent of the polymerization in the mold. The most obvious temperature to insert in the isothermal rate expression is the mold temperature, since the time averaged temperature during the 1.5–3.0 minute cure time will be very close to the mold temperature. Figure 6-39 provides a plot of the conversion as a function of time for the molding compound described by the kinetic rate expression shown in Equation (6-30). Results for the nominal mold temperature of 170°C and at other temperatures are provided. These results clearly show the initial rapid rate of reaction and the drop-off in reaction rate as the reactant is consumed. Higher mold temperatures do accelerate the reaction and reduce the time to an ejectable chemical conversion, but they also reduce the flow length according to the principles established in Section 6.4.2.3 and presented in Figure 6-26. Extending the degree of conversion in the mold past the point of ejectability is counterproductive because postcure, which is far less costly in terms of human and capital resources, serves the purpose of extending the cure to 100% conversion.

### 6.5.3 Gelation in the Gates (Gate Freeze-Off)

The gelation of the molding compound in the gates of the mold is a significant process event because after this the cavity is hydraulically isolated from the

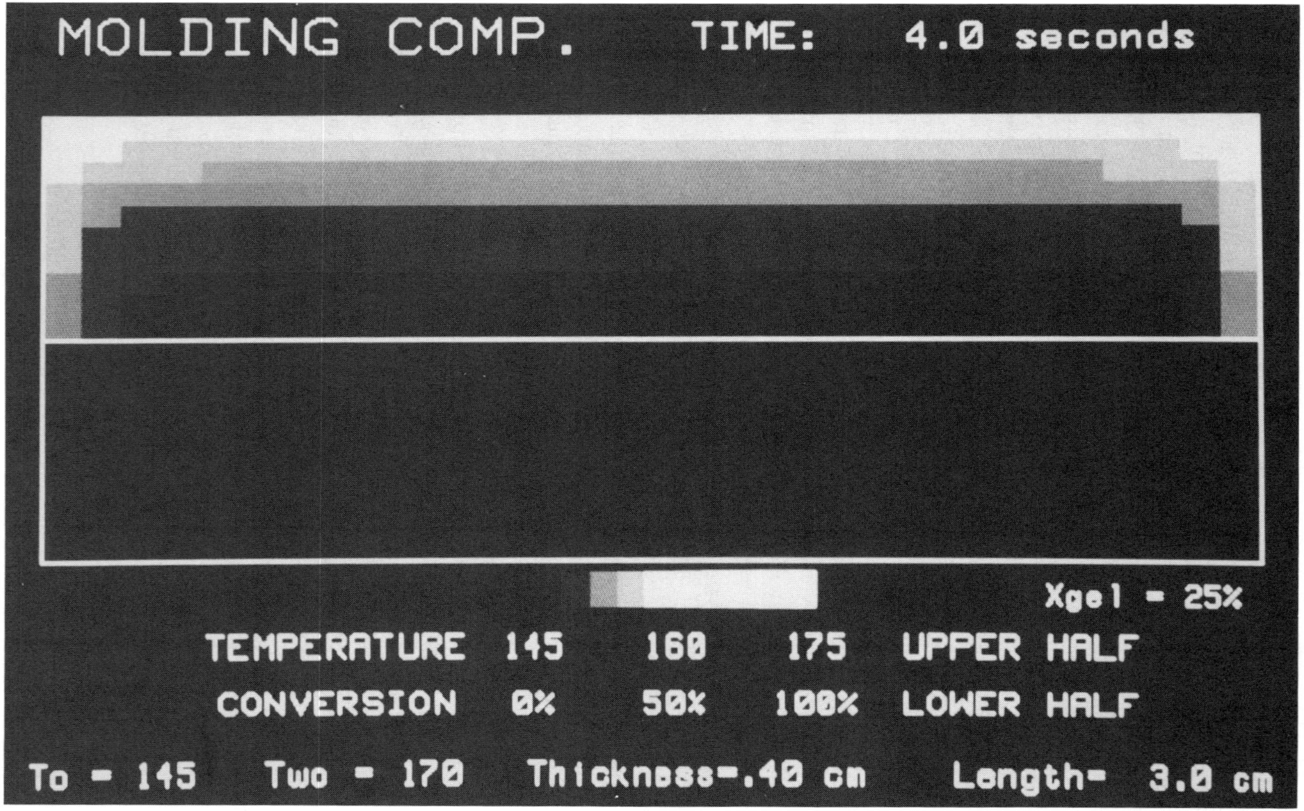

**FIGURE 6-38** A mapping of the conversion and temperature for a package molding compound during polymerization in a medium size DIP. The mappings are at a times of 4, 12, and 48 seconds after the cavity was filled. At each time, the cavity has been split on the centerline. The temperature is provided in the mapping above the centerline, whereas the conversion is mapped below the centerline. For both temperature and conversion, the mappings would be mirrored across the centerline with the assumptions and geometric model used. Neither leadframe nor IC device is included in this analysis. The initial condition is a uniform temperature of 145°C and a uniform conversion of 3%. The mold temperature is 170°C.

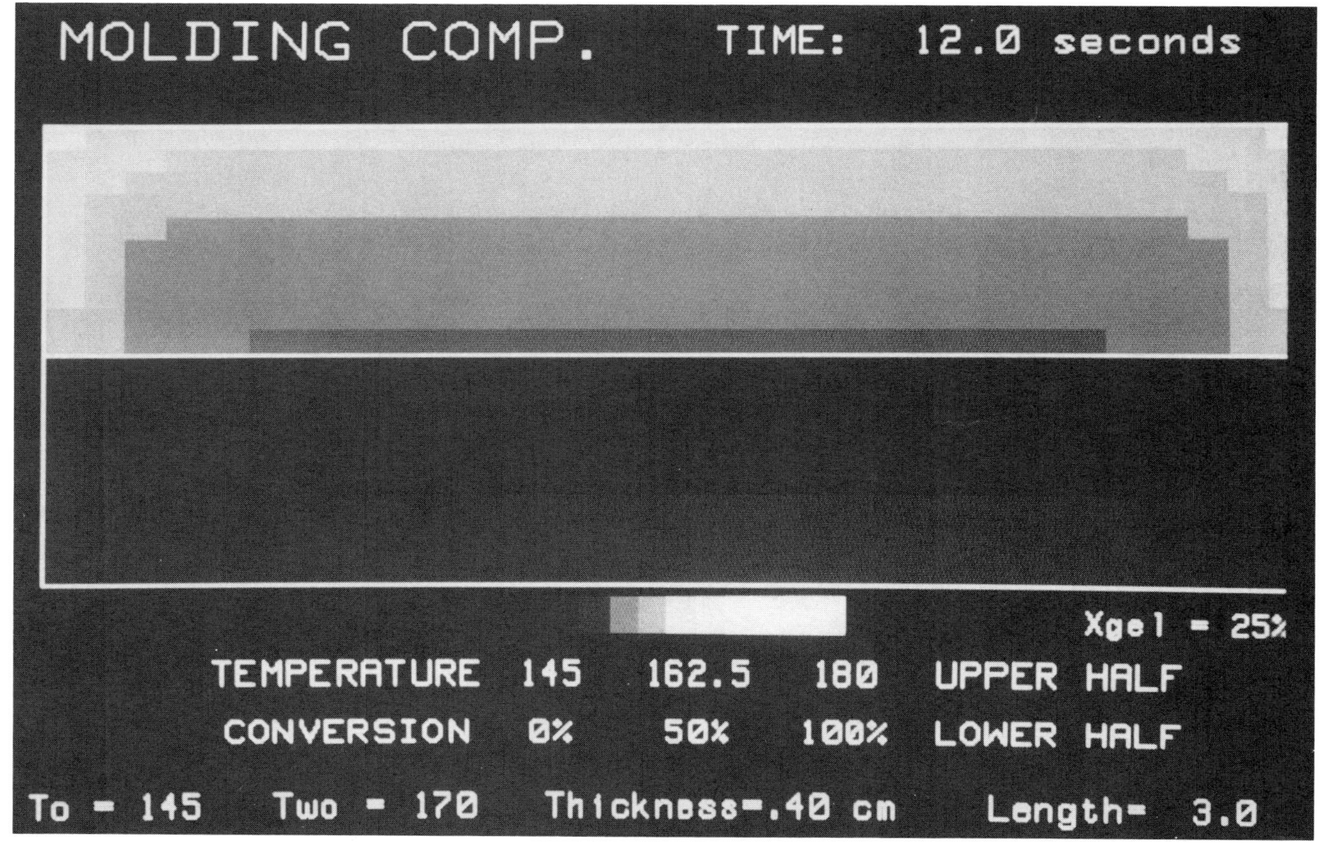


**FIGURE 6-38** *(Continued)*

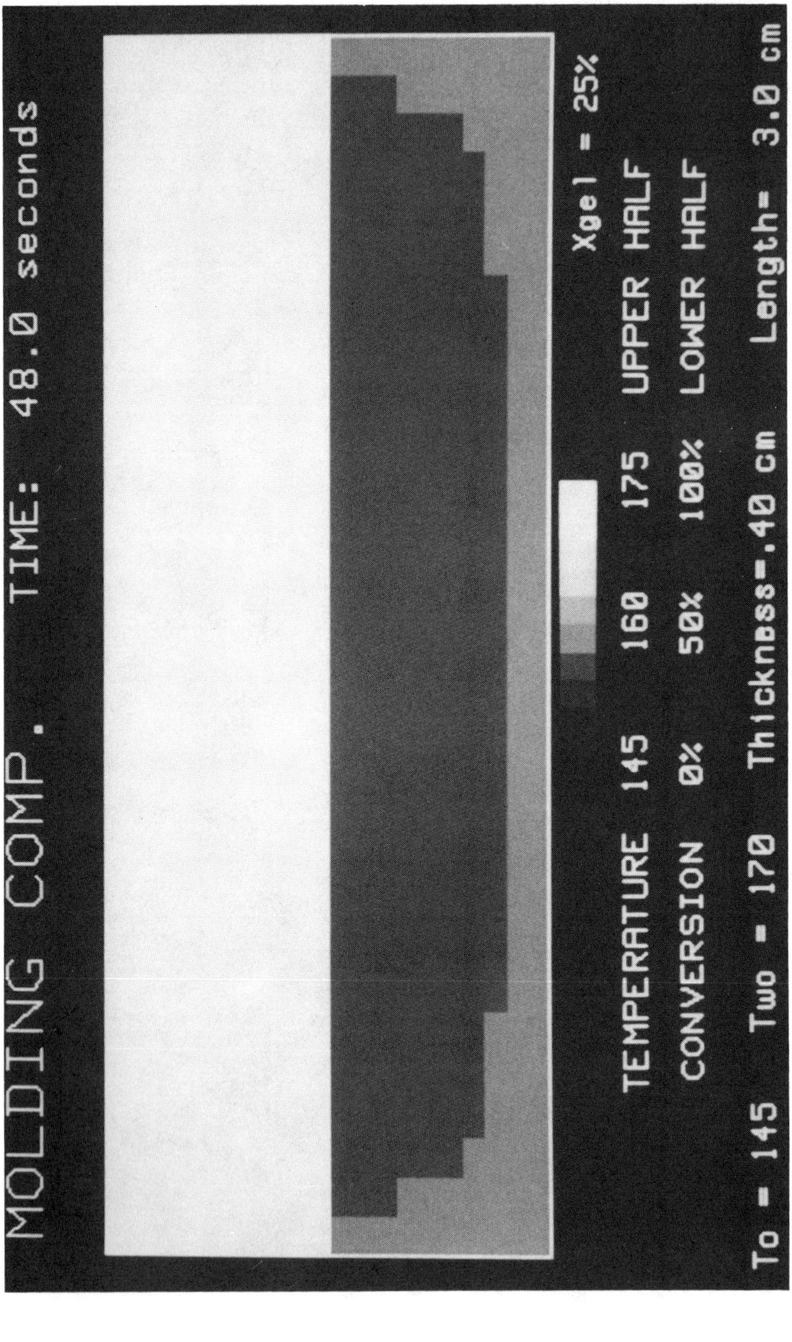

**FIGURE 6-38** *(Continued)*

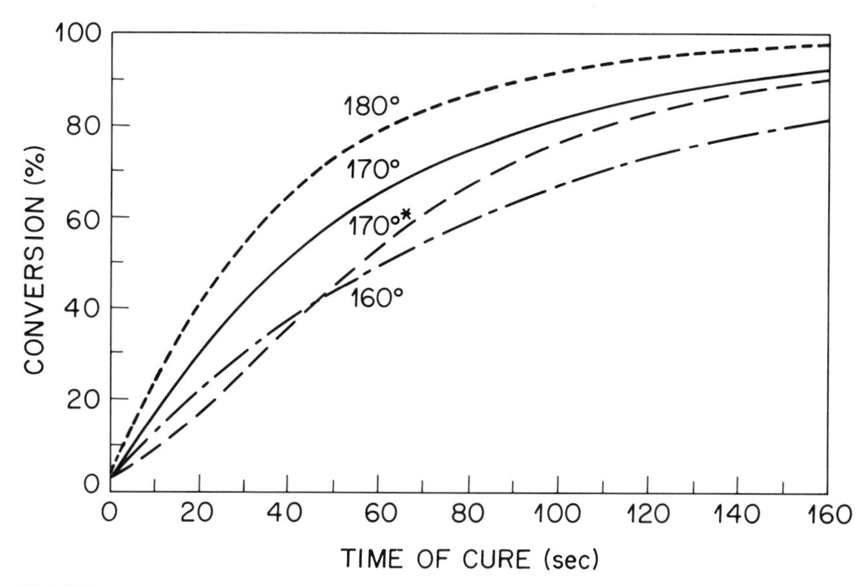


**FIGURE 6-39** The degree of conversion in the mold cavity under isothermal conditions plotted against time for three different temperatures of 160, 170, and 180°C. The curve denoted 170°* is not isothermal, but instead follows a time-temperature profile that emulates the temperature history in the cavity for a mold temperature of 170°C, including an initial temperature of 140°C and the mild exotherm, as predicted by the two-dimensional solution shown in Figure 6-38.

pressure source. This point is commonly known as *gate-freeze-off* despite the fact that this is a misnomer, since the material is hardened due to chemical crosslinking rather than temperature-dependent solidification as is the case in injection molding. Once the material in the gates has gelled, the packing pressure can no longer be applied through the gate. The most serious consequence of this phenomena occurs when the gates gel while other cavities are filling. The pressure during the filling stage is usually well below the packing pressure that is applied when the entire mold is filled. Cavities where the gates gel prior to complete mold filling never experience the packing pressure and they show greater porosity and a much higher incidence of voids, particularly small voids that would have otherwise been compressed. The problem is not hypothetical, but instead a genuine concern because the very narrow gate orifice heats the molding compound to the mold temperature immediately after the flow ceases. The molding compound in the cavities is generally at some lower temperature at this time and there is at least a 15 second delay before it reaches the mold temperature. Therefore, the gates have a significant head start on reaching gelation than the rest of the material in the cavities. Cure analysis to predict gelling in the gate can be confidently conducted under isothermal conditions at the mold temperature. The analysis should begin at the time the flow through the gate ceases. The implications of premature gate gelling and the steps that can be taken to eliminate it are discussed in Section 7.2.3.

### 6.5.4 Hot Hardness

The molded packages are ready for ejection when the degree of conversion reaches a level where the material has sufficient resilience and hardness to withstand the forces of ejection without significant or permanent deformation. The mold temperature is typically above glass transition temperature of the molding compound, so the material will be in a hard rubbery state at the time of ejection. Premature ejection prior to attaining the proper degree of chemical conversion causes a number of serious problems such as sticking in the mold, excessive deformation, cracking, warpage, and possible damage to the internal components due to the deformation. The degree of conversion at ejection will vary with molding compound, so no specific parameter can be stated. It is usually between 60% and 90% conversion, although deviations from this range are not uncommon. The parameter used to assess the ejectability of a part from the mold is known as the hot hardness, a term that signifies the hardness of the material at the mold temperature. The value is commonly measured by probing the molded parts while they still reside on the lower mold half with a hardness tester such as a durometer. If the hardness is tested at increasing time intervals of cure, a steady increase in hardness will be noted, followed by a constant plateau value. The parts can be ejected when this constant value plateau is attained as can be seen in Figure 6-40.

## 6.6 POSTMOLD CURING

Postmold curing is an integral part of plastic packaging because the molding compound does not reach 100% chemical conversion in the high production rate molding process. The reaction rate slows significantly as the conversion approaches complete conversion, making it counterproductive to expend capital and human resources to realize complete conversion in the molding tool. The usual procedure is to continue the reaction in the mold until the hot hardness reaches either a plateau value or a value that is suitable for ejection. The conversion at this point is in the vicinity of 60–90%, depending on the chemical kinetics of the specific molding compound. The glass transition temperature will not have reached its ultimate value and the mechanical toughness of the material may also be low. In many cases, the post cure operation is conducted immediately after molding as is shown in the process sequence provided in Figure 6-2. There are many other instances where it is preferable to conduct postcuring after code marking to eliminate an added heat cure cycle for the marking ink.

Postmold curing can be analyzed in the same manner as in-mold cure, with

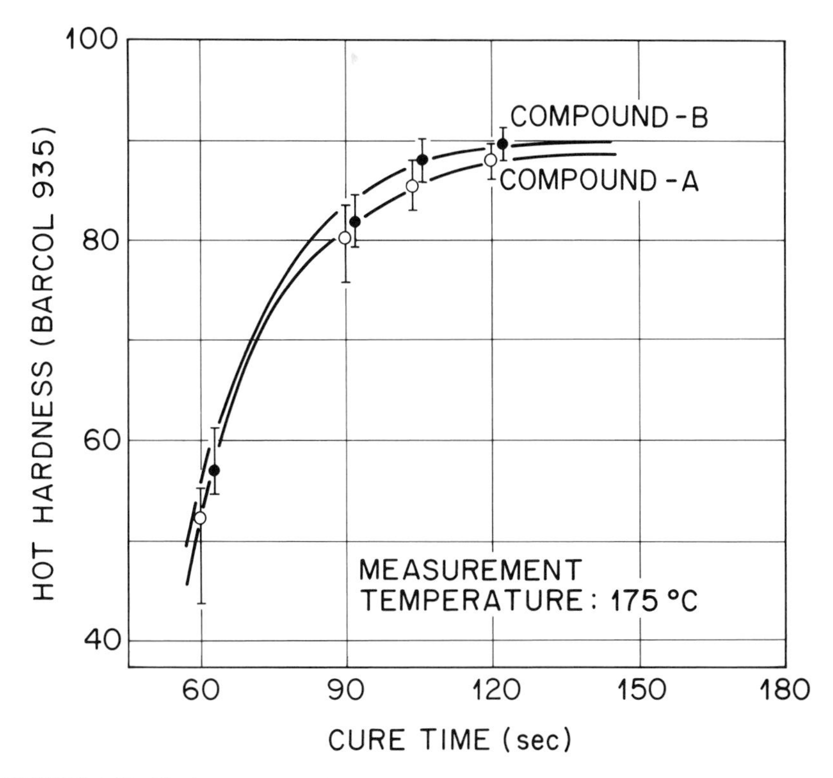


**FIGURE 6-40** The hot hardness of two similar grades of molding compound plotted against increasing time of cure in the mold. (Courtesy of Shin-Etsu Chemical Co., Ltd.)

the significant simplification of isothermal conditions at the postmold oven temperature. The most important consideration of postmold cure analysis is the development of thermomechanical properties with degree of cure. Matsuoka and coworkers [22] have published a treatment of the curing reactions of epoxy resins where they analyzed the cure process, the development of thermomechanical properties, and the effects of relaxation behavior. This work is particularly relevant since parts of it were performed on commercial packaging molding compounds. They show that there is a significant increase in the glass transition temperature as the conversion approaches 100% as shown in Figure 6-41 [22], indicating that proper post cure treatment is essential to realize the full material properties. Their results also indicate that glass transition temperatures well above the cure temperature can be achieved because relaxation phenomena occurring in the glassy state allow the reaction to continue at a significantly lower rate. Therefore, the reaction enters a relaxation-controlled region when the glass transition temperature reaches the temperature of cure as is shown

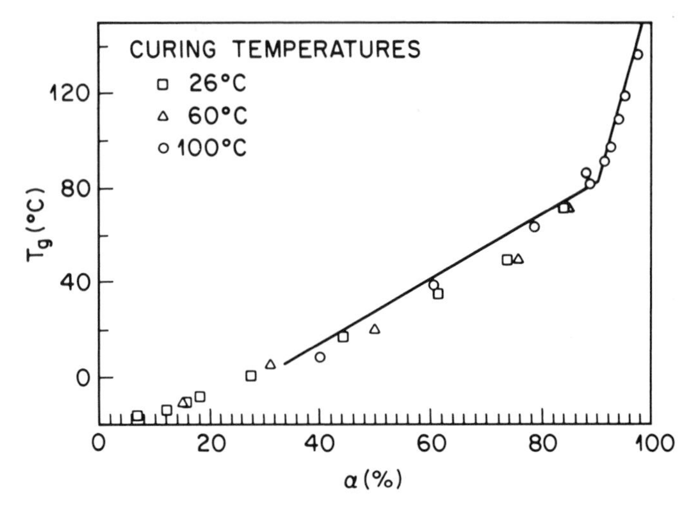


**FIGURE 6-41** The glass transition temperature of an epoxy resin plotted against the extent of reaction, α, for three different cure temperatures. The solid line indicates predictions of glass transition temperature from a thermodynamic property model. Note the steep rise in $T_g$ as the extent of reaction approaches 100%. (Reprinted with permission from *Macromolecules*, **22**, 4093–4098. Copyright 1989 American Chemical Society.)

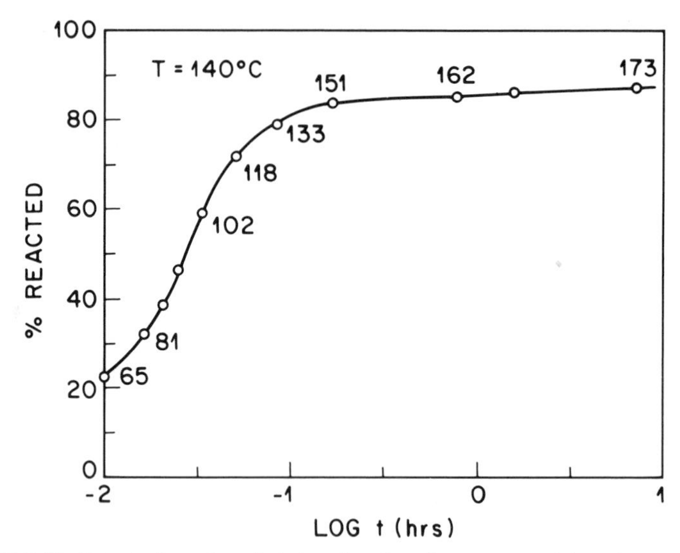


**FIGURE 6-42** Extent of reaction plotted against time for a commercial grade epoxy molding compound cured isothermally of 140°C. The numbers on the line indicate glass transition temperatures measured by DSC after quenching and then reheating each sample from 25°C. The shape of the line indicates that the reaction enters a relaxation-controlled region when the glass transition temperature reaches the temperature of cure. (Reprinted with permission from *Macromolecules*, **22**, 4093–4098. Copyright 1989 American Chemical Society.)

in Figure 6-42 [22]. These results have important implications for designing postmold cure treatments to achieve optimum thermomechanical properties.

## REFERENCES


1. L. T. Manzione, G. W. Poelzing, and R. C. Progelhof, "Experimental and Mathematical Treatment of Transfer Pot Temperature for Thermosets," *Polym. Eng. Sci.*, **28**(16), 1056 (1988).
2. L. T. Manzione, J. S. Osinski, G. W. Poelzing, D. L. Crouthamel, and W. G. Thierfelder, "A Semi-Empirical Algorithm for Multicavity Mold Balancing," *Polym. Eng. Sci.*, **29**(11), 749 (1989).
3. A. I. Isayev, K. D. Vachagin, and A. M. Naberezhnov, *J. Eng. Phys.*, **27**, 998 (1974).
4. T. J. Liu and C. N. Hong, "The Pressure Drop/Flow Rate Equation for Non-Newtonian Flow in Channels of Irregular Cross Section," *Polym. Eng. Sci.*, **28**(23), 1559 (1988).
5. S. Middleman, *Fundamentals of Polymer Processing*, McGraw Hill Book Company, New York (1977), p. 56.
6. L. T. Manzione, "Plastic Packaging of Microelectronic Devices," in *Chemical Engineering Education in a Changing Environment*, Engineering Foundation, New York (1989).
7. J. S. Osinski, L. T. Manzione, and C. Chan, "Thermal Runaway in Fast Polymerization Reactions," *Polym. Process Eng.*, **3**(1&2), 97 (1985).
8. R. B. Bird, W. E. Stewart, and E. N. Lightfoot, *Transport Phenomena*, John Wiley & Sons, Inc., New York (1960), p. 278.
9. S. Middleman, *Fundamentals of Polymer Processing*, McGraw Hill Book Company, New York (1977), p. 443.
10. Ibid., p. 431.
11. L. Graetz, *Ann Phys.*, **18,** 79 (1883); **25,** 337 (1885).
12. B. C. Lyche and R. B. Bird, "The Graetz-Nusselt Problem for a Power Law Non-Newtonian Fluid," *Chem. Eng. Sci.*, **6,** 35 (1956).
13. R. H. Perry and C. H. Chilton, *Chemical Engineers Handbook*, McGraw Hill Book Company, New York (1973).
14. E. N. Sieder and G. E. Tate, *Ind. Eng. Chem.*, **28**, 1429–1435 (1936).
15. A. Hale, Ph.D. Thesis, University of Minnesota, Chemical Engineering Department (1988).
16. N. P. Vespoli and L. M. Alberino, "Computer Modeling of the Heat Transfer Processes and Reaction Kinetics of Urethane Modified Isocyanurate RIM Systems," *Polym. Proc. Eng.*, **3,** 61–64 (1985).
17. A. Hale, M. Garcia, C. W. Macosko and L. T. Manzione, "Spiral Flow Modelling of a Filled Epoxy-Novolac Molding Compound," Soc. Plastics Engineers, ANTEC Papers, p. 796 (1989).
18. L. T. Manzione, Editor, *Applications of Computer Aided Engineering in Injection Molding*, Hanser Publishers, Munich, West Germany; distributed in the United States by Oxford University Press, New York (1987).
19. E. Broyer and C. W. Macosko, "Heat Transfer and Curing in Polymer Reaction Molding," *A.I.Ch.E. J.*, **22,** 268–276 (1976).
20. L. T. Manzione and J. S. Osinski, "Moldability Studies in Reactive Polymer Processing," *Polym. Eng. Sci.*, **23,** 576–585 (1983).
21. C. W. Macosko, *Fundamentals of Reaction Injection Molding*, Hanser Publishers, Munich, Vienna, New York (1989).
22. S. Matsuoka, X. Quan. H. E. Bair, and D. J. Boyle, "A Model for the Curing Reaction of Epoxy Resins," *Macromolecules*, **22,** 4093–4098 (1989).

# 7 Process Optimization

This chapter concerns the optimization of the molding process to improve plastic package molding productivity, yield and ultimate product reliability. The topics discussed include the identification and understanding of molding problems. Changes in molding compound, mold design and process parameters are the typical routes to optimization, although improvements in package design and assembly operations are not neglected. The concepts described in this chapter depend heavily on the concepts of the material, machinery and process that were developed in the earlier chapters, therefore the optimum benefit will be derived if this material has already been covered.

## 7.1 REDUCING THE FLOW-INDUCED STRESS

The flow induced forces are the most important contributor to yield loss during the molding operation. As such, they warrant a significant amount of attention in any packaging optimization study. The physical underpinnings of these forces were described in Section 4.2.1, where their magnitude was found to depend on the molding compound velocity and viscosity. For the usual case of a shear thinning molding compound, the viscosity is the more important component of the product. Therefore, the reduction of flow-induced forces depends on reducing the product of the molding compound viscosity and velocity at the time and at the conditions when it flows over the leadframe.

### 7.1.1 Predicting the Extent of Wire Sweep and Paddle Shift

Although it is generally preferable to minimize the flow induced stresses, it is also important to be able to predict the extent of wire sweep and paddle shift to guide the selection of molding compounds and process conditions. Often there are tradeoffs involved, and minimization of flow-induced stresses is not the only criteria in the optimization scheme. In many cases, the molding compound that provides slightly higher flow-induced stress, but improved shrinkage stress or other reliability properties is preferable, especially when the deformation caused by the higher-flow-stress molding compound is still negligible. The question is how to assess what degree of deformation can be expected and what level is acceptable. The second question is easier to answer and was discussed briefly in Section 4.2.1.2; the deformation to the wire bonds and leadframe should be insignificant. Essentially any discernible wire sweep and/or paddle shift found in prototype runs has the potential of becoming a serious molding problem due to the vagaries and inconsistencies of production molding. Conversely, no deformation at all is largely impossible. The flow of the molding compound generates forces and both the wire bonds and the leadframes have a finite mechanical modulus. There will be deformation in response to the force loading, and there may also be elastic recovery when the degree of deformation remains below the elastic limit for the materials. The situation for plastic package molding is more complicated than the simple mechanics alluded to here, however, since the medium applying the force is changing with time. Deformations below the elastic limit can be preserved in this case if the molding compound viscosity increases significantly during the force loading, or if the molding compound gels and hardens precluding substantial elastic recovery. Therefore, it is important to be able to predict deformations that result from specific molding compounds and specific process conditions.

There have been several attempts at quantifying the degree of wire sweep that occurs during the molding operation. An early report by the author [1] computed the viscous drag force for flow around a circular cylinder to estimate the force loading, and then computed the deformation of the wire based upon a distributed force loading on a cantilever using the mechanical properties of the gold metal. The length of the cantilever cylinder is that of the wire bond half length. The velocity gradient over the cavity half thickness provides the distributed force loading. Depicting the wire bond as a cantilever is probably a worst case scenario. Figure 7-1 shows a schematic illustration of this analysis. This work was the first to show analytically that the viscous drag forces generated during package molding could cause significant deformation, although

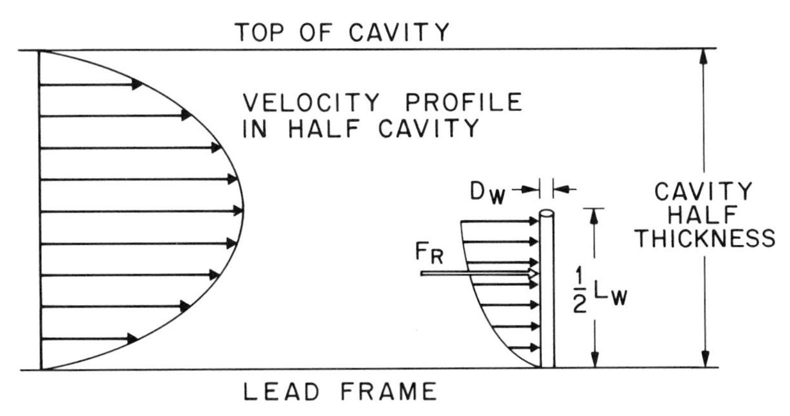


**FIGURE 7-1** Analytical treatment for predicting the degree of wire sweep that can be expected under specific material and process conditions. $F_r$ is the resultant force on the wire. This geometry represents maximum deformation.

the finding was that in the particular case investigated the deformation of the wire was small.

Another approach to analyze the wire bond deformation encountered during molding was offered by Nguyen [2]. In this method, the degree of deformation is quantified by the introduction of plastic hinges at the support ends of the bond and in the center of the span as is shown in Figure 7-2. The introduction of the hinges provides a means to represent and determine wire bond displacement. The hinge movement is elastic until the load exceeds the plastic load limit after which plastic deformation ensues.

The best approach to quantify wire sweep is to determine it experimentally. Taking precise measurements from enlarged X-ray photographs is tedious and will ultimately decrease the sample size that is investigated. A simpler but equally effective approach is to qualitatively rank order degrees of wire sweep, using an arbitrary scale where one extreme is no detectable wire deformation and the other is actual broken wires. A total experimental approach can become time consuming and expensive when there are a large number of molding compounds and process parameters to evaluate. A facilitated approach that combines both theory and experiment is to experimentally measure the degree of

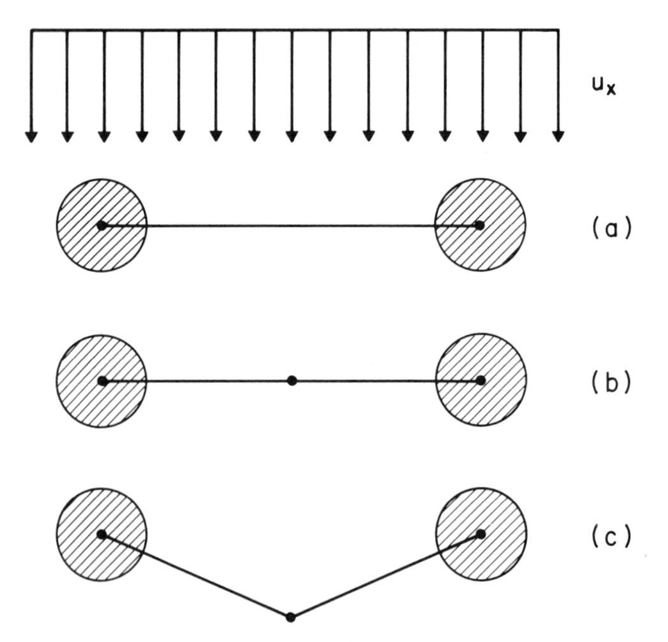


**FIGURE 7-2** Approach that introduces movable hinges to represent and quantify wire bond deformation during package molding: (a) introduction of hinges at end supports, (b) introduction of hinge in center of wire span, and (c) hinge deformations that simulate wire sweep. (From Reference [2]. Reprinted with Permission of the Society of Plastics Engineers.)

wire sweep for a reference case. An actual measurement on a case that shows some deformation is preferred. Assume the deformation is Hookean, that is it depends linearly on the force loading which scales to the product of viscosity and velocity as was described in Equation (4-8). As stated earlier, this particular expression is for a Newtonian fluid, so some error is introduced in its application. Use Equation (4-8) to derive a scaling coefficient that relates the degree of deformation to the resultant force loading, $F^*$. Compute the viscosity and velocity for this reference case and derive the scaling parameter as in shown in Equation (7-1):

$$D_{\text{wire}} = K_w F^* = K_w \frac{\eta V}{\ln\left(\frac{\eta}{V}\right)} \tag{7-1}$$

The effect of changes in viscosity and velocity on wire sweep can now be assessed once the $K_w$ coefficient has been established and verified. Although not acceptable for rigorous work, this is a reasonable way to address wire sweep problems.

Predicting the degree of paddle shift is probably more difficult than treating wire sweep. Certainly, there have been fewer reported analyses. Paddle shift and related lead finger displacement result from unbalanced pressure forces that occur during the flow over and through the leadframe. The magnitude of the lead finger deformation can be estimated by considering distributed forces acting along the wetted surfaces above and below the lead [1]. These forces counteract each other to some extent and would cancel if the position of the upper and lower flow fronts were matched and the cavity dimensions in each half were the same. The flow front positions are often mismatched, and this is often intentional so that the molding compound flows up through the wire bonds rather than across them which would be more damaging. The imbalance in the forces means that there is a resultant force on the lead finger which will cause a deformation as is shown in Figure 7-3. The extent of this deformation depends on the pressure forces behind the moving flow fronts which can be determined through Equation (6-4). The deformation of the lead finger can then be estimated through a mechanics analysis of the resultant forces [1], or a more generalized treatment which represents the distributed forces by the angle the force magnitudes make with the plane of the lead finger. The result of this analysis is presented in Equation (7-2). This relation, derived by E. Suhir of AT&T Bell Laboratories, can be used to estimate the degree of lead deformation for different molding compounds and process conditions. Paddle shift will also scale to this lead deformation since it arises from the same phenomena. Examining Equation (7-2), it is important to note the fifth-power dependence of deformation on the lead length, $L$, and the third power dependence on the lead thickness, $h$. These dependencies are what make it more difficult to mold larger packages

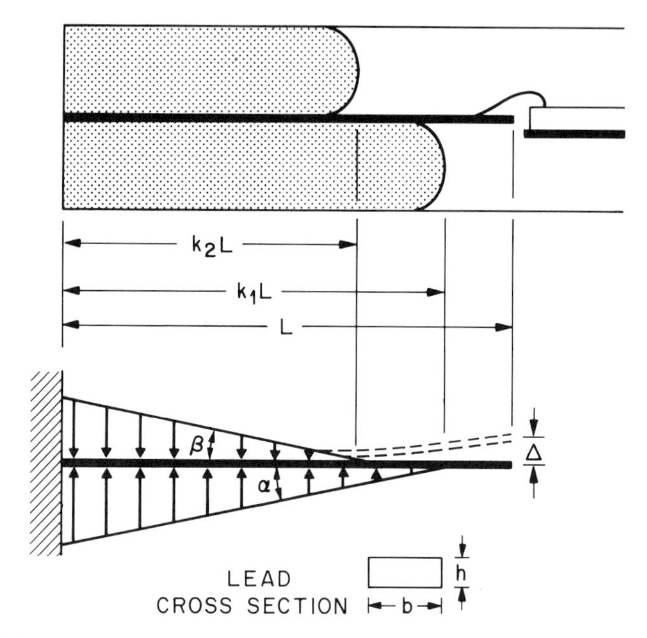


**FIGURE 7-3** Distributed forces on the lead finger of an integrated circuit leadframe during package molding for a specified mismatch in the location of the flow fronts, and the representation of the forces by the angle the distributed forces make with the lead. This representation is the basis of the analysis presented in Equation (7-2.)

with thinner leadframes:

$$\Delta = \frac{L^5}{120EI}\left\{5(m_1k_1^4 - m_2k_2^4) - (m_1k_1^5 - m_2k_2^5)\right\} \qquad (7\text{-}2)$$

where

$$I = \frac{bh^3}{12}, \qquad m_1 = \tan\alpha, \qquad m_2 = \tan\beta$$

### 7.1.2 Reducing Viscosity

The viscosity of the molding compounds used for plastic packaging is a complex function of shear rate, temperature and time. From our present understanding, the resin portion of the molding compound does not appear to have a long relaxation time, therefore its viscosity does not depend on shear history: all previous shearing and orientation relax in a time scale that is fast with respect to the time scales of the process. The filler particles can and will orient in a shear field and the dissipation of this orientation can impart a dependence on shear history, although this effect is probably small. Low flow-induced stress is achieved by selecting molding compounds that have a low viscosity at the temperature, time and shear rates of contact, and by adjusting the flow length and process parameters to assure that the lowest possible viscosities are realized.

**7.1.2.1 Molding Compound Selection** There has been a significant reduction in the viscosity of epoxy molding compounds used for microelectronics encapsulation in an effort to reduce the flow-induced forces and minimize flow induced yield loss as is discussed in the evolution of the molding compound discussed in Section 3.6.1.2. Nonetheless, some molding compounds have lower viscosity than others at various conditions of time, temperature and shear rate and in some cases it is important to identify the ones that have lower viscosity for the flow history and process parameters of the specific molding tool. These conditions can be stated generally, but the specific parameters will depend on the specific mold and package design. Figure 7-4 is a summary of the time/

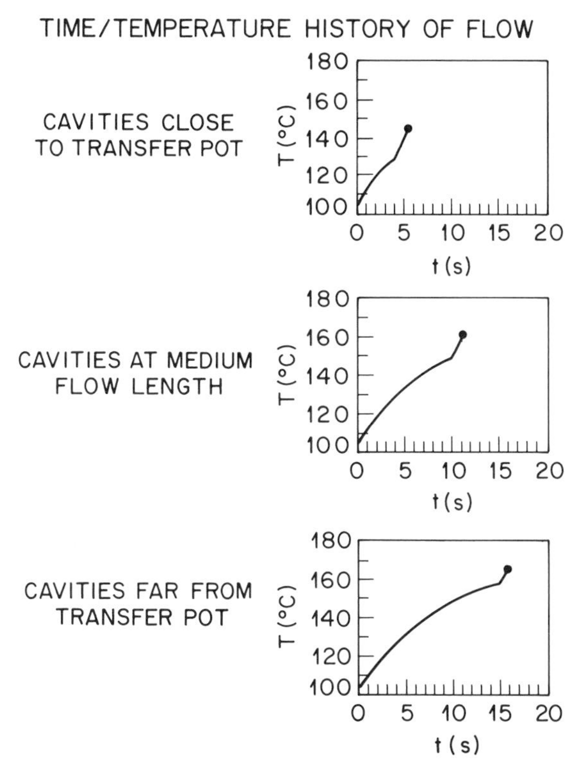


**FIGURE 7-4** Summary of the time/temperature history that represents the conditions of the molding compound as it contacts the devices. The smooth curve represents the temperature increase in the runner, whereas the break in the curve and straight line represent the heating through the gate.

temperature history that can be expected in most molding tools. Multiplunger tools are not included in this figure. In general, multiplunger molds are characterized by exceptionally short flow lengths and relatively low molding compound temperatures. The shear rates will be approximately the same or somewhat lower if longer cavity fill times are used in a multiplunger mold.

Figure 4-5 showed that the shear rates associated with flow-induced stress are relatively low compared to the variety of shear rates that are experienced throughout the process, whereas the shear rates associated with flow in the runner are in the vicinity of several hundred reciprocal seconds. Viscosity measurements undertaken with the aim of selecting molding compounds that will provide less wire sweep and paddle shift should therefore be conducted at low shear rates between 10–50 $sec^{-1}$. Many molding compounds can show high viscosity relative to other molding compounds at high shear rates, yet have lower viscosity relative to these same materials at the low shear rates that cause yield loss due to flow-induced stress. This behavior is illustrated in Figure 7-5, where the viscosities of several molding compounds are displayed. Molding compounds such as material A in Figure 7-5 can show good processability with regard to mold filling behavior and transfer pressure, but are likely to show

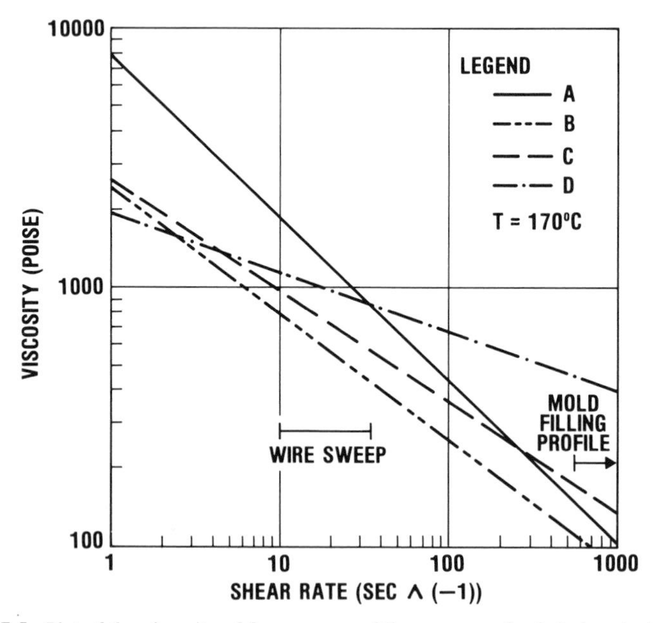


**FIGURE 7-5** Plot of the viscosity of four epoxy molding compounds plotted against shear rate. The shear rate regions that control wire sweep/paddle shift and those that influence the mold filling behavior are also shown.

more flow-induced deformation on devices with fragile leadframes and long wire bond spans. The behavior is closely tied to the shear thinning character of the material, represented here with the power law index, which controls the difference in viscosity between high and low shear rates. It is for this reason that flow length tests such as spiral flow are of limited value in assessing propensity for wire sweep. The spiral flow test, usually conducted at shear rates of hundreds of reciprocal seconds, is not a good match for the shear rate region that is responsible for flow-induced deformation and yield loss.

**7.1.2.2 Optimizing the Process Temperatures** The time-temperature history of the molding compound as it flows through the molding tool largely determines the viscosity as it contacts the device. The temperature history can be influenced by several design and material parameters that are readily altered. The effect of flow length was depicted in Figures 6-20 and 6-26. These results show that there is an appreciable flow time required before the molding compound attains a suitable temperature that will have sufficiently reduced its viscosity to provide low flow-induced stress. The required temperatures in the runner adjacent to the cavity are above 140°C. For the conventional-size runner cross sections used in Figures 6-20 and 6-26, the flow time to reach these higher temperatures corresponds to approximately 3–4 seconds. At conventional transfer rates, this flow time translates to a flow distance of 15 cm or longer. Many high-yield molding tools can and do have flow lengths below this value, but in cases where flow-induced stresses are an important consideration, a longer flow length should be considered. Higher temperatures can also be achieved by reducing the runner cross section, as was demonstrated in Figure 6-23 which showed heating in the runner for different hydraulic radii of the runner cross section. The price paid for the higher temperatures obtained with this approach is higher pressure drops to push the molding compound through the narrower flow channel as can be determined from Equation (6-4). An increase in mold temperature also increases temperature and decreases viscosity, but the latitude in this parameter is limited to probably not more than $\pm 5$°C, since it affects the cure kinetics and the flow length.

**7.1.2.3 Molding Within the Process Window** The concept of the processing window was introduced in Section 6.4.2.3 and Figure 6-26. It will be exploited in this section to provide a method for reduction of molding compound viscosity. A viscosity history of a single fluid element as it travels through the molding tool can be determined by applying the rheology of the molding compound to the time, temperature and shear history of the flow. The temperature profiles in the runner system provide a good starting point for temperature history. The position versus time history must include velocity changes that occur as the mold fills and as the material is diverted into the cavities. The viscosity history has a characteristic shape common to most thermoset processes conducted in heated molds. The initial viscosity is high since the initial temperature is low. The tooling and process parameters should be designed to prevent the molding compound from contacting the chips during this high viscosity phase. The viscosity drops rapidly as the molding compound heats during flow in the runner reaching a minimum value as the temperature converges on the mold temperature, the exotherm of the reaction being minimal. The viscosity resides in the vicinity of the viscosity minimum for some time because of the molecular dynamics of the polymerization. The viscosity then increases rapidly as the conversion approaches the conversion to gel because small increments of conversion contribute large increases in molecular weight since very large, highly branched molecules are joining. This steep increase defines the other end of the processing window.

The mold design and the process parameters should be configured so that all of the cavities are filled when the molding compound is within the process window. Many mold designs preclude meeting this criteria because of shortcomings in the design that cannot be overcome by modification of process parameters. Figure 7-6 shows the processing window and cavity contact intervals for a common mold design where the leadframes are located in close proximity

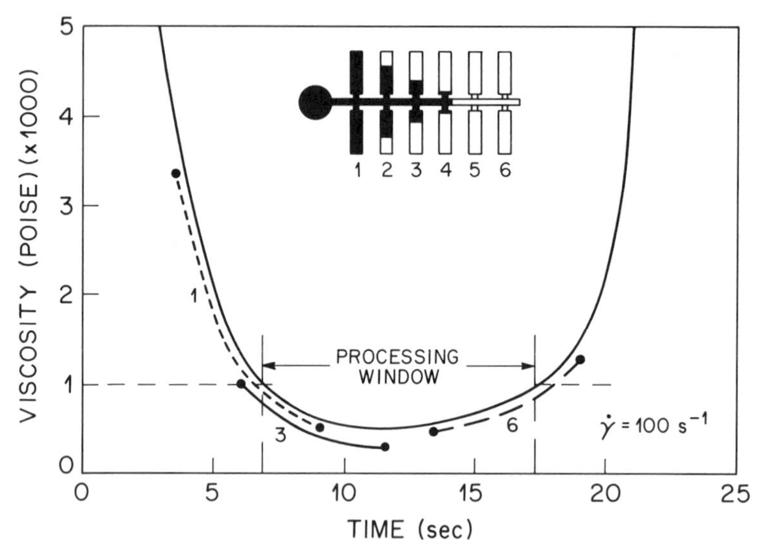


**FIGURE 7-6** Plot of the processing window and cavity contact intervals for a typical molding compound in a molding tool where the leadframes are in close proximity to the transfer pot as shown in the inset. The cavity contact intervals for the first, third and sixth cavity pairs, derived from a computer simulation of mold filling, are shown. The molding compound reaches the cavities close to the pot before it enters the processing window, whereas the contact interval for the sixth cavity pair approaches the far end of the window.

to the transfer pot providing a short runner residence and heat up time. The molding compound flows over the leadframes for approximately 5 seconds in an unbalanced mold and this is represented by an extended bar for several cavity sites on the six site leadframe. The molding compound reaches the first cavities on the leadframe strip before it has reached the low viscosity window. Mold mapping of failure locations often indicates lower yields in these cavities. Figure 7-7 shows the processing window and the cavity contact intervals for an improved mold design where there is a longer flow time prior to the molding compound reaching the cavity locations in an unbalanced mold. Although the cavity contact intervals are largely within the processing window for this case, the long individual cavity fill times push some to the edge of the window. Flow-induced damage may not be apparent in most instances, but operating close to the edges of the window makes the process more susceptible to the fluctuations and inconsistencies that are common in production operations. For this reason, it is preferable to have all the cavity contact intervals well within the processing window to assure the most consistent, trouble free operation.

There are several approaches to preventing the contact time interval from being outside of the processing window. One method is to simply enlarge the window. Some molding compounds will inherently provide a wider window

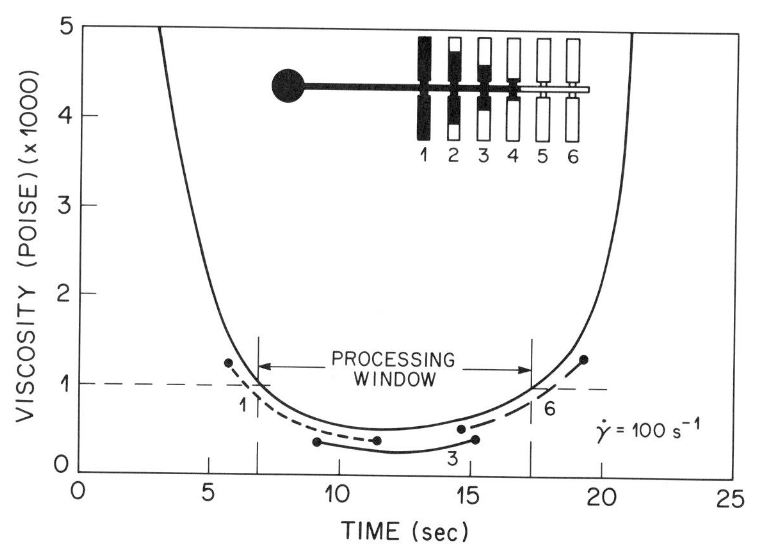


**FIGURE 7-7** Plot of the processing window for a typical molding compound in a molding tool where there is an extended flow length and flow time between the transfer pot and the cavities as shown in the inset. More of the cavities, and particularly the early cavities, fill within the process window in this case, but the wide spread in the contact intervals pushes some to the edges of the window.

because they reach acceptable viscosity at lower temperatures so that their process window begins after only 2–3 seconds of flow instead of the 6 seconds for the molding compound shown in Figures 7-6 and 7-7. One should select molding compounds with low viscosity at the relevant shear rates and temperatures as was described in Section 7.1.2.1. Another approach is to profile the velocity of the molding compound in the runner to prolong the flow time prior to reaching the chips, allowing greater heating and lower viscosity. There are tradeoffs associated with velocity profiling, however, that will be described in more detail in Section 7.1.3.1 which discusses programmed transfer rate. A higher mold temperature can also increase the temperature in the nearer cavities. This moves the onset of the window to shorter times but will also shift the closing of the window at the other end to shorter times; the overall breadth of the window remains the same or can decrease. Probably one of the most effective ways to contain cavity filling within the process window is to cluster the cavity contact intervals closer together through the use of balanced mold filling where all of the cavities fill uniformly. The mold fill profile, the processing window, and the cavity contact intervals for balanced filling are provided in Figure 7-8. The contact intervals are usually longer in balanced fill because the velocities in the cavities are reduced since the volumetric flow rate into the mold is distributed over all the cavities rather than a fraction of cavities. The intervals are all coincident in simultaneous filling which makes it far easier to contain all of them within the processing window. Means of achieving balanced filling and a more

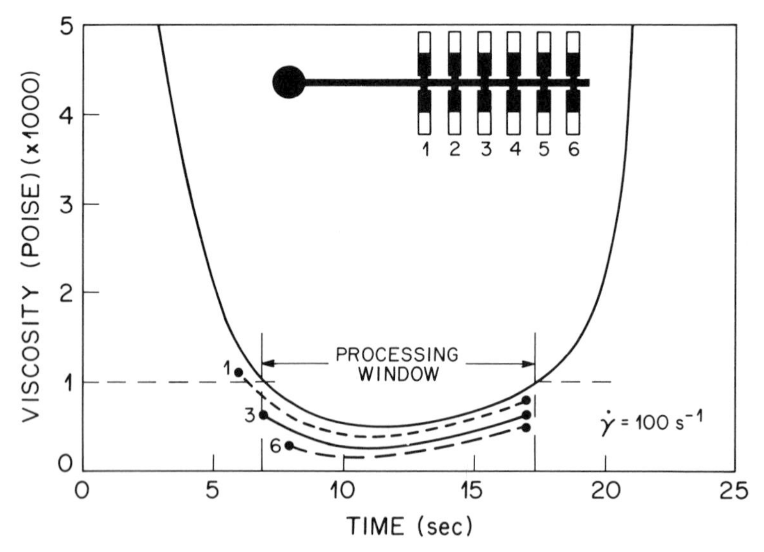


**FIGURE 7-8** Plot of the processing window and cavity contact intervals for a molding compound modified to provide a balanced mold filling profile. The mold fill profile is shown in the inset.

thorough evaluation of its benefits and drawbacks are described in Section 7.1.3.2.

### 7.1.3 Reducing Velocity

Velocity reductions have not received the same amount of attention as have reductions in molding compound viscosity. There are good reasons for this omission. According to Equations (4-4) and (4-8), the flow-induced stresses are a weaker function of velocity than viscosity. In addition, low-viscosity molding compounds are easier to understand and implement. Velocity reductions require active manipulation of mold design and process parameters according to a preconceived plan. Any molding tool changes are permanent and cannot be easily reversed to accommodate a different molding compound or solve a different problem. Nonetheless, velocity reductions are an important means of reducing flow induced stresses and improving molding yield, particularly when considering that viscosity reductions in epoxy molding compounds are bound to reach their limit before the need for additional flow-induced stress reduction dissipates.

**7.1.3.1 Programmed Transfer Rate** A programmed transfer rate can be used to lower the velocity of the molding compound in the cavities while still keeping the cavity contact intervals within the processing window. In the absence of a special device or feature for programmed transfer, it will be conducted at a constant pressure, although some presses may provide a constant transfer rate. At constant pressure, the ram movement and the molding compound velocity will slow as more of the mold is filled because the molding compound encounters more hydraulic resistance. The initial velocity in the approach runner and the velocities in the cavities closest to the transfer pot are all relatively high, whereas the velocities over the middle cavities are lower. The cavity velocities increase again when only the far cavities remain unfilled because all of the applied pressure and volume flow rate are now focused on the few remaining cavities. High flow-induced stresses are created in the cavities near and far from the transfer pot, whereas the middle cavities experience the lowest velocities and flow-induced stress. The far cavities can also have higher flow-induced stress caused by advanced cure. Mold mapping of open and short defects, a common flow-induced failure, often show lower yield in these near and far cavities. In mold designs with a short approach runner length, the high velocities in the near cavities are compounded by low temperatures and high viscosities, further lowering yields.

A programmed transfer rate can be used to overcome the intrinsic velocity profile. Some recent transfer press designs come with transfer rate profile capability. Older presses and new presses without this option can be retrofitted with a process controller to profile the transfer rate. The transfer time may be divided into anywhere from 5 to 10 increments for which an individual transfer rate can be set. Some units may offer the additional option of profiling either the transfer rate or the transfer pressure. The objective, however, remains the same which is to control and optimize the velocities in the molding tool. The specific transfer rate profile will depend on the specific mold design and the problem to be corrected. For flow-induced stress reduction in a conventional mold design with a medium length approach runner, the preferred transfer rate profile is slow at the start, then higher over the middle cavities, and slow again at the end of transfer when the last cavities are filling. The low initial velocities allow more time for the molding compound to heat in the approach runner, thereby lowering its viscosity. Low transfer rates at the end of the transfer keep the velocities in the last cavities to fill at an acceptable level when the entire volume displacement of the plunger is being driven into a few cavities. The higher transfer rates in the middle cavities are needed to keep the cavity contact intervals within the process window since a low velocity throughout the entire fill would prolong the mold fill time and push the contact intervals for the later cavities outside of the process window. This transfer rate profile is commonly known as a "reverse C pattern" because the pattern that is formed by the settings suggests a backward letter "C," as shown in Figure 7-9.

Different mold designs and molding compounds will require different transfer profiles. A molding tool with a long approach runner and a long flow length will require higher initial velocities, less of a velocity increase in the middle cavities, and a moderate velocity increase in the last cavities. The shape of the profile is approximately the same but the amplitudes are lower, providing a shape that is closer to a "close parenthesis" rather than a "reverse C". The overall effectiveness of transfer rate profiling also depends on the type of mold design. The instantaneous transfer rate applies to all parts of the mold, so it is preferable that the filling profile is balanced so that the flow fronts are in the same place in each of the chases. Transfer rate profiling is least effective when there are large differences between the approach runner lengths to the different chases of the tool, as is shown in Figure 7-10. The low-velocity period set for the short runner will expire well before the flow front in the long approach runner reaches the cavities. Conversely, an extra slow velocity in the short runner, needed to adequately heat the molding compound over the short flow distance, could overheat the molding compound in the long approach runner. Although some benefits can still be derived from transfer rate profiling, the optimum benefits will be obtained with balanced mold filling.

**7.1.3.2 Balanced Mold Filling** A number of the significant advantages of balanced mold filling have already been described. The velocity of the molding compound in each of the cavities is reduced because the total volume flow rate

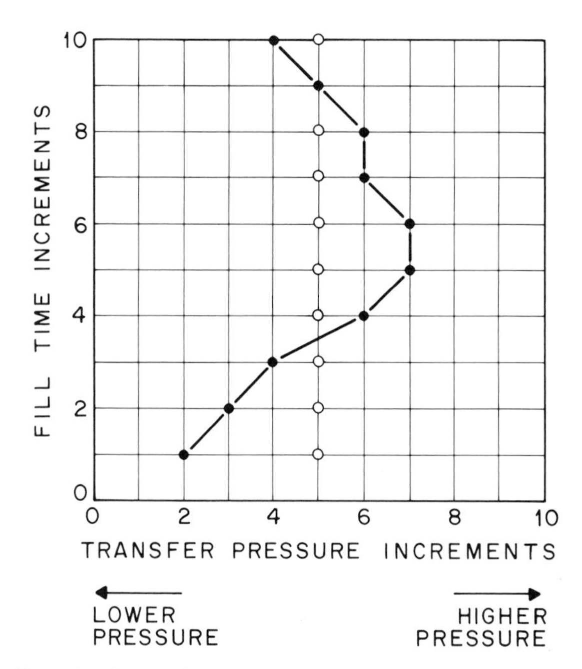


**FIGURE 7-9** Example of a transfer rate profile used often to realize reductions in flow induced stresses. The figure shows the placement of contact pins (solid circles) on a breadboard controller, a pattern that suggests a reverse letter "C". The open circles show the conventional constant pressure profile that provides the same overall mold fill time. The solid circles are connected for illustration purposes only.

into the mold is divided among a larger number of cavities. It is easier to keep the cavity contact intervals within the process window with a balanced mold because they are clustered together and the cluster can be moved to longer or shorter times through process parameter changes that do not shrink the process window. A balanced mold also increases the effectiveness of programmed transfer rate because the flow fronts reach the different sections simultaneously. The topic of this section is how to achieve balanced mold filling.

A conventional molding tool with uniform runners of a constant cross section and gates of uniform size will inherently fill nonuniformly because of two unrelated effects. The first effect is that the molding compound will reach and begin flowing into the cavities nearer the pot first, giving them a lead on cavities further downstream. In general, this advantage is slight and can be easily overcome. A more serious incongruity arises from the pressure gradient along the flow length. The pressure is highest at the transfer pot and decreases to atmospheric pressure (zero gauge pressure) at the advancing flow fronts. The pressure gradient is nonlinear because of the changes in volumetric flow rate as

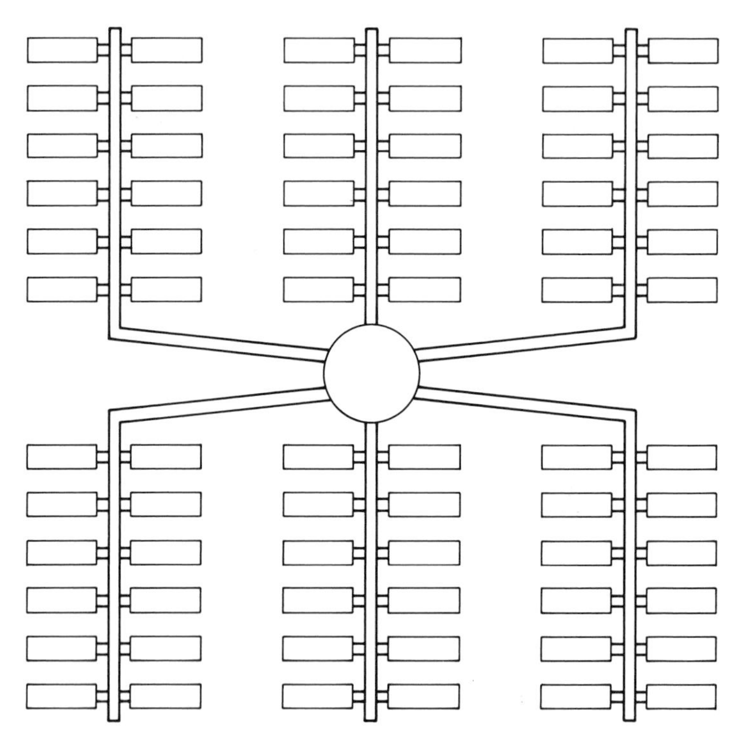

**FIGURE 7-10** Unbalanced molding tool that is not amenable to programmed transfer rate profiling because the molding compound will reach the different section of the mold at different times.

molding compound is diverted into the cavities, and the temperature, viscosity and velocity gradients that are inherent to the process. The gates that are closer to the transfer pot experience the higher pressure, and these cavities fill at a faster rate. Unlike the advantage of simple proximity to the pot, the pressure gradient is preserved throughout the entire mold fill time hence the later cavities do not catch up until the earlier cavities are full.

There are several approaches to balanced mold design and there are certain constraints that may be relevant. In this discussion we will maintain the constraint that the leadframe configuration must be preserved. One of the simplest means to achieve balancing is to arrange the cavities radially around a central transfer pot, but this compromises so much of the efficiency of the downstream and upstream semi-batch processes as to make the concept entirely unacceptable. Other more compatible options include manipulating the cross section area of the runner to control the volume flow rates and pressure drops, progressive alteration of the sub-runners connecting the runner to the gate opening, or modifying the gate sizes to control the volumetric flow rate through them. The degree of the modification plays an important role in choosing among the options, and this is where additional constraints apply. One of these is the need to maintain a uniform gate breakoff, both in appearance and in breakoff action. Excessively large gates will resist breaking and gouge a large chunk of the molded body leaving an unsightly gate vestige. Conversely, exceedingly small gates may be prone to clogging by filler or gel particles and prove to be troublesome. There are also limits on what alterations can be done on the runner and subrunners. Extra thick runners waste molding compound and require a long time to cure, thereby lowering productivity and material usage efficiency. Exceptionally small runners can be destabilizing because they heat the material to higher temperatures earlier in the process. The best way to choose among the options is to first model the mold filling profile for the specific mold configuration and molding compound in use to determine the degree of modification needed, then choose options that provide balanced filling yet satisfy all of the constraints with the minimum degree of gate and runner modifications.

**7.1.3.3 Velocity Surge During Packing** There is a sharp increase in pressure when the molding machine switches from the transfer pressure to the higher packing pressure. The high porosity of the molding compound can result in a very rapid densification on packing that causes a rush of molding compound into the cavity. The high velocities accompanying this rush can cause wire sweep. In addition, the rapid compression of any voids that may be present in the cavity can also cause local high velocities and wire sweep. A high incidence of wire sweep has been reported in cavities that also contained voids [3]. This type of flow-induced damage can be minimized by profiling the pressure transition between filling and packing. Minimizing the pressure increment can also be beneficial, but inappropriate in some cases where it results in insufficient packing pressure to properly densify the molding compound. Using a programmable process controller to profile the pressure transition is probably the best approach to minimize this problem. The specific profile will depend on satisfying the other transfer and packing pressure dependent criteria such as flash and resin bleed. These criteria are considered in Section 7.2.2, which describes packing pressure optimization.

### 7.1.4 Mold Filling Profile Modeling

A mathematical simulation of transfer mold filling is a valuable tool in optimizing molding tool design and performance. There is a burgeoning industry that has grown around the concept of computer aided engineering for polymer process operations [4], but the vast majority of these offerings concern the injection molding process. Several thermoset molding packages are available or under development [5, 6], and these can be made to accommodate the transfer molding of IC packages. In addition, there have been several published reports of simulating the filling profile in transfer molding with the objective of balanced mold filling [7, 8, 9, 10]. One of these published methods [7], co-developed by the author, will be reviewed below. The purpose of this review is twofold. The first will be to demonstrate the workings and results of a flow simulation program. The second purpose is to demonstrate the complexity of the undertaking, the simplifying assumptions required, and the deviations that these assumptions can cause.

The most rigorous approach to flow simulation in transfer molding would be to write three-dimensional energy, mass, and momentum balances for the flow, then solve this set of coupled partial differential equations over the flow path which would include all of the branches, sub-runners, and gates that make up a production molding tool. This approach is impractical, not only because of the effort involved in writing the program, but also in the computer time required to execute the code. The limited number of publications on this topic are convincing evidence that a sufficient amount of accurate information can be derived from a much simpler model. The model described herein is a one-dimensional network flow simulation that is semi-empirical in that it uses experimental measurements of pressure drop through the gates and runners to set the hydraulic resistances rather than attempting to derive them. Subsequent application of the model to different molding compounds need not repeat the pressure drop measurements but only modify the semi-empirical constants based on rheological testing of the material. Similarly, the model can be applied to different gate geometries by scaling the gate resistance based on the analysis provided in Appendix 4A.

The basis of the model is the assumption that in a branch flow the pressure drops between the flow fronts and the pressure source are all equal. The flow resistances among the different flow branches are likely to be different so the volumetric flow rate in each branch will vary with the hydraulic resistance encountered. Each flow path is divided into a large number of linear flow increments, typically 0.1 cm or less. The computation starts by assigning a suitable pressure in the transfer pot to drive the flow. Iterations begin at the flow front in the runner located a few centimeters from the pot to avoid a zero flow length instability. The gauge pressure at the flow front is zero. A small counter pressure or vacuum can also be accommodated. An initial volumetric flow rate is assumed to begin the calculation. Starting at the flow front and moving backward, pressure drops for flow in the runner are determined at each length increment using the relation shown below in Equation (7-3). The coefficient $K_R$ in this relation includes both the viscous and geometric contributions to the hydraulic resistance in the runner. It can be determined either experimentally from data as shown in Figure 6-13, or it can be predicted using an expression for pressure drop of non-Newtonian fluids in circular or rectangular channels such as provided in Equations 6-2 and 6-4. The exponent, $n_R$, is ostensibly the power-law index for the molding compound at the medium shear rates encountered in the runner (200–600 $sec^{-1}$):

$$\frac{\Delta P}{\Delta z} = K_r Q^{nR} \tag{7-3}$$

The pressure increments determined from Equation (7-3) are summed as the pointer moves backward toward the transfer pot. On reaching the pot, the summed pressure is compared with the applied pressure. If the pressures do not agree to within a specified tolerance, a new volumetric flow rate is chosen and a second iteration with this new flow rate is conducted down the length of the runner. The solution iterates in this way until the summed pressure agrees with the applied pressure to within the allowed tolerance. At this time, the flow fronts are projected forward a distance computed from the time increment and the volumetric flow rate that provided the agreement. The elapsed time is advanced by this time increment and the iterations are started again from the new flow front.

The locations of the gates are entered at the start of the program. When the pointer encounters a gate location, the algorithm computes the volumetric flow rate through the gate using the pressure sum up to that point with an expression as shown in Equations (7-4) which incorporates the hydraulic resistance of the specific gate, $R_i$, and the rheology of the molding compound at that point in the mold. The viscosity is represented by the coefficient $K_G$ and the power law index $n_G$. Note that Equations (7-3) and (7-4) allow the use of different viscosity coefficients and power law indices for flow through the runners and through the gates. In many cases, the rheology under these widely different conditions of deformation cannot be adequately represented by a single set of power law parameters. The coefficient $K_G$ can represent both the temperature and time dependencies of the viscosity, but only temperature dependence has been included [7]. The volumetric flow rate into the gate determined from Equation (7-4) is then subtracted from the flow rate in the runner and the computation continues upstream toward the transfer pot, repeating the above procedure each time a gate is encountered. An illustration of this solution is provided in Figure 7-11.

$$Q_i = \left(\frac{P_i}{K_{Gi} R_i}\right)^{1/nG} \tag{7-4}$$

The gate hydraulic resistance coefficients, which like the runner hydraulic resistance coefficients, include both the viscous and geometric contributions, can be determined experimentally as is shown in Figures 6-28 to 6-31. They can also be determined by a trial and error solution from the mold filling model

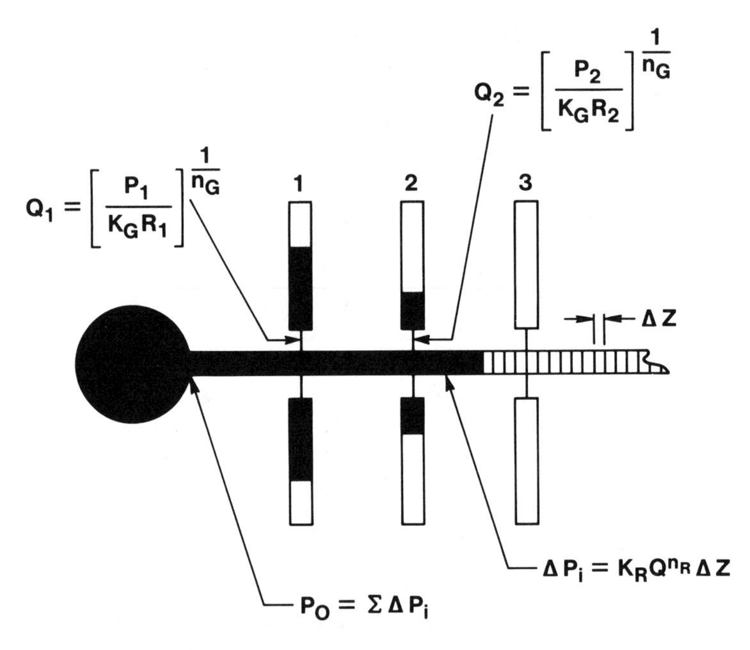


**FIGURE 7-11** One-dimensional mold filling simulation algorithm (From Reference 7. Reprinted with Permission of the Society of Plastics Engineers.)

if one or more filling profiles are known. The coefficient scales linearly with viscosity, but nonlinearly with gate geometry parameters such as the gate width, depth and land length. Similarly, the resistance does not scale linearly with volumetric flow rates since the molding compound is a shear thinning material. The functional dependence on these parameters can be derived from the gate pressure drop data provided in Figures 6-28 to 6-31. Yet another approach is to compute the $K_G$ coefficient from the gate geometry and molding compound rheology using the analysis provided in Appendix 4A.

The locations of the flow fronts in each cavity are determined by projecting the fronts in each time interval a distance computed from the volumetric flow rate into the cavity and the cross section area of the cavity. Flow resistance in the cavity can be assumed to be negligible compared to the resistance through the gate. A conditional statement terminates the program when all of the cavities are full.

The temperature dependence of the viscosity is incorporated into the viscosity coefficients $K_R$ and $K_G$ that appear in Equations (7-3) and (7-4). Either a WLF or Arrhenius temperature dependence function can be used as was discussed in Section 2.4.4. The expression would appear as shown in Equation (7-5) if an

Arrhenius dependence is employed. In this expression, the subscript $x$ denotes that either the gate or the runner coefficient may be treated in this way. The temperature at points along the runner at different times can be determined by employing one of the analyses outlined in Section 6.4.2.2. A simpler linear relationship between temperature and flow distance has also been used in mold filling simulation with satisfactory results [7]. For accurate modeling, the temperature used to determine the hydraulic resistance should be the temperature in the gate rather than the temperature in the runner. Methods for estimating gate temperature were described in Section 6.4.3.2.

$$K_x(T) = K_0\left(\frac{E_\eta}{RT}\right) \tag{7-5}$$

The time dependence of the viscosity (hydraulic resistance) can also be incorporated into the $K_R$ and $K_G$ coefficients. This dependence is much more difficult to include since the time that is used in the computation of the viscosity is not the elapsed time of filling, but the specific time-temperature history of the molding compound moving through the gate. For flow in the runner, it is acceptable to consider one entire length increment as a fluid element with the same time-temperature history. The same assumption can be applied to the molding compound that moves through the gates, but it is less justifiable since this material may be only a fraction of the total volume moving in a longer length of runner. Nonetheless, approximate treatments of this residence time distribution can be developed and used to determine the resistance. In many applications of transfer mold filling simulation, however, the time dependence of the viscosity can be neglected [7, 8, 10]. The viscosity increase due to polymerization is negligible during most of the mold filling time as was clearly shown in Figures 6-26, 7-6, 7-7, and 7-8. The temperature effects overwhelm the polymerization effects until near the end of the mold fill time. The mold filling profile is fully established at this late time, and in most molding tools the filling is complete before the viscosity begins to increase significantly. Therefore, the time dependence of the hydraulic resistance through the runner and through the gates can be neglected in most cases except where the fill characteristics late into the mold filling are desired. One example of an application where polymerization will have to be included is when the occurrence of incomplete fills are to be predicted.

An example of the results of this one-dimensional model for a conventional mold design with large gates is provided in Figure 7-12. The prediction is compared with experimental short shots in Figure 7-13. The agreement is in general very good. The largest discrepancies are the 10–15% over-estimation of the fraction of fill in the fifth and sixth cavities of the medium elapsed time and in the fourth cavity of the short elapsed time. It appears that the model is least accurate at low fractions of fill when the molding compound is just beginning

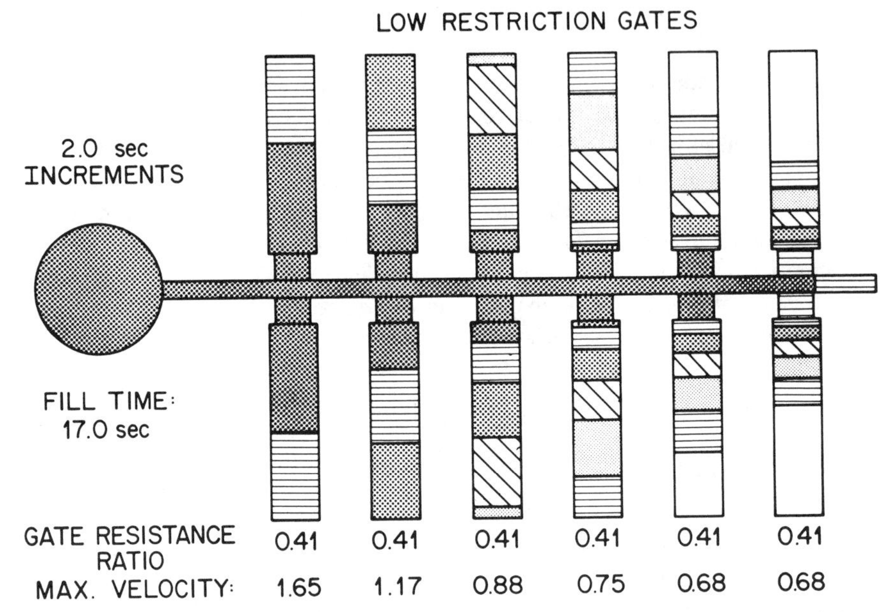


**FIGURE 7-12** Mold filling profile predicted from a one-dimensional network flow simulation. The mold is a conventional design with larger than normal gate flow areas. (From Reference 7. Reprinted with Permission of the Society of Plastics Engineers.)

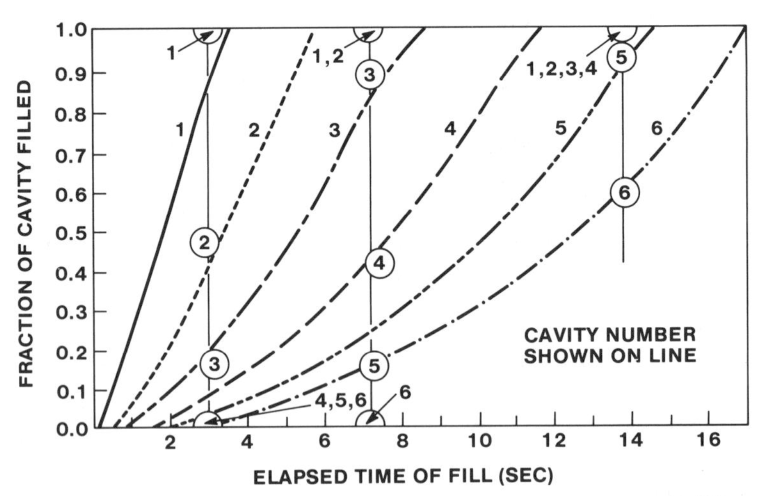


**FIGURE 7-13** Comparison of the predicted and experimentally determined mold filling profile for the profile shown in Figure 7-12. (From Reference 7. Reprinted with Permission of the Society of Plastics Engineers.)

to flow into the cavity. There is apparently some transient as the molding compound moves through the empty gate that is not accounted for in the model. Another discrepancy is in the under-estimation of the fraction of fill of the first cavity in the short elapsed time. The discrepancy in this case was attributed by the authors to under-estimation of temperature in the gate. A straight-line approximation to gate temperature was used which underestimated the gate temperatures of the first few cavities [7].

The modeling of branch flows may be required for some molding tools. This becomes important if there are severe imbalances among the chases in the mold so that some chases fill far ahead of the others, creating some of the same problems caused by cavity imbalance. Branches arise when a manifold design is used; a larger feed runner connects to several smaller runner arms. The network flow model described above can also accommodate branch flows. The pressure gradients in each arm can be different, but the pressure upstream of the branch point must be the same. The pressure summation is conducted separately in each runner arm with the stipulation that the pressures become and remain equal upon reaching the branch point, keeping in mind that the pointer moves backward from the flow fronts to the transfer pot. The volumetric flow rates of the individual runner arms sum in the feed runner section, although their contribution to the total volume flow rate in the feed runner may not be equal. This solution requires additional iteration since you have to iterate on all of the runner arm branches to obtain uniform agreement at the branch point. A simplified solution can be applied in the case where the configuration and hydraulic resistance encountered on each of the branches is approximately the same. In this case, it can be assumed that each branch contributes an equal share of the volumetric flow rate carried in the feed runner. Symmetric molding tools where the flow is fairly well balanced among the chases are amenable to this simplification.

### 7.1.5 Mold Filling Profile Dependence on Design, Process, and Material Parameters

It is difficult to predict the mold filling profile from the gate and runner geometries without the aid of an accurate flow simulation model or an extensive set of experiments. There are, however, a number of distinct trends that contribute to either greater or lesser degree of balancing. Imbalance arises from the pressure drop between the cavities on a leadframe strip, and from the fact that the molding compound simply reaches the nearer cavities first. Improved balancing, or closer approach to simultaneous fill, can therefore be attained by minimizing the effect of the pressure gradient between cavities. There are several ways of doing this which are reviewed in the following sections.

**7.1.5.1 Mold Filling Profile Dependence on Gate Geometry** Increasing the hydraulic resistance of the gates is an effective way of changing the filling profile. Gates with smaller cross section areas of flow require higher pressure to sustain a given volumetric flow rate, and therefore minimize the effect of the pressure gradient in the runner. The gates sizes do not have to be progressively smaller moving away from the pressure source to be effective. Smaller, uniform gates will improve balancing as is shown in Figure 7-14, particularly when coupled with a larger runner. With the proper temperature gradient in the runner, small uniform gates may provide a balanced filling. Completely simultaneous fill can be assured by using progressively sized gates. Smaller cross-sectional areas of flow are used closer to the pressure source to counteract the higher pressure at this end of the runner. The flow model can be used to determine the size of the gates using a trial and error approach. Another more approximate approach that can be used in the absence of a flow simulation program would be to fit the parameters for Equation (7-4) through a molding trial on an existing unbalanced design. By conducting a sequence of short shots it is possible to estimate the volumetric flow rate through each gate. Use the methods outlined in Chapter 6, Sections 6.4.2.2 and 6.4.3.2, to estimate the temperature and viscosity of the molding compound through the gates to set the $K_G$ coeffi-

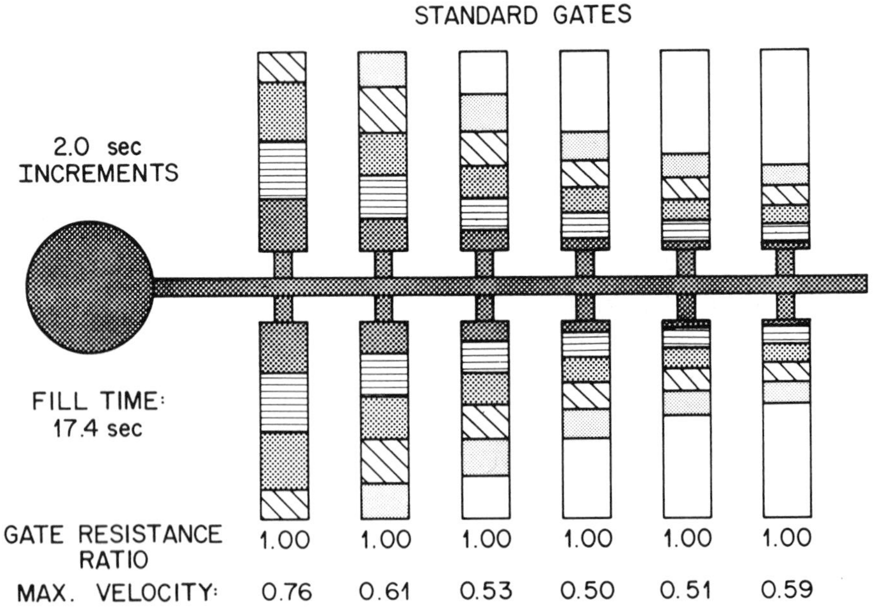


**FIGURE 7-14** Mold filling profile for uniform gates where the gate sizes have been reduced compared to the gate sizes used in Figure 7-12. A marked improvement in balancing is evident even though all other mold design and process parameters are the same.

cient using Equations (7-4) and (7-5). Equalize the volumetric flow rates through each gate by changing the resistance coefficient Ri to determine the resistance ratios of the gates. Then convert the resistance ratios to gate geometries by using the data for pressure drop versus gate dimensions presented in Figures 6-28 to 6-31. The data in Figures 6-28 to 6-31 can only be used in this way if the gate geometries and volumetric flow rates are similar to the gates in use. A more analytical and precise approach to selecting gate resistances using a mathematical treatment has been offered [7] and is provided in detail in Appendix 7A at the end of this chapter.

The mold filling profile for a set of progressively sized gates whose resistance was arrived at through the treatment described in Appendix 7A is provided in Figure 7-15 [7]. Satisfactory balancing is predicted. A molding tool incorporating these progressive gates was made and tested in a molding trial. A representation of the results is provided in Figure 7-16. The agreement is again good. The discrepancies are the same ones discovered previously: underestimation of fill in the first cavities and slight overestimation of fill in the middle cavities. The authors [7] again attribute these minor differences to underestimation of the gate temperatures and transients as the molding compound begins flowing into the empty gate and cavity.

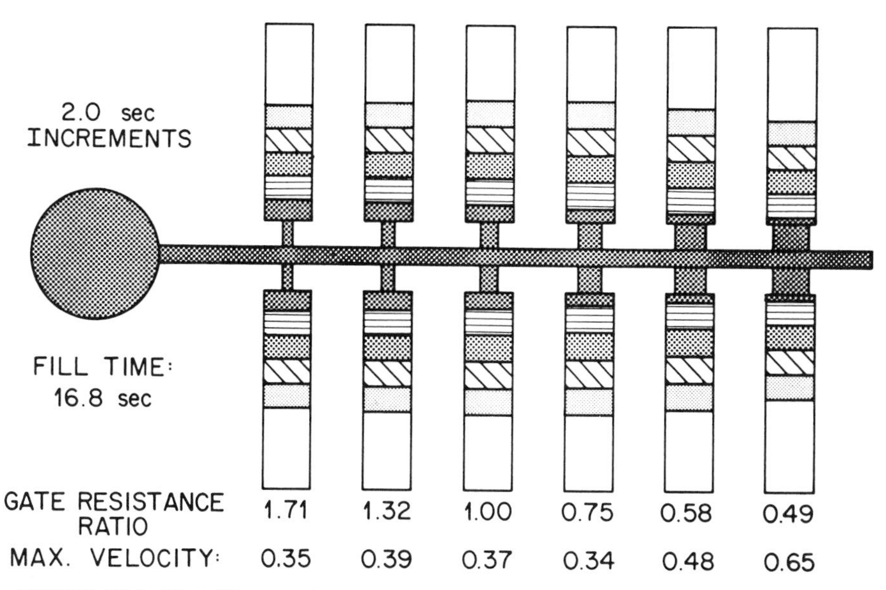


**FIGURE 7-15** Mold filling profile for progressively sized gates designed to provide uniform filling of all cavities. The gate resistances were determined through the mathematical treatment described in Appendix 7A. (From Reference 7. Reprinted with Permission of the Society of Plastics Engineers.)

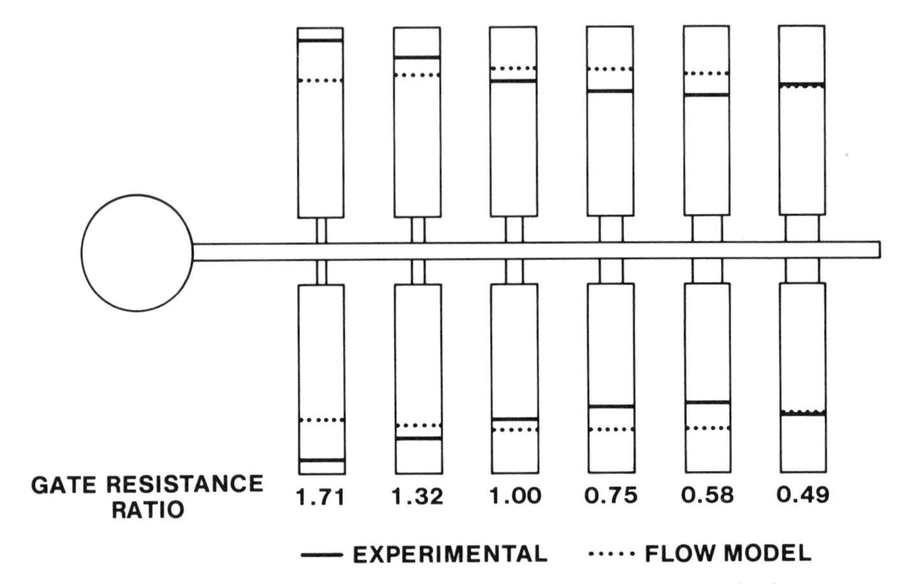


**FIGURE 7-16** Comparison of experimental and predicted flow front locations for the progressively sized gates intended to provide uniform mold filling as was shown in the simulation provided in Figure 7-15. (From Reference 7. Reprinted with Permission of the Society of Plastics Engineers.)

**7.1.5.2 Mold Filling Profile Dependence on Runner Geometry** Reducing the hydraulic resistance to flow in the runner by increasing the runner cross section reduces the pressure gradient in the runner and improves the uniformity of the mold filling. This is fairly effective with the only disadvantage being slightly greater molding compound usage. A large runner will reduce heat transfer to the molding compound resulting in a steeper temperature gradient in the runner. This provides lower temperatures, higher viscosity and higher hydraulic resistance at the high-pressure end of the runner which tend to nullify the inherent pressure gradient and further improve the balancing. This steep temperature gradient is not generally recommended because it can result in high viscosities and low yield in the cavities closer to the transfer pot, but there may be specific instances where it can be of use. A mold temperature increase or decrease is not effective in influencing the filling profile because it has a similar effect on the hydraulic resistance in the runner and in the gates, thus the relative difference in resistance is not altered.

*The Use of Tapered Runners* The feed runners and runner arms of a molding tool do not have to be of constant cross section. Tapering or step changes in cross-sectional area are feasible but not always prudent. Any step change in runner cross-section should be avoided. A step down to a smaller cross section causes an excessive pressure drop, whereas a step up will create air voids. Tapered runners have been used in transfer molding and other polymer operations to aid balancing and control pressure drop and heat transfer. In plastic package molding, the exact purpose and benefits of tapering are difficult to predict. The more common type of runner tapering is when the cross section area decreases in the flow direction [11]. Tapering in this way will increase the heat transfer and the temperature at the far end of the runner. This effect improves balancing because it reduces the resistance to flow through the gates in the later cavities. The other effect is the increase in pressure at the far end of the runner to push the molding compound into the smaller channel. The back pressure at the gates nearer to the transfer pot is increased and more molding compound is pushed into the nearer cavities, exacerbating the inherent imbalance of the mold. The degree to which these effects of higher temperature at the far cavities and higher pressure at the near cavities offset each other to provide greater uniformity of flow depends on all of the specifics of the molding compound rheology, mold design parameters such as the runner sizes and degree of tapering, and the process parameters. There are no generalizations that can be made from the physical phenomena of the flow, but it appears that in many designs based on conventional material, design and process parameters tapering does provide an improvement in balancing [11] although the effect is relatively small. Figures 7-17, 7-18, and 7-19 show the predicted mold filling profile derived from the one-dimensional network flow model for conventional mold design, a modified design with conventional gates and a tapered runner, and a modified design with smaller gates and a tapered runner. These results show that runner tapering alone cannot provide balancing with this molding compound/mold design combination, but tapered runners combined with re-

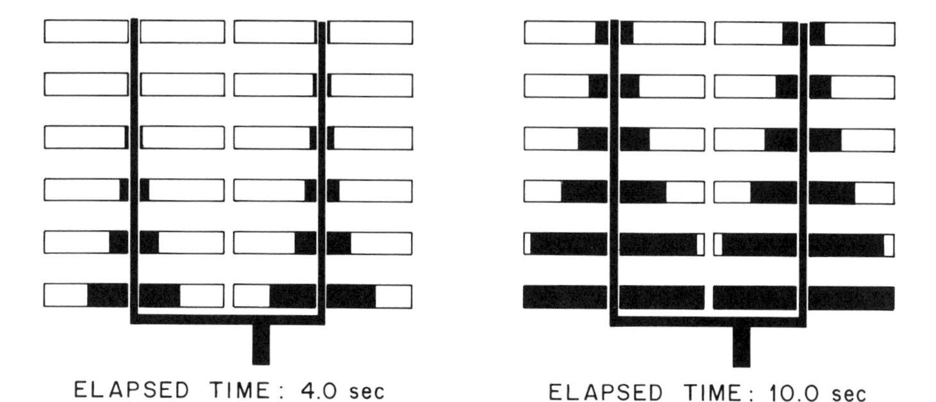


**FIGURE 7-17** Mold filling profile for one quadrant of a production scale molding tool. The tool design is conventional with uniform gate width of 0.500 in. (1.27 cm) with feed runner and runner arms of constant cross section.

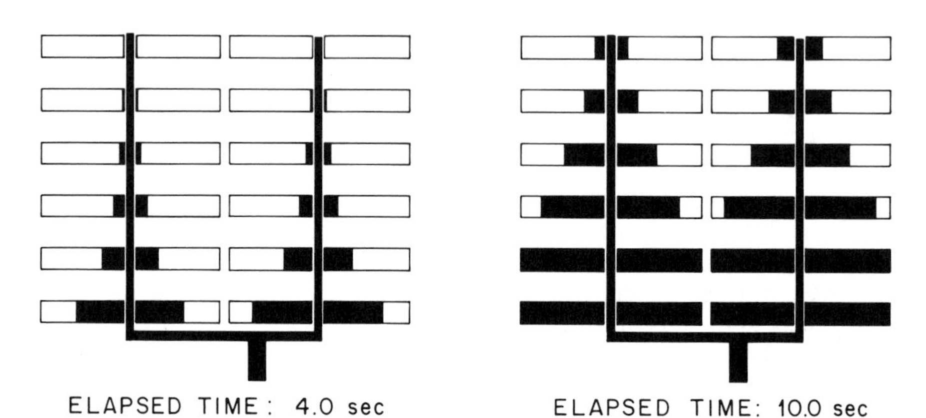


**FIGURE 7-18** Mold filling profile for one quadrant of a production molding tool. The tool design incorporates a tapering of the runner arm depths from 0.190 inch at the junction with the feed runner to 0.130 inch at the end of the arm. The gate sizes are the same as those used in the results shown in Figure 7-17.

duced size gates can provide a near balanced profile that may be acceptable in many applications.

**7.1.5.3 Mold Filling Profile Dependence on Process Parameters** It is difficult to generalize on the dependence of the mold filling profile on the process parameters because any specific modification in filling profile actually depends

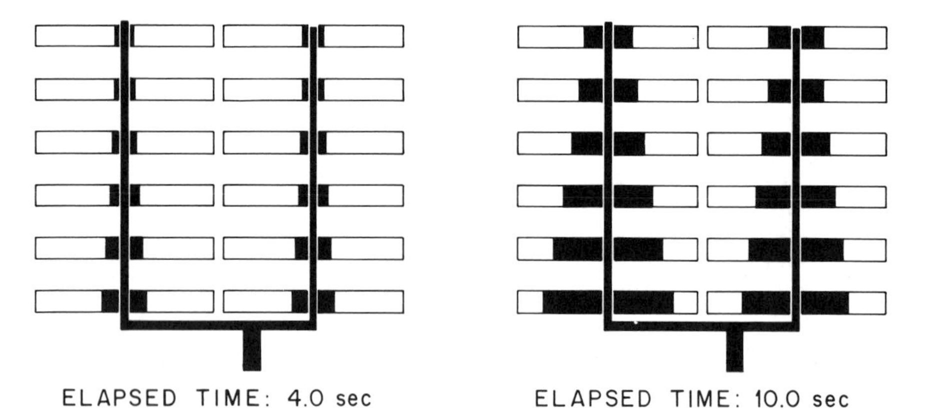


**FIGURE 7-19** Mold filling profile for one quadrant of a production scale mold. The tool design incorporates both runner tapering and smaller, uniform size gates to enhance the balancing of the filling profile. The width of the cavity gates was reduced from 0.500 inch to 0.250 inch, and the runner arm depths were tapered from 0.190 inch at the junction with the feed runner to 0.130 inch at the end of the arm.

on a complex interaction of a large number of material, design and process parameters. The best that can be accomplished in this discussion is to review some of the physical phenomena that influence the mold filling profile. Keep in mind, however, that the same process parameter change in a system using a different molding compound or mold design may have a completely different effect on the filling profile.

*Fill Time* The effect of filling profile on the fill time is difficult to generalize. Higher transfer rates increase the shear rates through the runners and gates, hence the mold filling will be most affected when the shear thinning index for flow through the runners and gates ($n_R$ and $n_G$) differ greatly. In cases, where these indices are close or the same the filling profile is largely insensitive to transfer rate, except for temperature effects. The temperature effects of a higher transfer rate are a lower temperature and higher viscosity at the cavities nearer to the transfer pot which would promote balancing. The effect of this temperature profile change on the mold filling profile is not great because the resistance to flow in both runners and gates is increased.

*Mold Temperature* The mold temperature has a minimal effect on the mold filling profile. It has equal influence on the flow resistance through the runners and the gates in most cases. For some molding compounds, however, the temperature sensitivity of the molding compound could be different at the very high shear and elongational rates encountered in the gates compared to the intermediate shear rates encountered in the runner. Figure 7-20 shows the small change in mold filling profile for both low and conventional mold temperatures.

### 7.1.5.4 Mold Filling Profile Dependence on Material Properties

*Shear Thinning Character* The shear thinning properties of the molding compound can have a profound effect on the mold filling profile. There is a significant difference in the deformation mode and deformation rate between the runners and the gates, therefore changes in the shear thinning character of the molding compound will have a significant effect on the relative difference in hydraulic resistance between gate and runner. It is for this reason that a balanced mold design should be considered to be molding compound specific. There is no assurance that a different molding compound would provide balanced mold filling in a design that worked well with the intended material. This is one of the principal disadvantages of employing precisely balanced design; switching to an improved molding compound could sacrifice balanced mold filling. Unfortunately, it is difficult to make generalizations about the precise relationship between uniform filling and shear thinning character. In this book, the shear thinning character of the molding compound has often been repre-

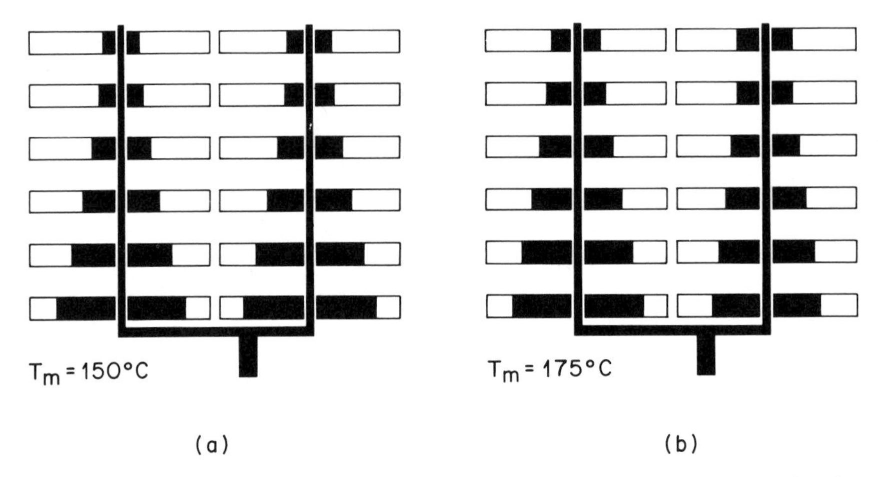


**FIGURE 7-20** Predicted effect of mold temperature on mold filling profile for a production size molding tool: (A) mold temperature of 150°C.; (B) conventional mold temperature of 175°C (same conditions as Figure 7-19). The transfer pressures were modified to achieve the same mold fill time in both cases.

sented by the Power Law model which uses a power law index, $n$, to quantify the extent of shear thinning behavior. Values approaching one are Newtonian in character and have a shear independent viscosity, whereas indices approaching zero have increasing shear thinning character. Most molding compounds are moderately shear thinning. Increasing the shear thinning character of the molding compound should provide a greater viscosity reduction in the gates than in the runners, but this is not unequivocal since the flow through the gates encompasses both shear and elongational deformations. In addition, changes in shear thinning behavior will likely be accompanied by other changes in rheological parameters such as the flow activation energy. Also, the specific changes in flow resistance through the gates and runners will depend on the specific gate and runner designs and process parameters. It is, therefore, difficult to make any generalizations about the effect of changes in non-Newtonian character on the mold filling profile.

*Temperature Sensitivity of Viscosity* Temperature sensitivity of the molding compound also has an effect on the mold filling profile. This parameter, reflected in the flow activation energy shown in Equation (7-5), changes the viscosity gradient in the runner. A lower activation energy means the molding compound viscosity is less sensitive to temperature and the viscosity gradient in the runner will be lower. A higher activation energy means greater temperature sensitivity and much steeper viscosity gradient; higher viscosity nearer to the pot and lower viscosities far from the pot. The steeper temperature and viscosity gradients tend to provide improved balancing, because it is more difficult to move molding compound into the near cavities, but this improvement is contingent upon all the other material, design, and process parameters.

*Fast Cure Kinetics* The use of a fast cure molding compound can have a profound effect on the mold filling profile. The more rapid cure of these materials means that the molecular weight and viscosity of the molding compound will be higher at the cavities far from the transfer pot. A simple increase in viscosity would not alter the mold filling profile significantly because the increased hydraulic resistance would be the same in runner and gates. The more usual case, however, is that a significant increase in molecular weight and viscosity changes the shear rate and temperature dependencies of the viscosity from the low molecular weight material. As an example, if the shear rate dependencies reflected in the $n_G$ and $n_R$ parameters of Equations (7-3) and (7-4) were the same at low degrees of cure and low molecular weight, it would not be surprising if they were different at advanced stages of cure. This is tantamount to the changes in power-law index examined above that can produce significant changes in mold filling profile. There is one caution associated with this discussion, though. If mold filling profile changes are caused by a simple increase in chemical reaction rate, as would be the case if simply more catalyst were added, then substantial viscosity increases are occurring during the mold fill stage. This is a precarious operating condition because significant viscosity increases do not occur until the molding compound is close to reaching the conversion of gel at which point all flow ceases. Therefore, the molding is very near to the edge of the operating window. There is a risk of short shots where all the devices are ruined, or the risk that several of the cavities may be filling outside of the optimum process window described in Figure 7-6, thereby sustaining high flow-induced stresses.

## 7.2 REDUCING VOIDS

Voids are a significant problem in plastic packaging of microelectronic devices. Voids in a semipermeable matrix such as the molding compound will condense moisture in liquid form, particularly when subjected to temperature and humidity cycles. The presence of ubiquitous salts, either leached from the molding compound or introduced during assembly and molding, can then form an electrochemical cell which greatly accelerates corrosion and shortens device lifetime. Voids on and around the device or the wire bonds are the most threatening, whereas voids far from the leadframe or the silicon may be benign. It is for this reason that voids are becoming a much greater problem than they were in the past. A larger fraction of package volume is consumed by the device and its wire bonds. There is also the growing popularity of small outline packages which place a large silicon device within a minimal amount of molded plastic. These trends increase the chances that if a void is in the molded body it will cause a failure, whereas in a larger package such as a 40-pin DIP there was a good chance that the void would be in a harmless location. There is probably less understanding of voids in transfer molding than any other molding problem. There are, however, several concepts that can be applied.

### 7.2.1 Reducing Entrained Air

There are significant quantities of air entrained in the transfer molding compound as it flows into the molding tool. This air is introduced from a variety of sources. One of the most obvious sources of entrained air is the transfer pot. The pot is fitted with a tight fitting piston that precludes the free egress of air and molding compound from the pot, forcing both into the runner system of the mold. There is a large volume of air in the transfer pot because the molding compound preforms fill only a fraction of the pot volume, albeit usually more than half. Some of this air vents out through the runner system and some slips past the plunger. But some fraction of air does remain in the pot after the molding compound is being forced into the runners and is forced out along with it. A photograph of entrained air in a transfer molding cull is shown in Figure 7-21.

One approach to eliminating this air is to minimize the air volume in the pot by making the preforms fill the pot volume as completely as possible. This entails using preforms whose diameter is as close to the pot diameter as allowable by handling considerations. Preform diameters that are 90% of the pot diameter are feasible. Operators can have difficulty inserting the preform stack in the pot when the preform diameter is too close to the pot diameter, so diameter ratios greater than 95% are tenuous. The softened preforms can hang up on the sides of the pot instead of reaching the bottom, causing production delays. Despite this difficulty, the concept is gaining acceptance because of accommodations made to handle the preform stack. One of the accommodations has been the use of a single preform of large diameter instead of using four to six smaller preforms whose misalignment can be one of the difficulties on insertion. The one preform also eliminates the seams and spaces between the preforms that are another source of air. Figure 7-22 shows a diagram of the conventional configuration and the single preform configuration. The difficulty in insertion of the larger diameter preforms can be alleviated by using automated preform handlers which can be more precise than human operators. In some instances, lower preheat temperatures are required to impart a stiffer consistency to the preheated molding compound to facilitate robotic handling. Lower preheat temperatures have implications in increasing flow-induced stress and molding tool wear.

**FIGURE 7-21** Transfer molding cull showing large air voids that result from air entrained in the transfer pot.

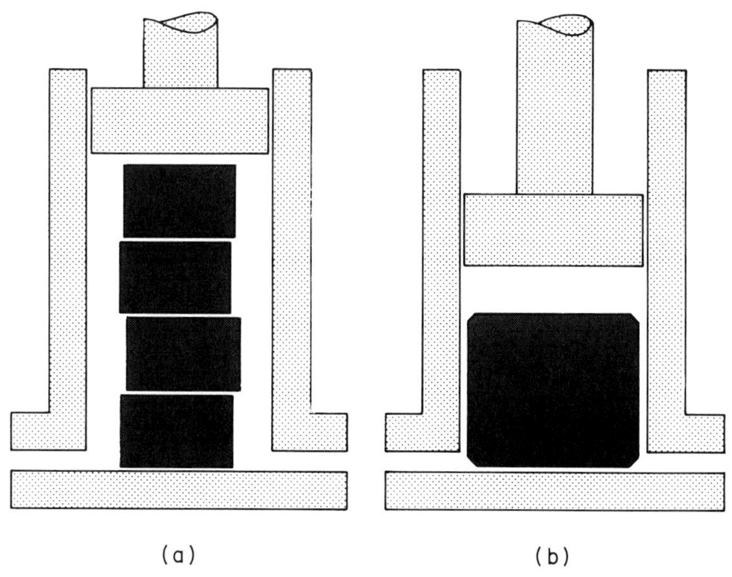


**FIGURE 7-22** Different configurations of preforms in the transfer pot: (a) the conventional configuration of multiple preforms of small diameter, and (b) one large diameter preform to minimize the air space in the transfer pot in an attempt to reduce the occurrence of voids.

### 7.2.2 Optimizing the Packing Pressure

The molding compound is a porous material when it flows at the low pressures characteristic of the mold filling phase of the packaging process. There is a continuous distribution of small, irregular shaped voids with sizes of the order of 0.5 mm (0.012 in.) or less. The application of the packing pressure, typically 600 psi (4.1 MPa) greater than the transfer pressure compresses the molding compound and eliminates the extensive porosity. Smaller, largely invisible microvoids may persist after packing as is shown in the micrograph provided in Figure 7-23, but these usually pose no reliability threat. They do, however, contribute to greater moisture permeability, so excessive microvoids are undesirable. Higher packing pressure will increase the molding compound density and reduce the number and size of microvoids, but some are exceptionally small crevices created by filler particle configurations that will be insensitive to the moderately higher pressures that can be used.

The packing pressure also plays an important role in compressing and eliminating the larger stochastic voids, those introduced by the air entrained in the transfer pot and runner, or caused by air trapped in the cavities or runners due to poor flow channel design and inadequate venting [11, 12]. The packing pressure does not eliminate voids through Boyle's Law alone, also known as *P-V-T* volume change, which states that the pressure multiplied by the volume and divided by the absolute temperature of a gas in state one, represented by $P_1 V_1 / T_1$, must equal $P_2 V_2 / T_2$. The packing pressure is at most four times greater than the transfer pressure, which means the void diameter decreases by a factor of the cube root of four which is less than 1.5. The packing pressure is effective because it initiates several different mechanisms in addition to the $PV/T$ volume reduction that all contribute to reducing the size and number of voids. These other mechanisms include pressure accelerated adsorption of air onto the free surfaces of filler and resin, pressure accelerated absorption and diffusion into the resin, local heating on volume reduction which increases the coefficients for adsorption, absorption and diffusion, and the break-up, migration and dispersion or larger voids into smaller voids and microvoids. Whereas the $PV/T$ reduction is immediate, the adsorption, absorption, diffusion and dispersion mechanisms are all time dependent. For this reason, it is imperative that the molding compound in each cavity experience the highest possible packing pressure for an appropriate time interval.

There are many material, design and process conditions that prevent the full applications of the packing pressure. One practical consideration is excessive resin bleed and flash in cases of high packing pressure. A poorly designed or fabricated molding tool, or an exceptional low viscosity molding compound, can exacerbate resin bleed, in some cases reaching the point where proper material density cannot be achieved because sufficient packing pressure cannot be applied. Employing a packing pressure profile instead of the conventional con-

**FIGURE 7-23** Photomicrographs of a fracture surface of the molding compound subjected to normal packing pressure. Magnifications of 1000× and 2500× are shown.

stant packing pressure can alleviate this problem in some cases [11, 12]. A packing pressure profile can be implemented through the use of a programmable process controller of the type that was discussed in Section 7.1.3.1. In the absence of such a device, the packing pressure is applied at a constant level for a specified time interval. The problem with a constant packing pressure is that applying high pressures for the entire packing interval aggravates resin bleed, which is a flow phenomenon where extent is proportional to time of applied driving force. The packing pressure can be optimized by using an exceptionally high packing pressure for only a short time to reduce the voids, and then reverting to some lower packing pressure to prevent excessive resin bleed [11, 12]. A summary of this discussion is provided in Figure 7-24.

### 7.2.3 Preventing Gate Freeze-Off

The packing pressure is applied through the gate of the cavity. If the molding compound in the gate reaches the conversion to gel before the molding compound in the cavity and before the application of the packing pressure, then the cavity is isolated from the packing pressure and will show greater porosity and greater occurrence of stochastic voids. It is not uncommon for this to happen. The molding compound in the gate is instantly raised to the mold temperature because of the rapid heat conduction across the very narrow throat of the gate, whereas the molding compound in the cavity requires about 30 seconds to reach the mold temperature. This gives the material in the gate a significant lead in conversion so that it reaches the conversion to gel far sooner. The problem occurs when the cavities nearer to the transfer pot fill a long time before the cavities far from the pot because of an unbalanced mold filling profile. The packing pressure is not applied until the entire mold is filled, providing a long time interval where the molding compound is static in the gates of the near cavities and rushing toward gelation. The packing pressure has not been applied because the molding compound is still moving under the lower transfer pressure into the far cavities. Thus, in an unbalanced mold filling profile, the first cavities to fill, typically those nearer to the transfer pot, will experience the packing pressure for the shortest time, if at all. Voids are often encountered at a higher rate closer to the transfer pot. This scenario is depicted in Figure 7-25 which shows the conversion in the gate and cavities at different positions in the mold.

The occurrence of voids may be reduced by using balanced mold filling [11]. In a balanced mold design (see Section 7.1.2.2), all of the cavities fill simultaneously, thus molding compound flows through all the gates until the time when all the cavities reach complete fill simultaneously. The amount of time between the point at which flow stops and the packing pressure is applied is thereby the minimum value for all gates, whereas in the unbalanced case shown in Figure 7-25 the time interval can be as much as 15 seconds, allowing enough

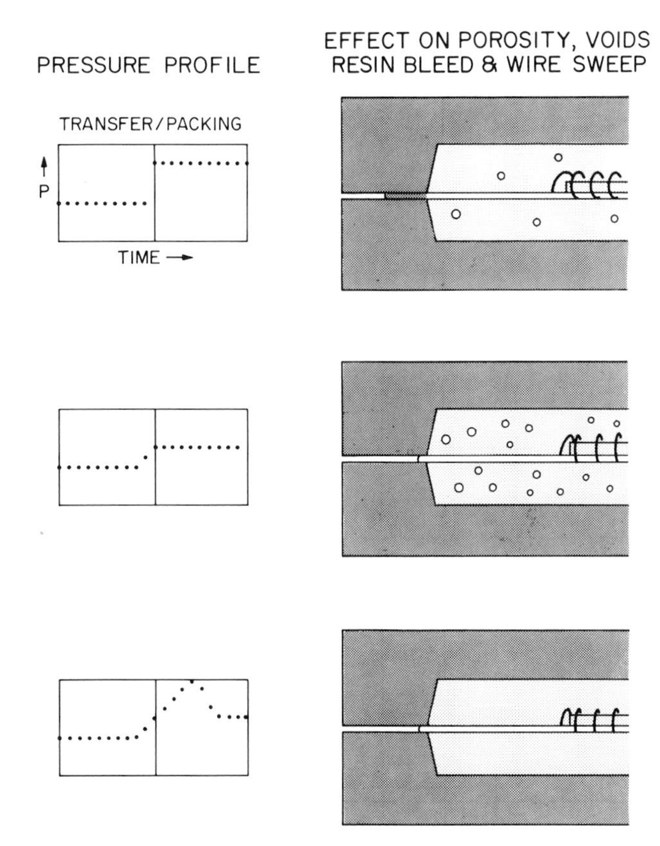


**FIGURE 7-24** The effects on molding compound porosity, resin bleed, and wire sweep for different packing pressure profiles: (A) (Upper) sudden application of constant, moderately high pressure which results in excessive resin bleed and wire sweep with some void reduction; (B) (Middle) gradual application of lower packing pressure which minimizes resin bleed but causes excessive porosity and minimal void size reduction; and (C) (Lower) a packing pressure profile that uses a gradual application of packing pressure and a short interval of high packing pressure to enhance void size reduction and reduce porosity but minimize resin bleed. (After Reference 12.)

time for the molding compound to reach the gel point. Gate freeze-off is prevented and porosity and voids are reduced when the time between flow stoppage and application of the packing pressure is minimal.

## 7.3 OPTIMIZING THE POLYMERIZATION

The polymerization in the mold plays an important role in optimizing packaging productivity and yield, and it also has ramifications for subsequent reliability.

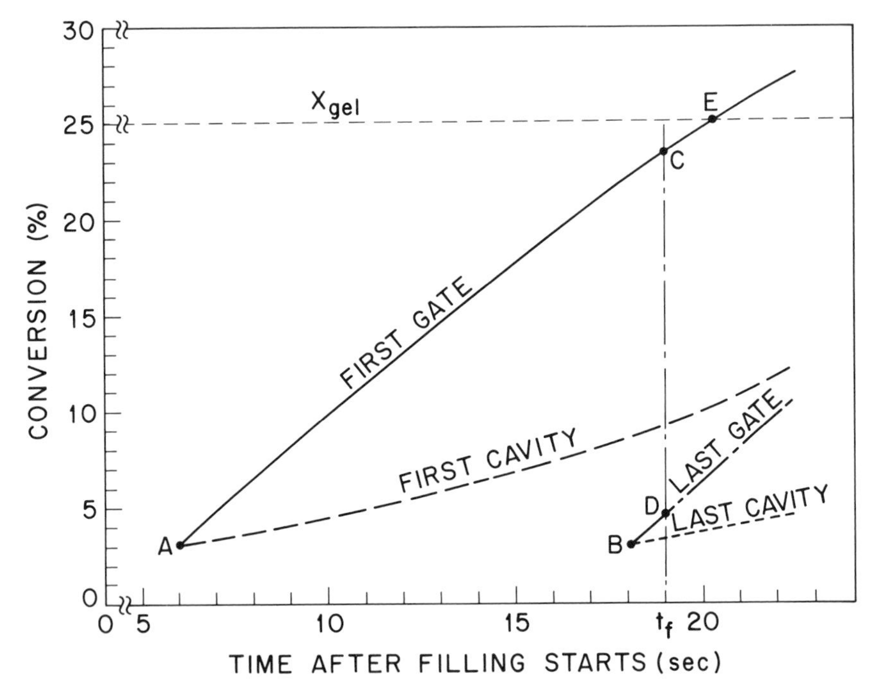


**FIGURE 7-25** The conversion history of molding compound in gates and cavities at cavity locations near and far from the transfer pot illustrating the potential for gelling of molding compound in the gates when a highly unbalanced mold fill profile exists. The letter designations of the points on the plot represent the following phenomena: (A) time at which the first cavity and gate fill, (B) time at which the last cavity and gate fill, (C) time at which the packing pressure is applied to the first cavity to fill, (D) time at which the packing pressure is applied to the last cavity to fill, (E) time at which the molding compound in the first cavity to fill reaches the conversion to gel, halting the application of packing pressure in the cavity.

### 7.3.1 The Importance of Moisture

Moisture has a profound effect on molding compound flow and cure behavior. Several of these effects are conflicting, so it is not simple to select the optimum moisture level for the preforms prior to molding. Moisture reduces the viscosity of the molding compound, so conditioning the preforms under some relative humidity is often beneficial to processability and reducing flow-induced stresses. The discussion in Section 5.2 concluded that the minimum relative humidity of the conditioning atmosphere be 20% R.H. Excessive moisture content, however, can cause excessive porosity and voids. Conditioning atmospheres above 50% R.H. were discouraged. Another serious consequence of moisture is the disabling of the cure mechanism, probably caused by a deactivation of the catalyst. In general, higher levels of moisture in the preforms reduce the cure rate and also lower the ultimate mechanical properties, particularly the glass transition temperature. The molecular changes induced by moisture present during the cure are not completely eliminated by subsequent drying or prolonged post-cure at elevated temperatures.

Moisture also has an undesirable effect on the coefficient of thermal expansion (CTE). Molding compound from preforms conditioned in a humid atmosphere can show a CTE that is significantly greater than the CTE for molding compound that was conditioned in relatively dry conditions. In addition, moisture in the preform and in the final molded package is likely to degrade the adhesion of the molding compound to the die and the leadframe, another reliability issue, as will be discussed in Section 8.2.4.

### 7.3.2 Optimizing the Mold Temperature and Uniformity

The uniformity of the temperature of the mold has important productivity and yield implications. The large transfer molds that are the mainstay of plastic packaging operations are subject to significant temperature deviations over the upper and lower mold surfaces and significant deviations of temperature with respect to time. These deviations result from a number of normal and abnormal conditions. The mold is heated through electrical resistance heater rods that pass through the body of the mold near to the molding surface. The surface of the rods is necessarily at a temperature above the intended mold temperature to ensure heat transfer to the molding tool that compensates for heat lost to the surroundings. The power to each individual heater is turned on and off in response to a feedback signal consisting of one or more thermocouples located in the upper and lower mold halves. The control mechanism can be either simple on-off control or proportional control where the power to the rod is proportional to the offset from the target temperature. Typically, the heaters in the upper mold half require lower power settings and less ''on'' time because of the heat rising off the lower mold half. Temperature gradients arise because there are a finite number of heater rods and a transient temperature gradient around each one. Also, unanticipated thermal loads, such as a draft on the molding tool or a lower ambient temperature can cause the surface temperature in sections of the mold to deviate significantly from the set-point. The use of only one or two feedback transducers compounds the problem because these few sampling points cannot accommodate the spatial and temporal changes that are occurring in different areas of the mold. The magnitude of these deviations can easily be 5–8°C over the surface of the tool.

The temperature gradients impact both the flow and cure of the molding compound. The molding compound temperature can be reduced by a few degrees if a long segment of runner is 5° or so below the nominal mold temperature. Although the consequences of this minor temperatures excursion are not overwhelming, the viscosity increase caused can make the difference in cases where the process is on the verge of flow-induced stress damage. The cure rate can also be adversely affected by temperature discrepancies. Low temperature will retard the cure in the affected cavities causing the hot hardness to lag behind resulting in ejection problems. Longer in-mold times are set to alleviate the hot-hardness deficiencies causing lower productivity. Conversely, local hot spots on the mold surface can cause premature flow seizure and gate freeze-off with the consequences being excessive voids and shot shots.

The molding process is optimized by assuring that the mold temperature is at the optimum set temperature, that the temperature is uniform over the entire mold surface, and that the profile is relatively free of transients. The steps that can be taken to improve temperature control over the mold include: (1) a maximum number of heater rods to improve temperature uniformity, (2) as many feedback thermocouples as possible, approaching one thermocouple for every one or two heater rods, (3) position the heater rods and the thermocouples as close to the mold surface as possible to minimize the temperature gradient between the rod and the surface, (4) use lower power settings on the heater rods which will promote longer "on" periods and less cycling during extended on-off cycles, (5) use a thermal insulating blanket on the molding tool to minimize the steep gradient at the side surfaces of the mold and to minimize the effect of drafts on the tool, and (6) with plate molds, use plate preheating to minimize temperature differences between the aperture plates and the rest of the mold. Some of these suggestions for temperature optimization are summarized in Figure 7-26.

### 7.3.3 Optimizing the Ejection Time

The ejection time is nominally set according to the molding compound supplier's recommendation. In many cases, however, it is advantageous to optimize the ejection time for the particular package design and molding parameters in use. There are no conversion or temperature criteria that can be established for ejection because the considerations in deciding when to eject do not correlate with any specific degree of conversion or temperature. These considerations mostly involve manifestations of low hot hardness which include mold sticking, package warpage, and ejector pin-induced deformation to the package. The measurement of hot hardness was described in Section 6.5.4. The material supplier can often provide a plot of the hot hardness of the molding compound obtained on a specific molded package or test specimen. This plot, an example of which was provided in Figure 6-40, shows a characteristic logarithmic growth which provides steep increases in hot hardness at early stages of cure with a constantly declining rate of hot hardness increase, leveling off at some plateau value that is characteristic of the mechanical modulus above the glass transition

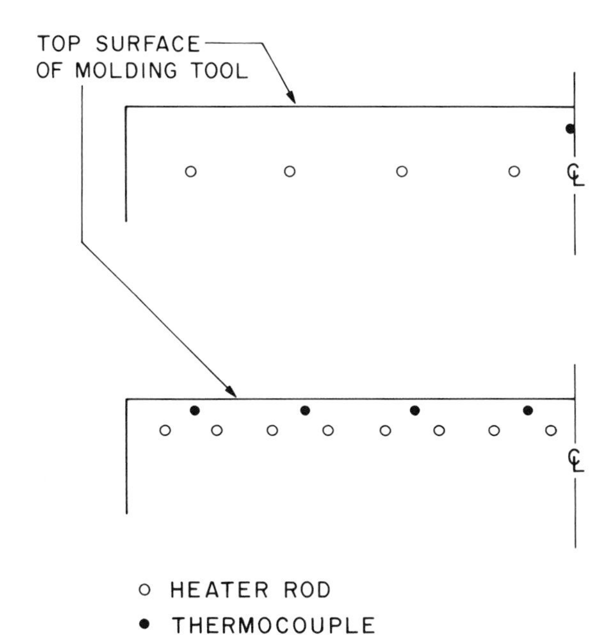


**FIGURE 7-26** Optimization of mold temperature stability and uniformity achieved through number and placement of the heating rods and thermocouples. The upper drawing shows a marginal configuration whereas the lower drawing shows a configuration that offers improved temperature uniformity and consistency.

temperature. The parts should not be ejected until the hot hardness has reached this plateau value. The material supplier determines the recommended cure time from these hot hardness data and from actual molding trials.

Optimization is conducted by realizing that additional in-mold cure time after reaching the hot hardness plateau is usually wasted since no further observable changes will occur, despite the continued increase in conversion on the molecular scale. The material supplier's recommendation may be too long in many cases. One reason is that the supplier may have wanted to keep the recommended cure time to a whole or half minute increment for simplicity. Another reason the recommended cure time may be too long is that the material supplier may have been conservative in recommending the ejection time to accommodate the various package designs that could be molded. In general, thicker molded bodies will show a slower approach to the limiting hot hardness plateau because of the slower ramp-up in temperature over the thicker cross section. Conversely, very thin packages will likely reach the hot-hardness plateau sooner because the molding compound reaches the mold temperature sooner. For these reasons, determination of hot-hardness of the molded body in the actual molding tool should be conducted in cases where high productivity is important. A cure time reduction of even 15 seconds is significant, particularly in very high-volume packages such as DRAMs, because it translates to a productivity increase of about 10%.

## 7.4 MINIMIZING ASSEMBLY PROBLEMS

The molding process is often the culprit in assembly problems that occur downstream of molding. The most prominent example of this coupling is in the trim and form operation. Other downstream processes such as ink marking and testing, as well as the overall perception of quality, depend on precisely molded packages.

### 7.4.1 Matching CTE of Leadframe and Molding Compound

Many downstream processes, and in particular the trim and form operation, are improved when there is minimal deformation of the leadframe caused by excessive mismatch in the coefficients of thermal expansion between the leadframe and the molding compound, and to a lesser extent between the die and the molding compound. Achieving minimum leadframe deformation is not simply a matter of selecting low CTE molding compounds because the process variables also influence the temperature history of the leadframe and the point at which its dimensions are constrained. In molding with a cavity chase mold, the leadframes are at room temperature when they are placed on the loading fixture and then moved onto the transfer mold where they register on locating pins, usually in the center of the leadframe length. The leadframes heat rapidly once they are in contact with the heated molding tool. The time interval over which they heat and expand freely is only a few seconds until the press clamps closed. There is additional heating during the 30 seconds it takes to preheat the molding compound and load it into the transfer pot, but the expansion in this case is constrained by the more than 100 ton clamping force of the press. When the leadframe expansion is constrained in this way, the leadframe will buckle between the clamp points which are effectively the edges of the molded body outline. This is more of a problem in quad packages that have leads on all four sides because the buckling is along the sides of the molded body that will be subject to the precision trimming of the dam bars between the leads. With DIPs the dam bars are all perpendicular to the buckling axis of the leadframe so it usually presents no problem. The problem, which is not as serious in cavity and multiplunger molds because the leadframe comes into direct contact with the heated mold, can be alleviated through proper compensation in the position of the cavities of the molding tool. This temperature compensation is most effective when the leadframe temperature at clamping is known and invariable.

The best way to assure the constancy of this temperature is to allow a longer time for the leadframes to reside on the mold prior to hard clamping. This is the soft close feature that was discussed as a press option in Section 5.4.

The problem is worse in plate molds (see Section 5.5.2 for a discussion of plate molds) because the leadframes are loaded onto the steel aperture plates whose temperature falls to near room temperature when cycled in and out of the press. Under normal operation, the aperture plates and leadframes would be hard clamped at a temperature well below the mold temperature because of the relatively long time required to heat the plates and the leadframes to the mold temperature. Yet another complication with plate molds is that the steel plates and the leadframes have different CTEs so that they will expand at different rates and to different final dimensions, with the precise temperature the leadframes reach when constrained by the clamping being uncertain. Again, the problem is much more severe in quad packages which have leads along the shrink axis. The problem can be effectively eliminated by using temperature compensated aperture plates, compensated to the mold temperature, and then ensuring that the plates and leadframes reach the mold temperature by preheating them outside the press in a convective preheat oven. Soft close heating is not as effective in plate molds because of the longer conduction distance through the aperture plates to heat the leadframe. The preheat oven is usually positioned next to the transfer press to minimize the travel and cooling of the heated plates into the press. In a production operation, it is advisable to have an oven that can accommodate at least three sets of aperture plates since the nominal preheat time for a set of plates is about 10 minutes. The general procedure is to load the aperture plates with the leadframes and then insert them into the preheat oven instead of in the molding press. The operator removes the set of plates that has the longest residence time in the oven and inserts that set into the transfer press, where it can be clamped immediately.

Additional buckling occurs after ejection in all three tooling options; cavity molds, plate molds and multiplunger molds, because the molding compound often has a higher CTE than the leadframe. The extent of the deformation is proportional to the degree of the disparity and the length of the structure, which is the edge length of the molded body. Therefore, the problem will become worse as package size continues to grow with higher lead counts. One partial remedy to this problem is to provide stress relief zones in the leadframe. One common design is to isolate the leads and dam bars from the outer rail of the leadframe so that the long rail can buckle extensively while producing only minimal distortion of the leads. The second remedy is to select molding compounds whose CTE approaches that of the leadframe. Some of the new ultra low stress molding compounds can have a CTE below that of a copper leadframe which is $16 \times 10^{-6}$ cm/cm-°C (16 ppM). In the case of a fairly large mismatch, the lead shrinks more than the molding compound meaning that delamination and voiding could occur on the lead surface. The shrinkage of the molding compound onto the lead is one of the major contributions to achieving good adhesion and preventing moisture ingress along the lead surface. Alloy 42, with a CTE of 4.3 ppM, cannot be approached by an epoxy molding compound.

### 7.4.2 Minimizing Package Warpage

Planarity of the molded body and of the leadframe is needed to assure high yield in the downstream processes of trim and form and code marking. Probably even more important, package planarity is a highly visible sign of packaging quality. Planarity issues become more serious with larger package outlines, particularly the high pin count PLCC and PQFP packages that can have package outlines approaching two square inches. The problem is also found on long aspect ratio packages such as 40 and 48-pin DIPs which use 0.100 in pitch and are over two inches in length. Larger dies within the package also appear to create more problems.

There are several apparent causes of package planarity problems. One is inherent asymmetries in the internal structure of the package which cause the entire package to warp to compensate for the disparities in the coefficients of thermal expansion. For example, a symmetric structure of copper sandwiched between two layers of epoxy molding compound of equal thickness will not warp when cooled from the mold temperature to room temperature even though it is subject to significant thermal shrinkage stresses. The symmetry balances the forces preserving the planarity of the structure. Conversely, copper bonded onto the surface of a slab of epoxy molding compound will warp. Warpage will also occur if the copper is within the slab of epoxy molding compound but not on the centerline. It is not always possible to maintain symmetry within a molded package, but is should be recognized that package warpage will be proportional to the degree of asymmetry. Similarly, any degree of paddle shift is likely to cause package warpage since it introduces asymmetry. In many cases, the degree of warpage of the molded body correlates with the degree of paddle shift.

The second consideration in package warpage is nonuniform cooling and solidification of the molded body after ejection from the mold. The molded body is at an elevated temperature of approximately 170°C at ejection and then rapidly thrust into room temperature. There will be large differences in the cooling rate between package surfaces that are exposed to air, placed in contact with metal, or placed in contact with other hot or cold plastic packages. For example, if the bottom of a package is placed on a metal surface with the top of the package exposed to ambient air, the surface with the metal contact will cool and shrink much more rapidly, essentially being quenched below its glass transition temperature. The top surface will cool at a slower rate and will reach the glass transition temperature at a much later time, thereby yielding to the shrinkage forces of the bottom part of the package. This will cause the package to warp to compensate for the greater shrinkage in the bottom half. This deformation may not be completely reversed when the top half ultimately reaches low temperature, because the bottom is rigid at this time and temperature, and it will not yield as easily as the top did when it was above the glass transition temperature. In this way, the deformation is semi-permanent. It can be removed by again heating the molded body to high temperature and then cooling it uniformly. Bear in mind, however, that this behavior also depends on the internal structure and symmetry of the molded body. A symmetric structure will not warp under symmetric cooling, but most package structures are asymmetric meaning that an asymmetric cooling profile will be required to achieve planarity. Finite element modeling of the shrinkage stresses and strains of the specific package geometry, discussed in Chapter 8, may be of some help in determining what the cooling profile should be to achieve a planar molded body.

## REFERENCES

1. L. T. Manzione, "Plastic Packaging of Microelectronic Devices," *Soc. of Plastic Engineers Ann. Tech. Conf. Preprints* (1984).
2. L. T. Nguyen, "Wire Bond Behavior During Molding Operations of Electronic Packages," *Polym. Eng. Sci.*, **28**(14), 926 (1988).
3. W. L. Hunter, "Association of Bonding Wire Displacement with Gas Bubbles in Plastic-Encapsulated Integrated Circuits," *Nitto Technical Reports*, Nitto Electrical Industrial Co., Ltd., Osaka (September 1987).
4. L. T. Manzione, Editor, *Applications of Computer Aided Engineering in Injection Molding*, Hanser/Oxford Univ. Press, Munich, New York (1987).
5. MOLDFLOW Australia Product Bulletin, MOLDFLOW Pty. Ltd., Colchester Road, Kilsyth, Victoria 3137 AUSTRALIA.
6. A. C. Tech Product Bulletin, Advanced CAE Technology, Inc., 31 Dutch Mill Road, Ithaca, NY 14850.
7. L. T. Manzione, J. S. Osinski, G. W. Poelzing, D. L. Crouthamel, and W. G. Thierfelder, "A Semi-Empirical Algorithm for Flow Balancing in Multi-Cavity Transfer Molding," *Polym. Eng. Sci.*, **29**(11), 749 (1989).
8. A. Kaneda, J. Saeki, M. Aoki, K. Otsuki and Y. Watanabe, in N. Suh, Editor, *First Intl. Polymer Processing Symposium Program*, M.I.T., Cambridge, MA (1977).
9. G. Astfalk, M. Saminathan, and D. L. Crouthamel, "An Algorithm for the Network Flow of Thermoset Materials," Soc. of Plast. Eng., ANTEC Papers (1987).
10. H. A. Ghoneim and S. Matsuoka, "Simulation of Balancing of Multi-Cavity Molds," presented at the Second Annual Meeting of the Polymer Processing Society, Montreal, Canada (April 1986).
11. A. Kaneda, "Transfer Molding Technology for Packaging of Large-Size LSI Chips," Semicon Conference Proceedings, Tokyo, p. 503, (November 15, 1989).
12. R. E. Farris and S. R. Krueger, "Process Control Technology for Plastic Packaging of Microelectronic Devices," Soc. Plast. Eng. Annual Technical Conference (ANTEC) (1987).

## APPENDIX 7A
## SELECTING THE GATE RESISTANCE FOR BALANCED FILLING*

The balancing of multicavity manifolds in thermoplastic injection molding is usually accomplished by varying individual diameters for runners of circular cross section, or by varying runner equivalent hydraulic radii for other geometries. The present case of microelectronic packaging is different in that balancing is to be achieved by varying gate resistance only, since the cavities are gated directly off of a common runner. It is the pressure drop along this runner during flow that causes unbalanced fill.

For the majority of cavity fill the entire runner will be full. Therefore, calculation of gate resistance will be based upon a full runner. The slight imbalance in fill that occurs before the runner is full is difficult to eliminate and unimportant, since the average cavity flow rate, and not the fill time, is the parameter to be equalized.

Consider the balancing of a 12-cavity chase of 40-pin DIPS as an example. If the flow rate in each cavity during fill is $q$, the flow rate in the runner leading to the 12 cavities is 12 $q$, and the flow rate in the runner segment leading to node $i$ in the mold is $[12 - 2(i - 1)]q$ (as shown in Figure 7A-1). The pressure drop across a unit runner segment can be represented by

$$\frac{\Delta P}{L} = K_R Q^{nR}. \tag{7A1}$$

For a runner segment of length $\Delta z$ separating nodes $i$ and $i + 1$, the pressure drop across this segment is then

$$\Delta P_{i,i+1} = K_R \Delta z Q^{nR} \tag{7A2}$$

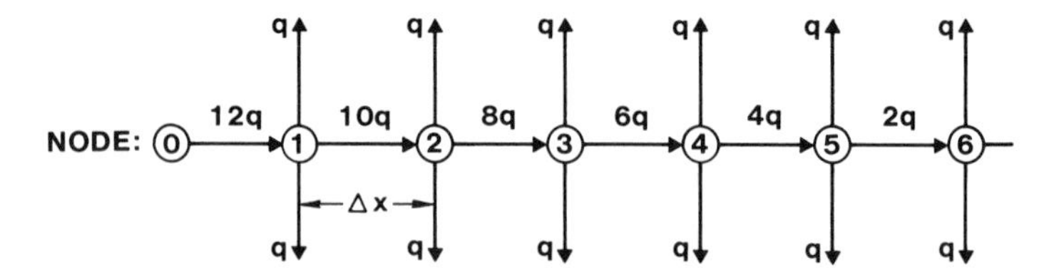


**FIGURE 7A-1** The flow-balancing algorithm.

*This analysis is derived from: L. T. Manzione, J. S. Osinski, G. W. Poelzing, D. L. Crouthamel, W. G. Thierfelder, ''A Semi-Empirical Algorithm for Flow Balancing in Multi-Cavity Transfer Molding,'' Polym. Eng. Sci., 29(11), 749 (1989).

where $\Delta z = 3.09$ cm in the present case. For $K_R = 0.0627$ (MPa/cm) $(\text{sec/cm}^3)^n$ and $n_R = 0.88$, the following relations can be written:

$$\begin{aligned} \Delta P_{12} = 0.194(10q)^{0.88} &= 1.472q^{0.88} \\ \Delta P_{23} &= 1.203q^{0.88} \\ \Delta P_{34} &= .939q^{0.88} \\ \Delta P_{45} &= .657q^{0.88} \\ \Delta P_{56} &= .357q^{0.88} \end{aligned} \tag{7A3}$$

Furthermore, the absolute pressure at each node is known to be:

$$\begin{aligned} P_1 = P_3 + \Delta P_{12} + \Delta P_{23} &= B_1 C q^{n_G} \\ P_2 = P_3 + \Delta P_{23} &= B_2 C q^{n_G} \\ P_3 &= B_3 C q^{n_G} \\ P_4 = P_3 - \Delta P_{34} &= B_4 C q^{n_G} \\ P_5 = P_3 - \Delta P_{34} - \Delta P_{45} &= B_5 C q^{n_G} \\ P_6 = P_3 - \Delta P_{34} - \Delta P_{45} - \Delta P_{56} &= B_6 C q^{n_G} \end{aligned} \tag{7A4}$$

where the rightmost term represents the flow through each gate. $B_i$ in this term is the geometry constant for the $i$th gate, representing the resistance to flow, $C$ is a material constant for a given molding compound at a given temperature and cure state, and $n_G$ is the power-law index for the molding compound. These parameters are also defined and used in Appendix 4A.

The latter two relate the shear rate experienced by the molding compound to its viscosity according to the power law:

$$\eta = C\dot{\gamma}^{n-1} \tag{7A5}$$

The product $B_i C$ is represented in Equation (7-4) simply as $K_G$. The constant $C$ is further resolved into an Arrhenius temperature dependence in the computer algorithm. Note that $K_R$ in Equation (7A1) could also be resolved into a geometric and a material constant. $n_G = 0.66$ for the molding compound used in this study.

Taking ratios with respect to gate 3 and substituting Equations (7A3) into Equations (7A4),

$$
\begin{aligned}
R_1 &= \frac{P_1}{P_3} = \frac{B_1}{B_3} = \frac{B_3 C q^{0.66} + 2.675 q^{0.88}}{B_3 C q^{0.66}} &&= \frac{B_3 C + 2.675 q^{0.22}}{B_3 C} \\
R_2 &= \frac{B_2}{B_3} = \cdots &&= \frac{B_3 C + 1.203 q^{0.22}}{B_3 C} \\
R_3 &= \frac{B_3}{B_3} = \cdots &&= 1.00 \\
R_4 &= \frac{B_4}{B_3} = \cdots &&= \frac{B_3 C - 0.939 q^{0.22}}{B_3 C} \\
R_5 &= \frac{B_5}{B_3} = \cdots &&= \frac{B_3 C - 1.569 q^{0.22}}{B_3 C} \\
R_6 &= \frac{B_6}{B_3} = \cdots &&= \frac{B_3 C - 1.953 q^{0.22}}{B_3 C}
\end{aligned}
\tag{7A6}
$$

$R_i$ represents the resistance ratios that must be determined. To do so, choose two of three things: (1) the flow rate $q$ to fill each cavity; (2) the pressure $P_0$ to transfer, or the pressure at any other node point; (3) a prescribed gate resistance (geometry) for one of the six gates around which to vary the others in reference. $q$ was chosen as 0.25 $cm^3/sec$, giving a 12 sec mold fill time in the present case. The effectiveness of the final design will be a weak function of $q$ due to the apparent difference between the material power law indices in the runner and the gates, which explains the presence of $q^{0.22}$ rather than $q^{0.0}$ in Equation (7A6).

To avoid calculating resistance values that are impractical, such as very small or very large gates, a practical intermediate was chosen as a reference for gate 3. This reference gate has an entrance angle of approximately 20°, a width of approximately 0.660 cm, and an exit depth of 0.031 cm. These dimensions would yield a $B$-value of approximately 3500 $cm^{-3n}$ (see Appendix 4A). If $C = 786$ Pa–$sec^n$ for the present molding compound under process conditions (see Appendix 4A); then $B_3 C \cong 2.75$ MPa $[sec/cm^3]^n$. This is the value of $K_G$ for this gate discussed previously. Substituting into Equations (7A6), we can solve for the resistance ratios:

$$
\begin{aligned}
&R_1 = 1.71 \qquad R_2 = 1.32 \qquad R_3 = 1.00 \\
&R_4 = 0.75 \qquad R_5 = 0.58 \qquad R_6 = 0.49
\end{aligned}
\tag{7A7}
$$

These values of $R_i$ combined with $B_3 = 3500$ and a runner described by the constants in Equations 7A3 should then yield balanced fill. This range of gate resistances can also be practically achieved in an actual mold. This treatment assumes isoviscous fill ($C$ = constant), therefore the results are not exact, but the resistances can be used as a starting point in the nonisoviscous flow model presented herein. As mentioned previously, however, these constants are realistic in that they were obtained from an actual curing system under process conditions. If enough is known of the average temperature and cure history of the material at each node, a $C_i$ constant could be calculated at each node to account for nonisoviscous fill, before calculating $R_i$. Note that these numbers depend upon the specific runner used; a runner with a different $K_R$, or pressure drop per unit flow length for a given material, will yield different results.

By the nature of Equations (7A6), it can be seen that gates with an overall high resistance ($B$-value) require a less extreme range of resistance ratios. Such gates would require higher pressures for a given flow rate, however, and very small gates may cause fill problems. Similarly, a large runner reduces the ratio range as well since the constant $K_R$ is then smaller. The constants outlined above, however, are based upon practical, workable gate and runner dimensions.

# 8 Package Reliability

Quality must be an overriding consideration in package design and manufacture. A poor quality package can compromise other advantages of device performance such as speed, low cost, and small size; features that may have been acquired at considerably more cost than those associated with achieving high package quality. The factors that affect quality of molded plastic packaging largely concern device performance and reliability. The performance metrics of the package are more a function of the package design and material selection rather than the manufacture of the package, so this topic will not be covered extensively here. Reliability, on the other hand, is intimately related to package manufacture, although this is often overlooked.

## 8.1 FAILURE RATE ANALYSIS

All packages will ultimately fail; it is the rate of failure that is the discriminating factor. The number of failed devices in a population will continue to increase with time, but the rate of failure is not constant and how the rate changes with time encompasses a wealth of information on the origin and severity of a reliability problem. If the failure rate of a population of packaged devices is plotted against time, the typical failure rate distribution is as shown in Figure 8-1. This is the characteristic "bath tub" curve [1] that is common to most integrated circuit devices. The initial high rate of failure is most often attributed to intrinsic device deficiencies, also known as "infant mortality." The percentage of devices lost to infant mortality is more strongly affected by wafer fabrication rather than package design and manufacture. This suggests that these failures actually represent marginal devices that passed multiprobe testing but could not survive the rigors of assembly, molding and burn-in. The burn-in and test cycle conducted after package manufacture is intended to eliminate this relatively large number of premature failures.

The second region of a typical failure rate curve is the flat portion which shows a relatively low and fairly constant rate of failures. The level of this constant rate region is largely package dependent. It is in this region that the devices experience high numbers of on-off cycles and extended use periods of full thermal loading. Intermetallic growth and corrosion failures related to moisture permeation under the passivation layer, moisture ingress along the interfaces, package cracks or chip cracks are the common causes of failure during this period. This is the time when the product is in the customer's possession

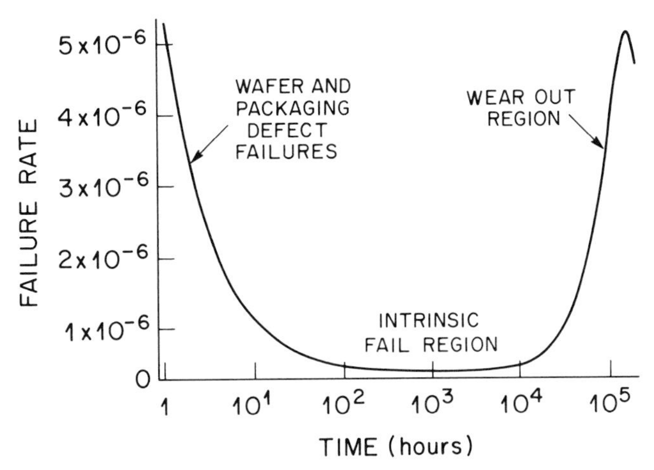


**FIGURE 8-1** Failure rate versus time showing the characteristic bathtub shape. (After Reference 2.)

and well before the expected lifetime of the system. A relatively high failure rate here is responsible for system failures and the attendant perception of poor quality. Performance-driven applications where reliability is a high priority such as military and medical apparatus have often turned to ceramic packaging to reduce the failure rate in this region. The advantage in reliability enjoyed by ceramic packages has been reduced significantly, however.

The final region of the curve is the segment of rapid upturn in failure rate as the packaged devices reach their expected lifetime. Mechanical problems such as cracking and inordinate stresses are unlikely to have an important influence on the time of this upturn since those problems would have taken their toll much earlier. The typical cause of this wear-out behavior is the accumulated effect of slower corrosive processes. Although the occurrence of the upturn itself is normal, no one expects a color television or a personal computer to last 30 years, the time of the upturn should exceed the expected lifetime of the system to satisfy the customer's quality expectations. In a large multichip system, the expected lifetime of the individual components should exceed the expected lifetime of the system by several times to assure relatively trouble free operation. The dependence of system performance on device reliability was discussed previously in Section 1.4.2 and Table 1-2.

The fraction of a population that fails within a given time interval $dt$ can be represented by a function $f(t)\,dt$. The cumulative number of failures is then represented by the integral of this function over the time period of interest, which is $t_1$ in the following relations. (*Note*: Some sources use a dummy variable of integration in place of time to avoid confusion with the time variable of the integration limit; $t_1$ is the upper limit of the integration in the following discussion.)

$$F(t) = \int_0^{t_1} f(t)\,dt \tag{8-1}$$

The reliability function is then defined as the fraction of the population that is surviving at time $t$:

$$R(t) = 1 - F(t) \tag{8-2}$$

The instantaneous failure rate, also known as the hazard rate, is then the failure rate $f(t)$ normalized to the fraction surviving $R(t)$:

$$h(t) = \frac{f(t)}{R(t)} = \frac{f(t)}{1 - F(t)} \tag{8-3}$$

Typical failure rates for ICs are so low that a scale factor is often used to translate the rate to a number that is easier to handle and compare. The most common failure rate unit is known as a FIT (*F*ailure un*IT*), which is defined as 1 failure per $10^9$ device hours of operation. Typical target values for ICs used in multichip systems are of the order of 10–15 FIT. Different failure distributions representing different failure mechanisms can now be described through different $f(t)$ functions.

### 8.1.1 Weibull Distribution Function

In a Weibull distribution function, the failure rate varies as a power of time. The cumulative number of failures at a given time is represented by the relation:

$$F(t) = 1 - e^{-(t/\lambda)^\beta} \tag{8-4}$$

The utility of such a distribution is that it describes the failure rate in terms of two parameters, $\beta$ and $\lambda$. $\lambda$ is known as the characteristic lifetime because it is the time at which the fraction of devices that has failed reaches $(1 - 1/e)$, which has the numerical value of 0.632. $\beta$ is known as the shape parameter which characterizes how the failures distribute about the pivot point at $\lambda = 1.0$, where the cumulative number of failures must equal 63.2% of the starting population by definition. When $\beta < 1.0$, the failure rate decreases with time, representative of the early failure period of infant mortality. $\beta = 1.0$ yields a constant failure rate typical of the second segment of the bathtub curve where the device is in use. $\beta > 1.0$ represents an increasing rate of failure characteristic of the wear out period of the device lifetime. The limitation of the Weibull distribution is that it cannot represent all the regions of device failure behavior with a single shape factor. Combining relations with different constants is one approach to minimizing this limitation. A distinct advantage of this two parameter model, however, is that the dependence on the parameters can be linearized so that data can be analyzed efficiently with a graphical method. The axes for this linear representation of the Weibull distribution are derived by taking logarithms as shown below [2]:

$$1 - F(t) = e^{-(t/\lambda)^{\beta}} \tag{8-5}$$

$$\ln(1 - F(t)) = -(t/\lambda)^{\beta} \tag{8-6}$$

$$\ln\left\{\ln \frac{1}{1 - F(t)}\right\} = \beta(\ln t - \ln \lambda) \tag{8-7}$$

Plotting the failure rate $F(t)$ in the form of $\ln\{\ln[1/(1 - F(t))]\}$ yields a straight line whose intercept at $\ln\{\ln[1/(1 - F(t))]\} = 0$ defines $\lambda$, the characteristic lifetime. The slope of the line is the shape factor $\beta$. Figures 8-2 [2] and 8-3 [2] show Weibull plots for several characteristic lifetimes and several shape factors. Real data are unlikely to provide straight lines so the parameters must be estimated with linear regression techniques. The utility of such distributions, however, is the ability to quantify, characterize and compare failure rates so as to suggest failure mechanisms and track improvement strategies.

### 8.1.2 Duane Plotting

Duane plotting [3] is often applied to obtain a quick estimate of device failure rates when the number of failures is small. Duane plotting begins with the premise that the average failure rate can be represented by some power of time.

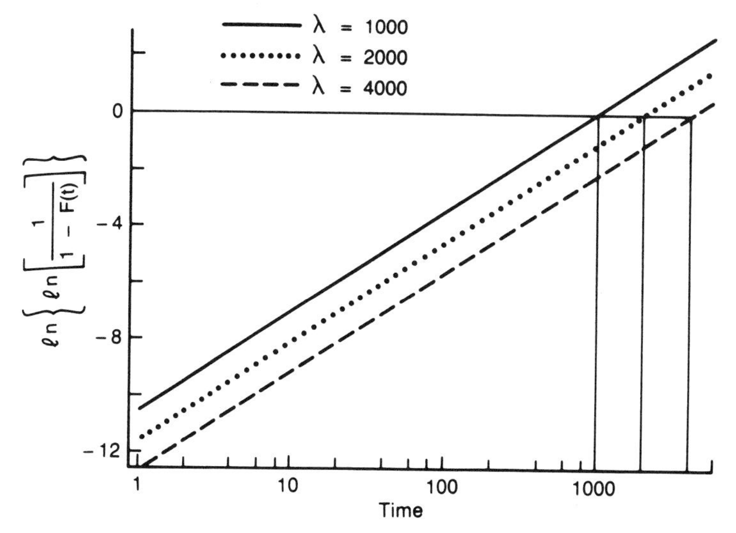


**FIGURE 8-2** Weibull plot showing the failure rate for several different characteristic lifetimes represented by the variable λ [2].

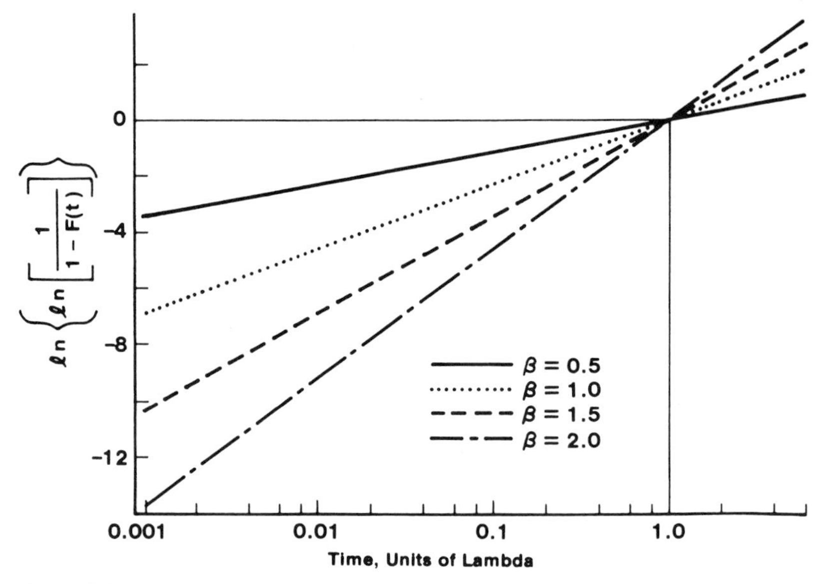


**FIGURE 8-3** Weibull plot showing the failure rate for several different shape factors of the distribution represented by the variable $\beta$ [2].

$$\text{Average Failure Rate} = \frac{F(t)}{t} = Kt^{-S} \tag{8-8}$$

and

$$F(t) = Kt^{1-S} \tag{8-9}$$

$$f(t) = K(1 - S)t^{-S} \tag{8-10}$$

When the log of the average failure rate is plotted against log time, the data should fall on a straight line with negative slope $S$ and shift parameter $K$. The failure rate is $h(t) = f(t)/(1 - F(t))$, but since $F(t)$ is $\ll 1$ the failure rate is approximately equal to $f(t)$:

$$h(t) = K(1 - S)t^{-S} \tag{8-11}$$

or

$$h(t) = (1 - S)(\text{Average Failure Rate}) \tag{8-12}$$

Users of the Duane plot must beware of extrapolating $h(t)$ too far from the initial failure zone where $F(t) \ll 1$ may no longer be valid. This type of extrapolation will result in a severe underestimation of failure rate in the steady state and wear-out failure regions [4].

### 8.1.3 Log-Normal Distribution

Another type of distribution function used to analyze failure data for IC devices is known as the log-normal distribution. In this approach, the logarithms of the time to failure are considered to follow some normal distribution. The algebra for this distribution method is more complicated than the Weibull or the Duane plot, so only the results are provided here. (A thorough review of statistical treatment of reliability data is available [5].) The cumulative distribution function $F(t)$ is expressed in terms of the median time to failure $t_{50}$ which is further defined as an exponential function of the single parameter $\mu$.

$$t_{50} = e^{\mu} \tag{8-13}$$

A further scaling parameter is then defined from $t_{50}$ and an equivalent timescale $t_{16}$ when 15.866% of the devices have failed:

$$\sigma \cong \ln\left(\frac{t_{50}}{t_{16}}\right) \tag{8-14}$$

The cumulative distribution function is then:

$$F(t) = \frac{1}{\sigma\sqrt{2\pi}} \int_0^{t_1} \exp\left\{-\frac{1}{2}\left(\frac{\ln t - \mu}{\sigma}\right)^2\right\} \tag{8-15}$$

The cumulative distribution function for the log normal distribution is not as amenable to numerical computations as the Weibull or the Duane plot, nor is it straightforward to convert log-normal data to the failure function $h(t)$ or the more convenient expression of $h(t)$ known as FITs. Goldthwaite [6] developed a graph to convert from the log-normal parameters $\sigma$ and $t_{50}$ to the instantaneous failure rates $h(t)$ and FITs. This plot, known as a Goldthwaite plot, is shown in Figure 8-4.

## 8.2 PACKAGE FAILURES

Package related failures are often associated with corrosion caused by excessive thermal loads or moisture ingress due to package cracks, delamination, or passivation layer cracks. The package and passivation layer cracks are mostly caused by excessive thermal shrinkage forces. In addition, device performance can be affected in the absence of specific mechanical failures. Device parameters such as small signal current gain in bipolar devices and transconductance in MOS devices are most affected by large mechanical stresses [7]. One means to reduce stress-induced parameter changes and failures is to actively manage thermal shrinkage forces in the package design, materials selection, and process parameters.

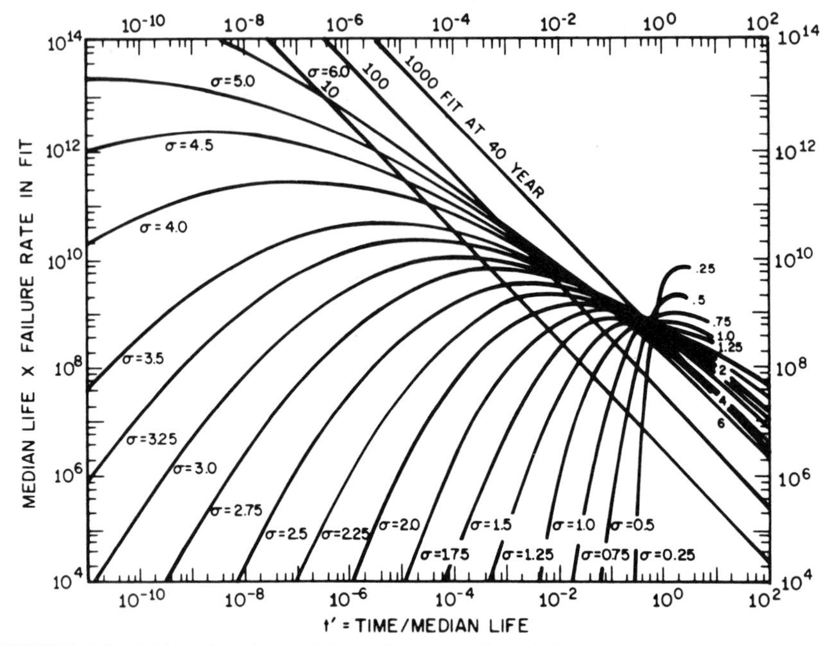


FIGURE 8-4 Goldthwaite plot, which provides normalized failure rate versus normalized time for the log-normal distribution. Lines of constant FIT rate are straight lines of slope − 1 on this plot [4].

### 8.2.1 Shrinkage Stress Management

The shrinkage stresses in a molded plastic package result from the disparities in the coefficients of thermal expansion among the various materials that are contained within the package. These materials are all in intimate contact, thereby transmitting the forces across the interfaces in a way that cannot occur in a premolded plastic or ceramic package where the device often resides in a vacated cavity. The magnitude of the thermal shrinkage forces can be estimated through the following stress-strain relationship which shows the dependencies on the coefficients of thermal expansion, the tensile modulus and the temperature excursion.

$$\sigma = K \int_{T_1}^{T_2} \frac{\left(\alpha_p(T) - \alpha_i\right)}{\dfrac{1}{E_p(T)} - \dfrac{1}{E_i}}\, dT \tag{8-16}$$

In the above expression, the $p$ subscript denotes the polymeric based molding compound and the $i$ subscript denotes the inorganic material which can be the silicon device or the metal leadframe. $K$ is a geometric constant for the specific design. The temperature limits represent the full extent of the temperature excursion which for most temperature cycling tests are likely to extend past the glass transition temperature of the molding compound. Both the coefficient of thermal expansion (CTE) and the tensile modulus of the molding compound change significantly at the glass transition temperature so it is important that the complete temperature dependencies of these two parameters are known. The expansion coefficient and modulus of the inorganic material can be assumed to be temperature independent. In most cases of high modulus inorganics such as metals, silicon or ceramic, a simplification can be made by realizing that $1/E_i$ is much smaller than $1/E_p$, so the inorganic material term has very little effect on the magnitude of the denominator. A further assumption can be made by realizing that both the CTE and the tensile modulus of the molding compound are relatively constant throughout the glassy region, change significantly at the glass transition temperature and then remain relatively constant through the rubbery region. This behavior obviates the need for complete temperature dependent data for the CTE and the tensile modulus, replacing these dependencies with single values for glassy and rubbery regions. The integral is then approximated with a difference equation which has terms for the stresses generated above the glass transition temperature in the rubbery region (denoted with an $r$ subscript) and in the glassy region (denoted with a $g$ subscript):

$$\sigma = K(\alpha_{pg} - \alpha_i)E_{pg}(T_g - T_1) + K(\alpha_{pr} - \alpha_i)E_{pr}(T_2 - T_g) \quad (8\text{-}17)$$

Yet another consideration in gauging shrinkage-induced forces is the concept of an anchoring temperature, the temperature at which the molding compound effectively exerts force on the internal structure of the packaged device. In most cases, this temperature is taken to be the glass transition temperature based on the assumption that stresses incurred from the high temperature endpoint to the glass transition temperature are insignificant. This assumption permits dropping the second term of Equation (8-17) to provide the form used in most comparison studies:

$$\sigma^* = (\alpha_{pg} - \alpha_i)E_{pg}(T_g - T_1) \quad (8\text{-}18)$$

In this relation, the geometric constant $K$ has been dropped. The computed quantity is no longer the actual stress but rather a stress parameter denoted by $\sigma^*$. The stresses, represented to different degrees of rigor in Equations (8-16) to (8-18), can and do cause damage to the microelectronic device. This damage manifests itself in several forms that are summarized in Figure 8-5. Each of these types of damage are discussed in the following subsections. In each case, the material and process aspects of the failure are highlighted. Surprisingly,

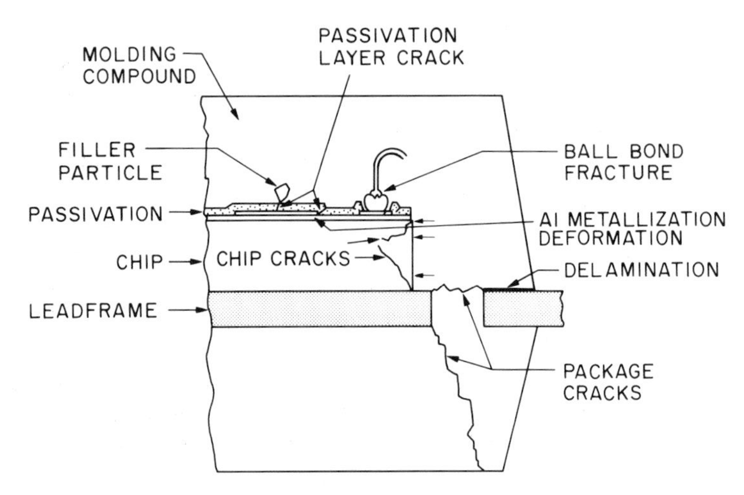


**FIGURE 8-5** Different mechanical stress-induced failures that can be found in a plastic packaged integrated circuit device.

many thermal shrinkage stress problems are affected by the processing, particularly the flow of the molding compound into the cavity and its subsequent cure. Although the stress parameter defined in Equation (8-18) is a metric of stress for a particular combination of materials, it does not include design parameters and interfacial phenomena. Although this relation can be useful for ranking molding compounds, it is unlikely to be of much value in predicting which materials will fail.

### 8.2.2 Package Cracking

Package cracking results when the stresses generated by the constrained shrinkage of the molding compound exceed the strength of the molding compound. The magnitude of the shrinkage forces depend on the package geometry, the CTEs of molding compound, leadframe and die, and the temperature excursions as related in Equations (8-16) to (8-18). Although this relation is adequate to monitor the degree of the stresses and to guide the selection of materials, it cannot indicate if the package will crack in a specific application. Mechanical analysis using a numerical technique such as finite element analysis (see Section 8.4) can provide a mapping of the stress levels throughout the package, but it is still difficult to predict the occurrence of cracking. The difficulty results from the inability of mathematical techniques to adequately represent catastrophic and stochastic phenomena such as cracking failure. Finite element analysis is discussed in greater detail in Section 8.4, whereas the intent in this section is to describe the design and process considerations in minimizing the propensity of cracking.

The presence of stress concentration points within the package significantly increases local stress levels. A severe stress concentration can magnify essentially any stress to above the strength of the molding plastic. In molded plastic packages, the most common stress concentration points are the edge of the chip and the edge of the die support paddle as is shown in Figure 8-5. In the event of cracking, a stress concentration point is invariably at the crack origin site as can be seen in the micrograph shown in Figure 8-6. Any radius that can be added to these edges will reduce the degree of stress concentration and thereby improve the crack resistance. The manufacture of the leadframe influences the edge sharpness to the point that it may be worth investigating whether a particular stamping or etching process provides a duller corner. There is not much that can be done about the sharp edge of the silicon device, although different saw and blade combinations may provide less sharp edges than others. Yet another form of stress concentration derives from poor adhesion along the molding compound interfaces with the die and the leadframe. With complete adhesion between the molding compound and the leadframe or die, there is less stress concentration at the corners because of the deformation and strains in-

**FIGURE 8-6** Micrograph depicting a crack originating at the stress concentration point at the sharp corner of the bottom surface of the paddle support. (Courtesy of A. Lustiger, AT&T Bell Laboratories).

duced over the adhered length. The adhesion provides a horizontal restraint to the horizontal shrinkage stresses for the geometry depicted in Figure 8-5. In the absence of adhesion, there is no horizontal restraint along the surface of the die, so the forces are transmitted directly to the edge as is shown in Figure 8-5. The magnitude of the stresses can be several times greater in the case of partial or complete loss of adhesion as was shown in Figure 4-17 [8]. This underscores the importance of making adhesion a selection criteria for molding compounds as was discussed in Section 4.3.2.

Mechanical interlocking can also be used to prevent loss of adhesion with the molding compound. This is often realized by including holes in the leadframe that mechanically lock the molding compound to the frame with material that flows into the hole. The most common position for these leadframe holes is in the short lead stubs that are just inside the dam bars as is shown in Figure 8-7. These pins prevent lateral movement of the molding compound over the lead that would break the adhesive bond leading to stress concentration and rapid moisture ingress. Small tabs on the sides of the leads are also used to promote mechanical locking of the lead into the molding compound as is also shown in Figure 8-7. The tabs are more common on thin leads which are not wide enough to accommodate a hole. Another possible location for interlock holes is on the bottom surface of the die attach paddle. Delamination is a common problem along this surface resulting in cracks at the pad edges.

**8.2.2.1 Package Cracking Due to Surface Mount Attachment** There is a type of package cracking that is caused specifically by surface mount attachment

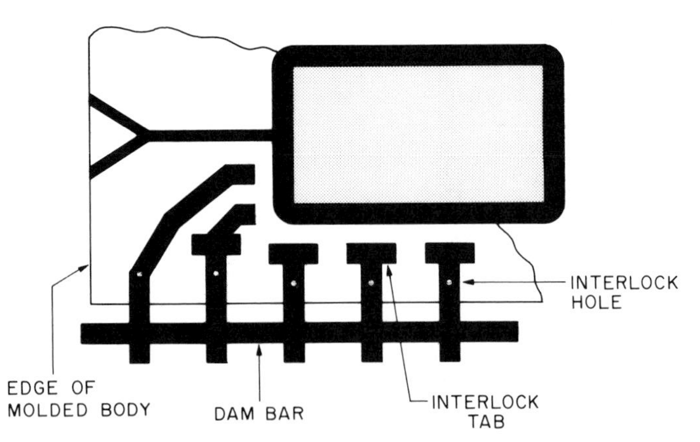


**FIGURE 8-7** Interlock tabs and holes in the leads intended to provide improved attachment of the molding compound to the lead, thereby preventing delamination.

of the component to the circuit board [9, 10]. Brief summaries of the attachment options for surface mount components are presented below:

1. IR Reflow. This consists of prolonged exposure of the circuit board with components to high temperature induced by absorption of IR radiation. It is a high-productivity method since all components on the board are attached simultaneously. The peak temperature varies with thermal mass distribution.
2. Vapor Phase Reflow. Exposure of circuit board with components to high temperature solvent vapor, usually a chlorofluorocarbon (CFC). It is also high productivity since all components on the board are attached simultaneously.
3. Solder Dipping. The entire circuit board and components are dipped in molten solder. High productivity, since all components on board are attached simultaneously.
4. Laser Heating. A laser beam, typically a YAG laser, is used to heat only the lead attach area. Low productivity since it is conducted one lead at a time.
5. Thermode Pulse Heating. A heated metal block formed to the lead attachment pattern contacts the leads of an individual package. A current pulse may also be used to rapidly reflow the solder. Productivity is low because it is a one component at a time attachment method.

Several of these methods expose the entire molded body to the solder temperature for periods of 30 seconds to 5 minutes. The wave solder technique used with through-hole packages exposes only the leads of the package to the solder temperature, so the introduction of surface mounting was the first time that the entire molded body was elevated to such temperatures, yet there have not been significant changes in the molding compounds to allow them to withstand this exceptional thermal loading. Solder dipping imparts the greatest stress on the molded body because it provides for the most rapid heating being a liquid immersion process. The other global heating techniques of IR reflow and vapor phase soldering also impart significant thermomechanical stresses. The last techniques, laser heating and thermode heating, impart the least thermal-induced stress since they are equivalent to the localized heating experienced in wave solder attachment of through-hole packages. Unfortunately, these localized heating techniques have far lower production rates than dipping, IR, or vapor phase reflow so that they are not commercially important at the time of this writing.

The mechanism of package cracking on reflow solder attachment is related to moisture absorbed within the molded body, both dispersed throughout the volume and adsorbed at the interfaces. This water is absorbed soon after manufacture if the packages are not kept in a humidity controlled environment. The molding compound is permeable to moisture even in the absence of cracks or delaminated interfaces. On heating to the reflow temperature above 200°C, the plastic package softens rapidly since it is at least 30–40°C above its glass transition temperature. The absorbed water, some of which may be in the form of clustered water in the pores of the molded body or at the interfaces, vaporizes and expands. A particular problem has been the delamination of the molding compound from the bottom of the die support paddle as is shown in Figure 8-8 [9]. This expansion severely distorts the package outline, popping it out and inviting the common label of "popcorn problem." Cracking can occur in the expanded state with the tearing of the molded body at the lower corners of the pad support, or it can crack on cooling and shrinking back to its normal dimen-

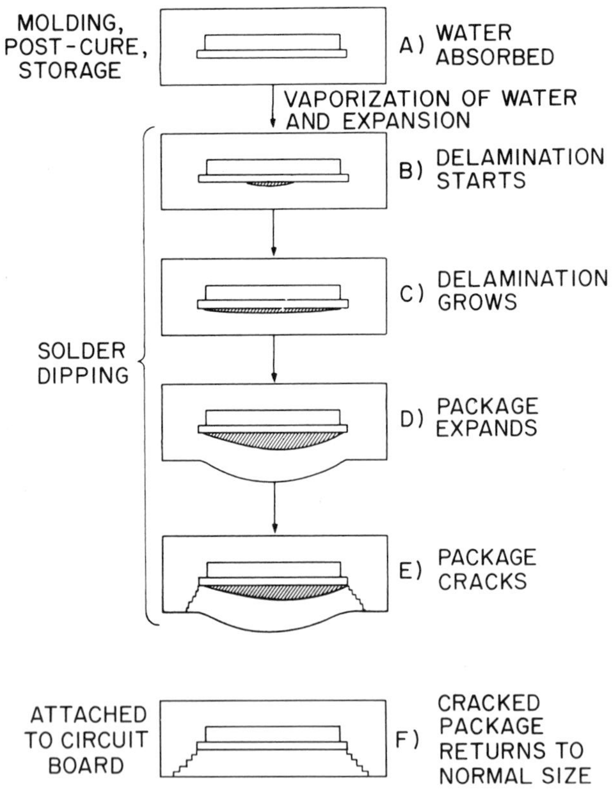


**FIGURE 8-8** Mechanism for moisture-induced package cracking on surface mount attachment. (After Reference 9.)

sions. The lower pad surface is completely delaminated in most cases, effectively concentrating the thermal shrinkage forces on the paddle edges.

There are several approaches to relieving reflow solder cracking problems. One simple solution is to store the molded devices under desiccant until the time they are attached to the circuit board. Although this is feasible, it is generally undesirable because it imposes significant restrictions on component handling in the factory and reduces efficiency. Drying of the components just before attachment could also relieve the problem, but it is also difficult to implement. Approaches that prevent the problem through modification of molding compound are preferred because they impose little or no restriction on the downstream circuit board assembly steps. One approach is to reduce the moisture absorption of the molding compound. This is difficult because of the inherent hydrophilic character of epoxy materials which can absorb as much as 3% by weight of water. A material with a significantly lower glass transition temperature absorbs less moisture than one with a higher $T_g$ [12] because it has less free volume. The fraction of free volume in a polymeric material changes with temperature but is the same for all materials at their glass transition temperature.

Another approach is to significantly increase the glass transition temperature of the molding compound until it nears the solder reflow temperature. Epoxies are capable of reaching this level of glass transition temperature so this approach is also feasible. The problem is in maintaining the low stress properties that are also important in surface mount components. The stress level goes up as the anchor temperature (glass transition temperature) increases as was shown in Equation (8-18). Yet another approach would be to keep the glass transition temperature below the solder reflow temperature to minimize stress, but increase the modulus of the molding compound in the rubbery region so that the material does not sustain enough vapor-induced deformation to damage the package. Increasing the strength above $T_g$ would prevent tearing as well. Improving the adhesion of the molding compound to the bottom die pad surface is an effective route to preventing the delamination there and preventing damage to the package [11]. Although no clear choice for a remedy has emerged at the time of this writing, it appears that most molding compound vendors are using one or more of the above approaches to alleviate the problem.

### 8.2.3 Passivation Layer Cracking

Passivation layer cracking [13, 14] occurs when the stresses imposed by the shrinkage of the molding compound exceed the strength of the fragile passivating material. It is difficult to determine whether phosphosilicate glass (P-glass) or silicon nitride is more prone to this problem because there have not been comparative studies published on the same die in the same package. Passivation layer cracking is often associated with ball bond lift-off and ball bond shearing because of the close proximity of these structures to the passivation layer damage on the edges of the chip. The use of a compliant cover coat material such as silicone rubber significantly decreases the stresses transmitted to the passivation layer, but the use of this material is not without complications. The adhesion between the silicone material and the epoxy molding compound is poor and there could be a cavity between the two materials due to the much higher coefficient of thermal expansion of the silicone as was depicted in Figure 3-11. This loss of adhesion concentrates the shrinkage stresses at the chip corner where the edge of the passivation layer is often exposed. Higher incidence of package cracking, passivation layer cracking and ball bond shearing can result when this type of stress concentration exists.

Okikawa et al. [13] have conducted a finite element solution of the stresses within a molded plastic package which includes separate element layers for the metallization and the passivation layer. A comparison of the mechanical properties of the materials at the interface between the chip and the molding compound is provided in Table 8-1. The passivation and metallization layer thicknesses examined were similar to those found on actual devices; passivation thickness of 1 micron and metalization thickness of 1 micron. Computation of the shear stress on the passivation layer showed zero stress at the center, rising non-linearly to a maximum at the chip corner as is shown in Figure 8-9 [13]. This analysis was unique in that it included the effect of the underlying metallization. The probability of cracking is enhanced as the width of the aluminum line under the passivation layer increases. The line widths at which the strength limit of the passivation layer is exceeded for the different positions on the chip are also shown on Figure 8-9. It is apparent from these results that the occurrence of passivation layer cracking increases with chip size and width of the underlying aluminum metallization lines, as well as the other stress related parameters such as the difference in the coefficients of thermal expansion, the molding compound modulus, and the temperature excursion. As in most ther-

**Table 8-1. Mechanical Properties of the Materials at the Device Interface [13]**

| *Material* | *Tensile Modulus* $kg/mm^2$ | *CTE,* $ppM/°C$ | *Poisson's Ratio* |
|---|---|---|---|
| Molding compound | 670–1400 | 18–26 | 0.30 |
| Silicon | 19200 | 2.6 | 0.30 |
| Aluminum | 600 | 23.6 | 0.30 |
| SiN | 6500 | 5.0 | 0.30 |
| P-Glass | 3000 | 5.0 | 0.30 |

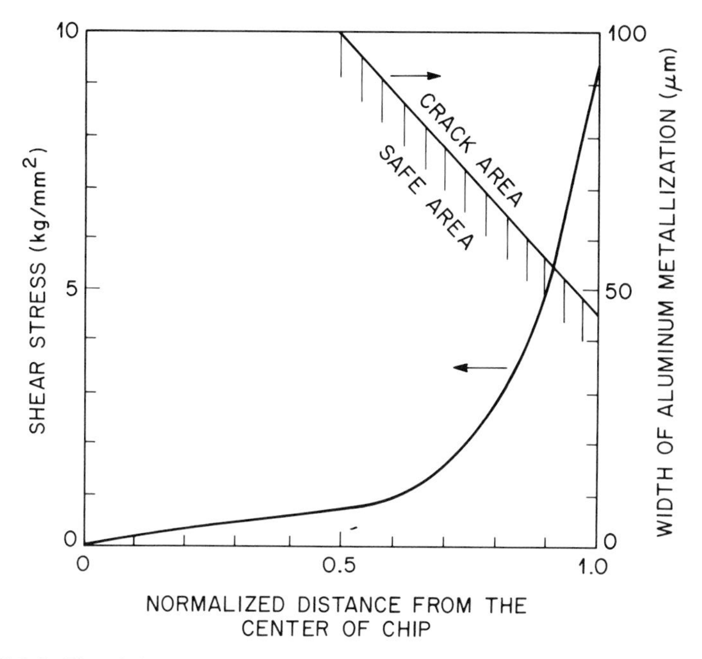


**FIGURE 8-9** The relation between stress over the chip and distance from the chip center. Plotted over these results are the widths of the aluminum metallization where the stresses in the passivation layer exceed the upper crack limit of the passivation. (After Reference 13.)

momechanical stress issues, repeated temperature cycling promotes a greater incidence of cracking.

Several reports [13, 14] have found that the mechanical behavior of the passivation layer can be treated through a simple approximation based on Euler's equation. This analysis assumes that passivation layer cracking can be modeled as the buckling of a long column. The stress required to cause buckling is:

$$\sigma_p = \frac{n\pi^2 E}{L/r^2} \tag{8-19}$$

In this relation, $L$ is the width of the passivation layer, $r$ is its radius of gyration, $E$ is its modulus of elasticity, and $n$ is a constant. If the shape of the passivation layer is a rectangular slab whose width is greater than its thickness, the radius of gyration can be written in terms of the thickness $t$ alone [14]:

$$r = \frac{t}{2\sqrt{3}} \tag{8-20}$$

Substituting Equation (8-20) into Equation (8-19) then yields an expression for the stress required to cause cracking in a passivation layer of thickness $t$ and length $L$:

$$\sigma_p = \frac{0.0835K\pi^2 Et^2}{L^2} \tag{8-21}$$

Okikawa et al [13] reports a value of $K = 4$ from experimental observations of passivation layer cracking, whereas Edwards et al. [14] report a value of $K = 10$ derived from computer simulation work. In general, Equation (8-21) shows that the propensity for passivation layer cracking, indicated by a lower value for $\sigma_p$, is greater with wider and thinner passivation layer dimensions. Thicker, narrower passivation layers are more resistant to cracking.

### 8.2.4 Delamination of Interfaces

Delamination at interfaces between the different materials within the package is a major cause of moisture ingress and subsequent premature package failure. The principal areas for delamination are along the leads extending from the chip to the edge of the molded body, and along the die surface itself. Delamination on the die is particularly troublesome because it also causes significant stress concentration at the edges of the die, as well as deformation of device features due to the unconstrained movement of the molding compound. In some cases of severe separation, the delamination can be detected by X-ray of the edge of the molded body. In most other cases, special analytical tools such as scanning acoustic tomography (SAT) [15] have to be employed. SAT is a non-destructive evaluation method that uses acoustic imaging to show delaminated regions. An example of SAT evaluation of the back surface of the paddle support in a large PLCC package is provided in Figure 8-10. SAT can be used to monitor the growth of a delaminated region with continued temperature cycling. Delamination is most likely to start at the corners of the chip where the stress concentrations exist. The chip edge represents a physical and mathematical discontinuity which can amplify the force to many times the bulk values because of the severely reduced area upon which the force is applied. Present molding compounds can often withstand hundreds of temperature cycles without showing delamination. Once delamination has begun, the line of stress concentration follows the adhesion edge on the chip surface since this is the first constraint to molding compound movement encountered. The progression of delamination with temperature cycling is illustrated in Figure 8-11 [8].

### 8.2.5 Metallization Deformation

The aluminum metallization on the device can be deformed by the shrinkage forces imposed by the molding compound [16, 17] as is shown schematically

**FIGURE 8-10** Scanning acoustic tomography (SAT) image of the back surface of the paddle support for a 100 PLCC package. The image shows delamination over a large fraction of the paddle area. (Courtesy of A. Lustiger, AT&T Bell Laboratories).

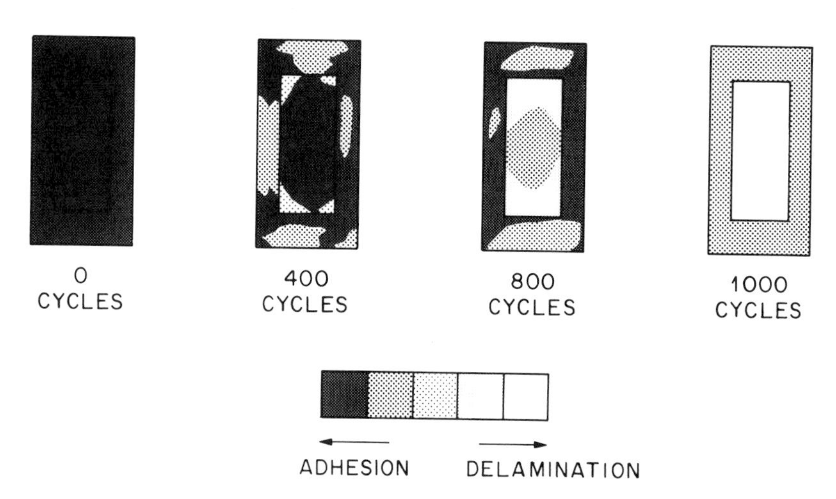


**FIGURE 8-11** Mappings of the adhered and delaminated regions on the surface of a large chip with increasing number of temperature cycles. The results are drawn from the color plates provided in Reference 8.

in Figure 8-12. Damage most often occurs at the edges of the chip, furthest from the zero tangential stress point in the center. The mechanism of the deformation is the relative movement of the molding compound over the device surface due to disparities in the coefficients of thermal expansion. The full extent of this deformation in the absence of any constraints can be surprisingly large. An estimate can be determined by multiplying the difference in the coefficients of thermal expansion of silicon and molding compound by the temperature difference and one half of the length of the chip as is shown in Equation (8-22) below. In this relation, $\alpha_p$ denotes the CTE of the molding compound, whereas $\alpha_{Si}$ denotes the CTE of the silicon device.

$$\begin{aligned} \Delta L &= (\alpha_p - \alpha_{Si})(175° - 25°)(L) \\ &= (15 \times 10^{-6}/°C)(150°C)(3000\ \mu) = 6.8\mu \end{aligned} \tag{8-22}$$

Aluminum line deformations of the order of several microns are often found, although it would be unusual to achieve the entire theoretical limit shown in Equation (8-22). This is because there are some constraints that operate, and these provide the best route to minimizing the problem. The molding compound

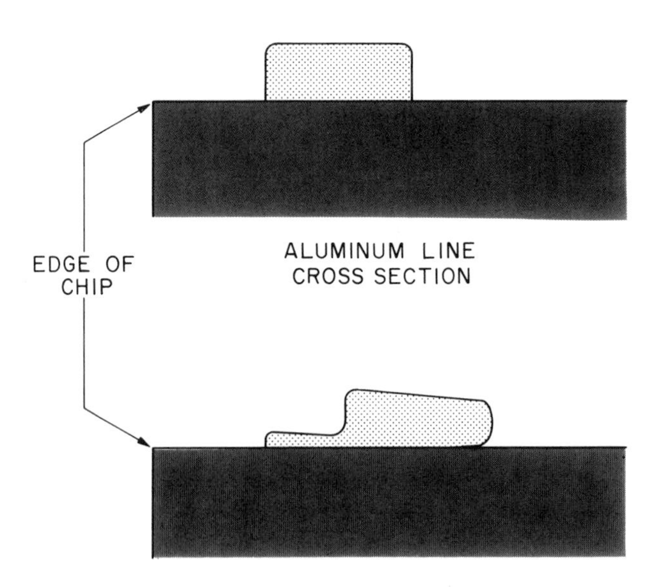


**FIGURE 8-12** Aluminum metallization deformation that can be sustained during shrinkage of the molded plastic package.

is constrained in its relative movement over the chip by its adhesion to the chip surface, therefore the maximum unconstrained deformation shown in Equation (8-22) is rarely observed. Loss of adhesion between the die surface and the molding compound allows relative movement and thereby promotes damage to the aluminum. Package cracking is also often associated with aluminum line deformation for the same reason. Repeated temperature cycling degrades the adhesion and causes micro- and macro-cracking, which allow the lateral movement that results in deformation. Therefore, molding compounds with superior adhesion to the internal materials of the package will provide lower aluminum line deformation. Covering the chip with a compliant material such as silicone rubber can effectively eliminate aluminum line deformation since this compliant layer sustains all of the deformation induced by the molding compound.

The degree of metallization deformation can be quantified using the mechanical properties of the materials involved. One report [17] uses a semi-empirical approach to show that the degree of deformation $\delta$ can be correlated with the distance from the center of the chip $r$, the maximum temperature of the test $T_{\max}$ and the number of temperature cycles $N$:

$$\delta = \left(r - 1.91 \exp\left\{-3.8 \times 10^{-8} N \exp\left(7.2 \times 10^{-2} T_{\max}\right)\right\}\right)\beta \quad (8\text{-}23)$$

In this relation, $\beta$ is a metric of the average stress level on the device, which can be derived experimentally through a stress transducer or estimated from the following relation:

$$\beta = \int_{T_{\min}}^{T_{\max}} (\alpha_p - \alpha_{Si})\, dT \quad (8\text{-}24)$$

Equation (8-24) should not be applied to molding compounds and device geometries other than the ones for which it was derived. It is provided to illustrate the dependence of deformation on molding compound properties, test conditions and the number of test cycles.

### 8.2.6 Ball Bond Liftoff, Shearing, and Fracture

Open circuit failures are often caused by loss of contact between the ball bond and the bonding pad on the chip, a phenomena known as *liftoff*. Liftoff can be the result of either tensile or shearing forces induced by thermal-induced stress. Open failures can also occur by fracture of the ball bond or the attached wire. These failures can occur during molding as a manifestation of flow-induced stress, or as a failure mechanism after molding. Molded-in stresses, thermal-induced stresses, and temperature cycling can fatigue the metallic bond and result in fracture failure that appears in any of the configurations shown in Fig-

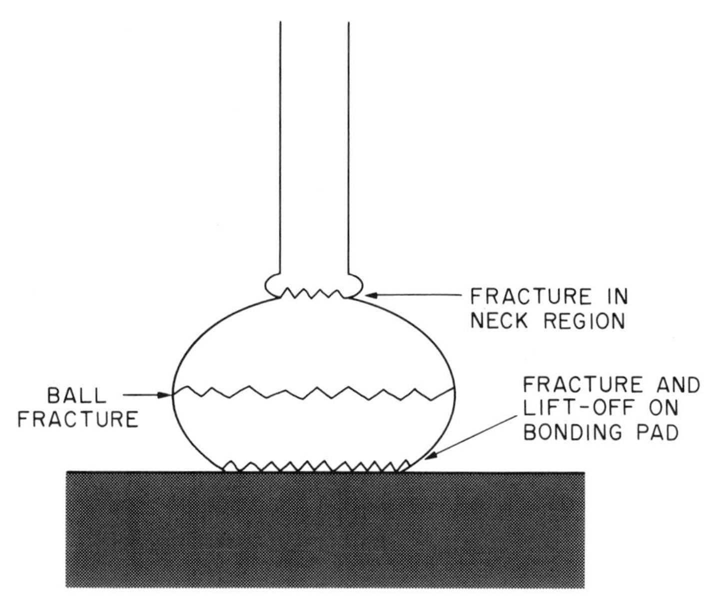


**FIGURE 8-13** Types of ball bond failures that can be caused by excessive flow-induced and thermomechanical stresses.

ure 8-13. A micrograph of a device with a single ball bond liftoff failure is shown in Figure 8-14. As in the case of metallization deformation and passivation layer cracking, the presence of package cracks or delamination in the vicinity of the ball bond greatly enhances the chances of incurring a ball bond fracture or liftoff. These failures are largely a manifestation of the thermal shrinkage forces within the molded package, and can be minimized by reducing these stresses as was described in the previous sections of thermomechanical-induced failures. These include closer matching of the coefficients of thermal expansion among molding compound, leadframe and device, minimizing the anchor temperature, and optimizing the design to reduce the stress concentration sites and the stress gradients. In addition, molding compounds that have excellent adhesion and high strength can help reduce mechanically-induced ball bond failures.

### 8.2.7 Test Chip Evaluation of Thermomechanical Stresses

A test chip can be used to assess the degree of thermomechanical-induced stress within the molded package. Test chips from different manufacturers will be configured differently but all will include some type of strain gauge structure to measure stress levels. Other features can include structures to monitor metal-

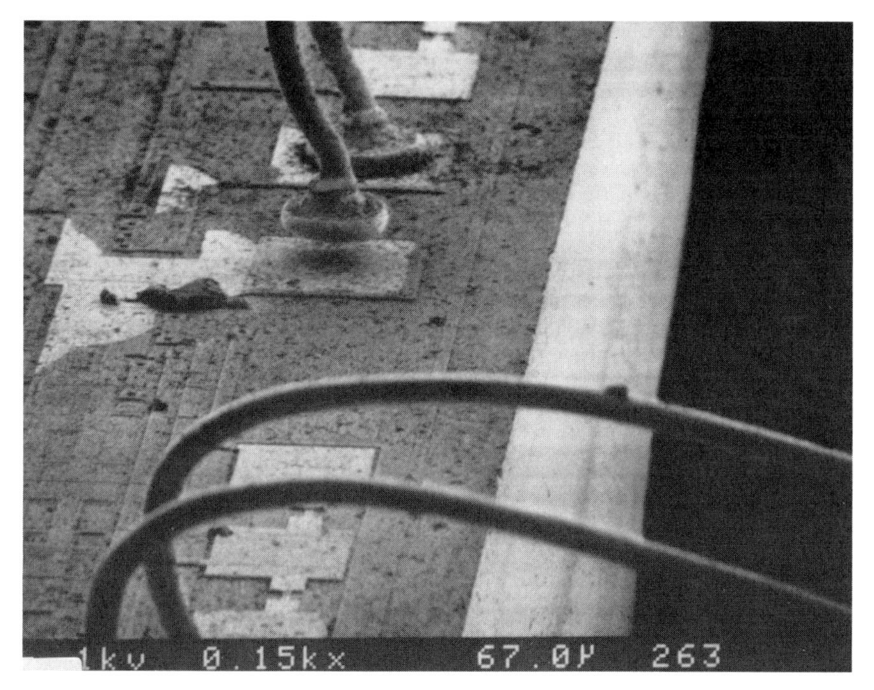


**FIGURE 8-14** Micrograph of a device that has a single ball bond liftoff failure induced through temperature cycling. The micrograph was of the bare surface of the device after the molding compound was removed by de-encapsulation. (Courtesy of A. Lustiger, AT&T Bell Laboratories.)

lization movement, passivation layer cracking and the effects of moisture. With careful calibration the stress levels that induce package cracking, passivation layer cracking and metal line movement can be determined. The issue with using test chips to evaluate molding compounds and package designs is the question if live chips are preferable. It is unlikely that this issue will ever be resolved. Test chips offer a standardized test that can be used over several generations of molding compounds and package designs. Live chips probe the effects of packaging on state of the art feature sizes and structures, they can uncover heretofore unknown packaging problems, and their reliability is the ultimate goal of any evaluation program.

### 8.2.8 Effect of Molding Parameters on Thermomechanically-Induced Failures

Several of the thermomechanical failures described above are affected by the molding parameters. Vestiges of the flow-induced stresses sustained by the wires and the device during cavity filling are retained in the molding compound long after molding and disadvantage these structures in service and in accelerated testing. The implications are that the molding process and molding parameters can be manipulated to improve component reliability. Several types of failures and the associated molding phenomena that affect them are listed below:

*Paddle Shift* The symmetry of the package is compromised in the presence of paddle shift causing the stress gradients to be compressed and increased. The chip is moved closer to the surface of the package, shortening the moisture diffusion length. The displacement of the paddle stretches the wire bonds and places greater stresses on the ball bonds, improving the chances for ball bond fractures. Paddle shift is most often realized with the paddle edge away from the gate sustaining the greater displacement. A mapping of the defects would indicate a higher incidence of failure in the pin numbers away from the gate when this type of paddle shift is affecting reliability. There are many other molding related issues that also bias the failure mapping away from the gate, so this defect profile is not associated exclusively with paddle shift.

*Wire Sweep* The wires often remain stressed after sustaining the flow-induced stress during molding, placing significant stresses on both the ball bond on the chip and the wedge bond on the lead. The term wire sweep is usually used to denote visible wire deformation, typically a lateral movement in the direction the molding compound flowed through the cavity. Under this lateral deformation, both types of bonds can develop kinks at the attachment point as the wire is pulled off the axis of the bond. This type of deformation is most often noted on the sides of the chip facing the gate. A failure rate bias on these pin numbers could indicate excessive wire bond stress imparted by the mold filling. There can be enhanced failure rates due to flow-induced stresses on the wire bonds in the absence of noticeable deformation. This occurs when the molding compound stresses the wires along the axis of the ball bond by flowing up through the wires as intended. The molding compound billows out the wires like wind filling a sail, but in some cases the stresses can be excessive and cause immediate or latent ball bond lift-off. The incidence of these failures is often greater away from the gate. The origin of this bias is explained in the next section.

*Defect Location Bias Away from the Gate* Defect mapping can indicate that failures are biased either near or away from the gate. Biases near the gate are often attributable to high molding compound velocity and jetting that occur before the flow field has had an opportunity to stabilize and spread over the entire flow cross section. A bias of high defect density away from the gate is more difficult to understand. One explanation may be that the chip and paddle are at a lower temperature than the sides of the cavity and the leads. The cavity walls are at the mold temperature and the leads are tightly clamped to these surfaces. The chip and paddle are not as well thermally interconnected since the pad assembly is supported by anywhere from two to four exceptionally thin cantilever beams which span the entire distance to the center of the package. The pad assembly has a large thermal mass compared to the thin leads requiring a much longer time to heat. The consequence of this physical configuration is that the chip and paddle can be at a lower temperature than the cavity and leadframe, particularly if the leadframes are not preheated adequately in a soft-close interval or in a preheater. The molding compound can heat to near the mold temperature when it flows through the gate and along the leadframe, but on encountering the colder pad assembly the temperature of the molding compound begins to drop and its viscosity increases. Therefore, the molding compound that flows up through the wires on the gate side of the paddle has a lower viscosity than the molding compound flowing through at the far side of the paddle as is illustrated in Figure 8-15. This results in greater flow-induced forces on the wires away from the gate.

## 8.3 THERMAL MANAGEMENT

The corrosion processes responsible for many of the failures found in IC packaging are chemical processes which are temperature dependent. Higher storage

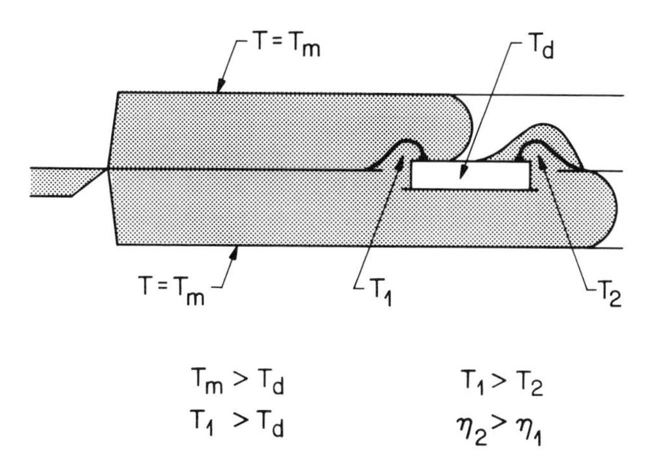


**FIGURE 8-15** Flow effects that can promote a higher defect density away from the gate of the cavity for the situation when the die and paddle assembly did not reside in the heated mold long enough to reach the mold temperature. The molding compound temperature drops and its viscosity increases as it flows around the colder die and paddle, placing greater stress on the wires on the downstream side of the chip. In this figure $T_m$ is the mold temperature, $T_d$ is the temperature of the die and paddle assembly, and $T_1$ and $T_2$ are temperatures of the molding compound at the points indicated.

and use temperatures will accelerate the reactions and shorten the time to failure. Higher temperatures also have a significant effect on the operation and lifetime of the transistor junctions. Increasing the temperature of the junction results in an increase in the reverse voltage, which causes a further increase in power dissipation. This cycle continues until the temperature difference between the transistor junction and its environment reaches some level where the heat transfer, which is proportional to the temperature difference, can match the rate of heat production. Obviously, in cases of poor heat transfer the equilibrium temperature can be high. The effect of thermal stress on transistor junctions was reviewed by Lang et al [18] and Anderson et al. [19]. Higher operating temperatures increase the failure rates of essentially all components within an IC device as depicted in the listing of acceleration factors for hybrid circuit elements shown in Table 8-2 [20, 21, 22]. Overall, the failure rate of a typical IC increases with an exponential dependence that is illustrated in Figure 8-16 [22, 23] which shows the thermal acceleration factor for digital bipolar devices. Other types of integrated circuits show the same overall behavior. The conclusion from these data is that it is important to maintain low operating temperatures to promote extended device and system lifetime, and that the acceleration factor can be highly nonlinear for some device features such as intermetallic bonds.

### 8.3.1 Heat Transfer Analysis

In many cases it is necessary to conduct heat transfer analysis to ensure that the operating temperature at the chip surface does not exceed the extended lifetime limit for the particular device elements; this temperature is considered to be 75°–85°C for most silicon ICs. Several simple approaches to heat transfer analysis will be discussed here to introduce the issues and dependencies in thermal management. The introduction and widespread use of more sophisticated finite

**Table 8-2. Thermal Acceleration Factors for Various Hybrid Circuit Elements [20, 21, 22]**

| *Circuit Element* | *25°* | *50°* | *75°* | *100°* | *125°* |
|---|---|---|---|---|---|
| Au–Al ball bond | 1 | 4 | 20 | 100 | 1200 |
| Au–Au wedge bond | 1 | 1 | 1 | 1 | 1 |
| Al–Al wedge bond | 1 | 1 | 1 | 1 | 1 |
| Thick-film resistor | 1 | 2 | 3 | 4 | 5 |
| Chip capacitor | 1 | 1.5 | 2.5 | 6 | 25 |
| Low-power trans. chip | 1 | 3 | 9 | 27 | 70 |
| High-power trans. chip | 1 | 2 | 6 | 18 | 54 |
| SSI circuit (25 gates) | 1 | 1.8 | 9 | 41 | — |
| LSI circuit (100 gates) | 1 | 1.8 | 9 | — | — |

—indicates no data available.

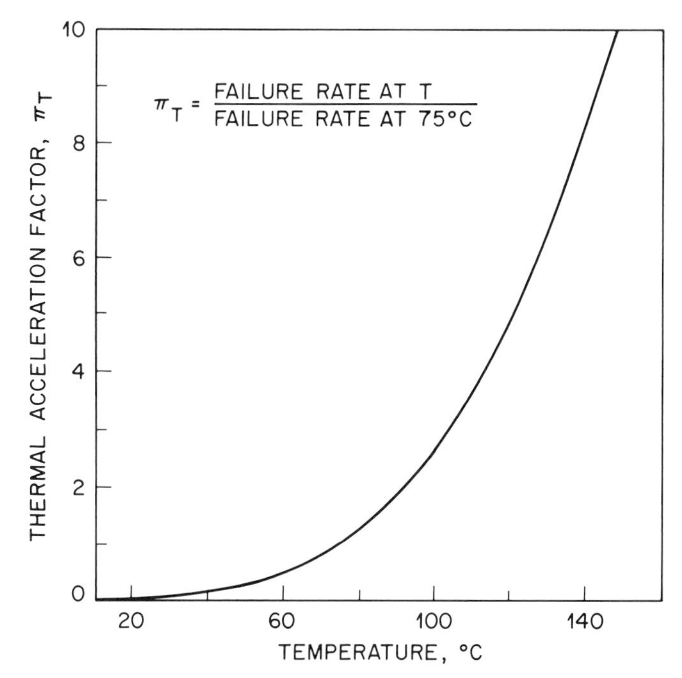


**FIGURE 8-16** Thermal acceleration factor for failure of bipolar digital devices. (Reprinted with permission from Hemisphere Publishing Corporation, NY, *Thermal Analysis and Control of Electronic Equipment*, 1983 by A. D. Kraus and A. Bar-Cohen, [22].)

element methods for thermal analysis has largely supplanted these simpler, approximate methods for estimating operating temperatures and optimizing thermal management, however.

Heat transfer from the surface of the chip to the outer surface of the plastic package is dominated by conduction through the molding compound and along the leadframe. This type of heat transfer is amenable to analysis since convective transport through fluid motion is not a consideration. The heat transfer from the surface of the package to the environment does involve convective transport and is therefore more difficult to treat rigorously. A simple approach is to represent both the conductive heat transfer through the package volume and the convective heat transfer to the ambient atmosphere as coupled resistances in series [24]. As an example, Figure 8-17 shows the cross section of a plastic packaged device represented as a composite wall where each section of the wall has a specific thermal conductivity and thickness. The first surface is the surface of the chip which has a heat production rate of $q$. the bulk ambient temperature $T_b$ is known, the temperature gradients throughout the package and at the surface of the device are sought.

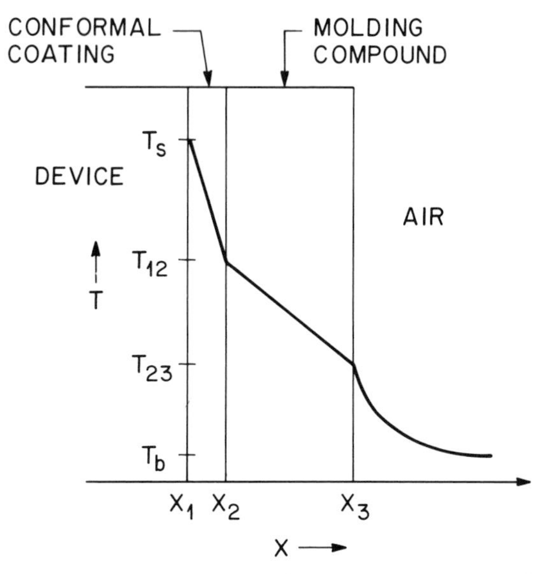


**FIGURE 8-17** Heat transfer through a plastic molded package as conduction through a composite wall where each section of the wall has a separate thickness and thermal conductivity. There is a heat generation term $q$ at the surface of the device and convective heat transfer represented by a heat transfer coefficient at the surface of the package.

At steady state, the heat flux through all sections of the slab and from all interfaces will be the same, allowing us to write:

$$q = q_{12} = q_{23} = q_{34} \tag{8-25}$$

The heat flux across any section of the composite can be related to the product of the temperature gradient through the section and the thermal conductivity:

$$q = k_i \frac{dT}{dx} \tag{8-26}$$

The heat flux from the interface with the environment can be related to the temperature gradient across the interface through the heat transfer coefficient $h$:

$$q = hA\Delta T \tag{8-27}$$

Combining Equations (8-25) to (8-27) provides the following relation which can be used to illustrate the dependence of chip operating temperature $T_s$ on the heat production rate of the chip, the thicknesses of the package layers, their thermal conductivities, the interfacial heat transfer coefficient, the area for convective transport and the ambient temperature $T_b$:

$$T_s - T_b = q\left(\frac{1}{hA} + \frac{x_1 - x_2}{k_{12}} + \frac{x_2 - x_3}{k_{23}}\right) \tag{8-28}$$

This relation includes most of the primary considerations in thermal management of plastic packages. The device operating temperature $T_s$ depends on the ambient temperature that is available: lower-temperature environments promote lower operating temperatures, hence the use of refrigeration units in some high-density packaging schemes. It is also inversely proportional to the thermal conductivities of the components of the package; high thermal conductivities of the molding compound and any other chip coating promote lower operating temperatures for a given thermal loading. This underscores the need to develop high thermal conductivity molding compounds that can accommodate chips with higher heat outputs. The issue of thermal management of new high power chips will be further addressed in Section 9.1.2. Equation (8-28) shows that the device operating temperature is proportional to the thicknesses of the slab sections: thinner packages can dissipate more heat than thicker packages and this dependence is also linear.

The chip surface temperature is inversely proportional to the heat transfer coefficient on the outer surface of the molded body. This coefficient depends on the velocity and velocity profile of the medium moving past the surface, its physical state (liquid, solid, boiling or condensing), and its own thermal properties. As an example of the effect of heat transfer coefficient on heat dissipation, Figure 8-18 [22] shows the approximate temperature differences attainable as a function of surface heat flux for several different types of convection. Both free convection, where the fluid motion is driven by gravity effects alone, and forced convection, where the fluid motion is driven externally, are considered in this plot. These different modes of convection correspond to different heat transfer coefficient values. Table 8-3 provides a listing of approximate heat transfer coefficients for these different convection modes. The last term in Equation 8-28 that is instructive for our purposes is the area term. A low heat transfer coefficient can be directly compensated for by increasing the area of heat transfer. In many cases this is accomplished by attaching a high surface area radiator to the package itself. This radiator or heat fin assembly, two of which are shown in Figure 8-19, should be made of a good thermal conducting material to bring the high temperatures out far onto the fin surfaces. Low-thermal conductivity materials support a greater temperature gradient along the length of the fins and thereby reduce the overall fin effectiveness since the temperature difference between the fin and the environment is reduced. The heat dissipation needs of the most advanced, high-density interconnections often go beyond what can be attained with heat radiators, high-thermal-conductivity molding compounds, and optimized designs. These systems, such as large mainframe computers, supercomputers, and digital telephone switches, require far more active thermal management strategies which include the high heat transfer coefficients attained with forced air or liquid cooling. A well known example of a thermally optimized design that brings high-thermal-conductivity material in contact with the device and uses enhanced fluid cooling to remove the heat from these materials is the

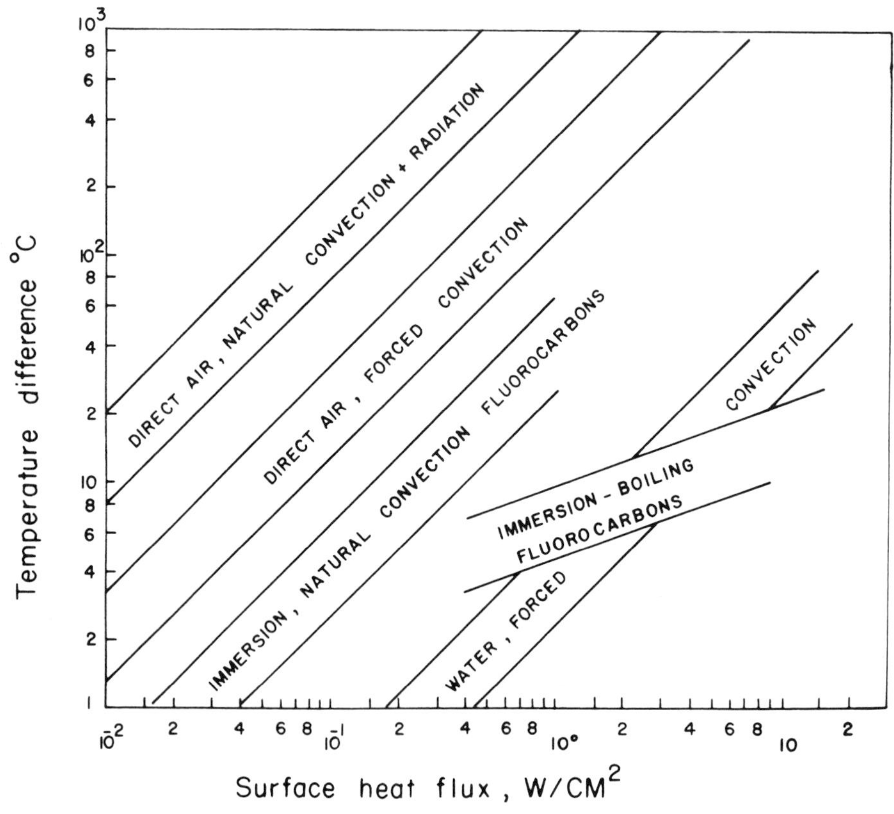


**FIGURE 8-18** Approximate temperature differences sustained for a given surface heat flux for several free and forced convection modes. (Reprinted with permission from Hemisphere Publishing Corporation, NY, *Thermal Analysis and Control of Electronic Equipment*, 1983 by A. D. Kraus and A. Bar-Cohen [22].)

## Table 8-3. Heat Transfer Coefficients for Natural and Forced Convection Modes. (After Reference 24.)

| *Convection Mode* | *h, cal/cm²-sec-°C* |
|---|---|
| *Free Convection* | |
| Gases | 0.000001–0.00006 |
| Liquids | 0.0003–0.002 |
| Boiling water | 0.003–0.056 |
| *Forced Convection* | |
| Gases | 0.00003–0.0003 |
| Viscous liquids | 0.00014–0.0014 |
| Water | 0.0014–0.028 |
| Boiling water | 0.0028–0.28 |

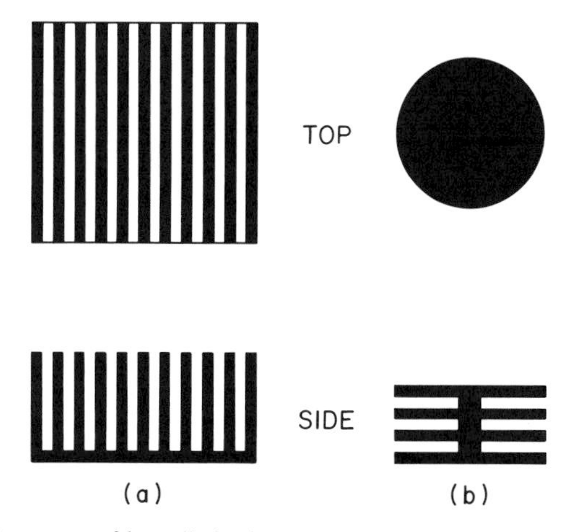


**FIGURE 8-19** Two types of heat dissipating radiators that would be attached to a package to increase the surface area of heat transfer and lower the device surface temperature: (a) directional radiator where the convective air flow must be parallel to the fins, (b) nondirectional radiator where the flow direction is indeterminate.

I.B.M. thermal conduction module [25], a drawing of which is provided in Figure 8-20.

Equation (8-28) also shows the dependencies and relationships of the individual terms to the overall heat transfer. This relation shows that thermal heat transfer through a package is analogous to an electric circuit in series in that the resistances add in series. The implication of this behavior is that a single low heat transfer term will limit the overall heat transfer despite high heat transfer properties of the rest of the system. As an example, switching to a high-thermal-conductivity molding compound may have an insignificant effect on the device operating temperature if it was not the rate-controlling term. The presence of a compliant conformal coating on the device can effectively limit the heat transfer to the point that other package features such as high-thermal-conductivity molding compound or forced convective cooling cannot overcome it. This can often be assessed by comparing thermal resistances of the individual components of heat transfer. The resistances are taken to be the terms inside the parentheses in Equation (8-28). The thermal conduction terms are $x_i/k_{Ti}$, where $x_i$ is the thickness of the specific material, such as the molding compound or the conformal coating, and $k_{Ti}$ is the thermal conductivity of this material. The interfacial heat transfer resistance is then $1/hA$, indicating that the magnitude of heat transfer by this route is inversely proportional to the product of the heat transfer coefficient and the area of heat transfer. Increasing the area through the addition of a radiator may provide enough additional heat transfer to avoid a more costly change in the type of convection used in the system.

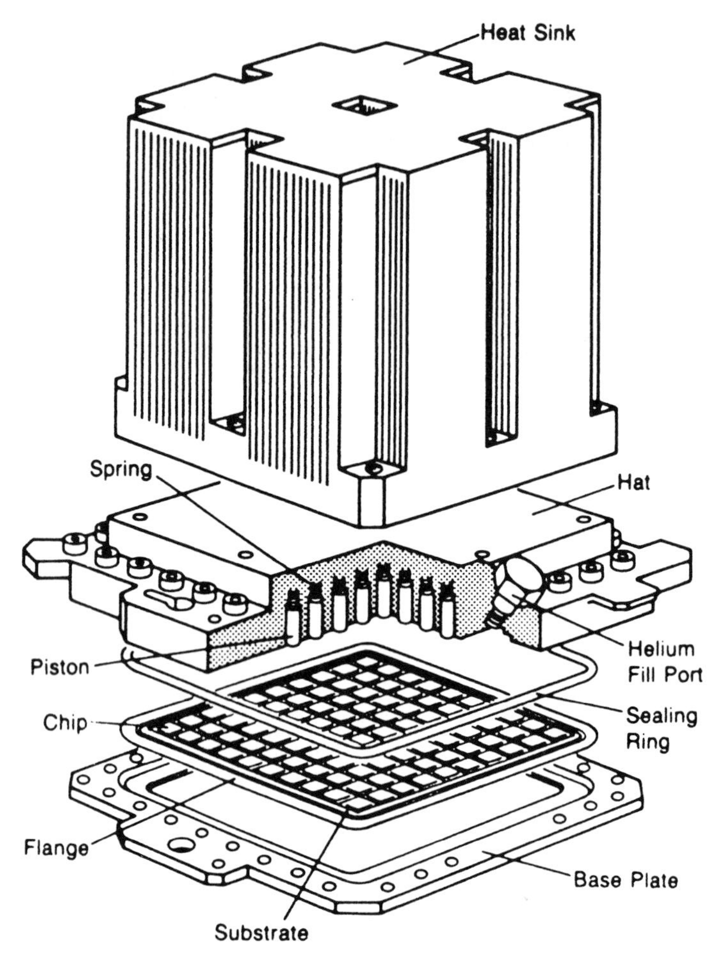


**FIGURE 8-20** I.B.M. 3081 Thermal Conduction Module (TCM), a high-density interconnection and thermal management configuration [25].

Referring to Table 8-3, however, it is apparent that changes in the nature of the convective heat transfer make orders of magnitude changes in the heat transfer coefficient which has a large impact on the overall heat transfer. Comparing resistances in this way allows the designer to identify which parts of the package are most limiting to the overall heat transfer rate and to then address corrections to these rate determining effects.

The simple analysis shown above does not account for the heat loss through the leadframe, a significant heat transfer mechanism in many cases. The leads approach close to the chip and paddle assembly and tap into the high temperatures that surround the device encased in the relatively low thermal conductivity molding compound. The metal leads have a significantly higher themal conductivity than the molding compound so they are effective conduits to conduct heat away from the chip, the only limitation being their small cross sectional area of heat transfer. Copper is a superior heat conducting leadframe material to Alloy 42 and has enjoyed a resurgence of use since chip power levels have mandated greater heat dissipation. The limitations to the amount of heat that can be moved through the leadframe are the thin cross section of the paddle support beams, the fact that these beams do not make direct contact outside the package, and the limited heat transfer through the molding compound and wire bonds that thermally connect the leads to the chip.

Although the one-dimensional solution described above is useful for assessing the dependencies of heat transfer on the material and design parameters it cannot be applied for more rigorous design work, particularly if there is a question that the device operating temperature may approach the limit for reliable performance. A better estimate can be obtained through a two-dimensional or three-dimensional solution of the energy balance that was discussed in Section 6.4.2.2.1. A numerical solution is usually preferable, particularly when there are several separate materials and interfaces present. This approach can address a more detailed package geometry including the leadframe and other internal features such as polyimide tapes and conformal coatings. The 2D solution is relatively simple to prepare and there are several resources that either provide these algorithms or thoroughly describe their development [26]. Three dimensional solutions are far more tedious to develop and require significantly more computer time to execute. The principal limitation of both 2D and 3D finite difference solutions is that only limited geometric complexity can be analyzed. These techniques are largely restricted to rectilinear shapes with few geometric complexities, asymmetries or anomalies. A finite difference solution with more than a few materials, structures and material interfaces becomes exceptionally complicated to write. Finite element analysis is much more appropriate when a more rigorous approach to package geometry and operating parameters is warranted. These methods and their application to both thermal and mechanical analysis are described in the next section.

## 8.4 FINITE ELEMENT ANALYSIS OF THERMAL AND STRESS GRADIENTS

Finite element analysis (FEA) [27] has emerged as a major tool in the promotion of improved reliability due to its unique capabilities in analyzing the thermal and mechanical gradients in complicated composite structures [13, 14, 28, 29, 30]. A finite element solution is one in which the domain of the solution is divided into smaller regions called elements. The shape of the elements can be triangles or quadrilaterals in two dimensions, and tetrahedra and pentahedra in three dimensions. Different types of elements can be mixed within the same solution as long as they are of the same dimension and all of the domain is covered without any gaps. The collection of elements over the entire domain of the solution is called the mesh. Meshing of the domain is a labor intensive task if done manually. Most commercial FEA packages, however, are now available with mesh generation algorithms that greatly simplify this procedure. Operator intervention is still needed in the form of mesh refinement which improves the resolution of the solution in areas that are of particular interest. The governing equations for the solution are transformed into algebraic equations which are an approximation of the governing equations now called the element equations. The solution within each element is approximated by an algebraic expression which constrains the functional form of the solution to within a set of unknown coefficients. Element equations are derived by substituting this expression into the governing equations. The element equations are identical in form for all elements of the same geometric type. Different element equations are required for elements with different shape. The loads on the domain, either external or internal, are specified by forcing terms that are also included in the element equations where applicable. The set of element equations is often written in the convenient matrix form:

$$\begin{pmatrix} K_{11} & K_{12} & K_{13} & \cdots & K_{1j} \\ K_{21} & K_{22} & K_{23} & \cdots & K_{2j} \\ K_{31} & K_{32} & K_{33} & \cdots & K_{3j} \\ \vdots & \vdots & \vdots & \cdots & \vdots \\ K_{i1} & K_{i2} & K_{i3} & \cdots & K_{ij} \end{pmatrix} \begin{pmatrix} a_1 \\ a_2 \\ a_3 \\ \vdots \\ a_i \end{pmatrix} = \begin{pmatrix} F_1 \\ F_2 \\ F_3 \\ \vdots \\ F_i \end{pmatrix} \tag{8-29}$$

The coefficients $K_{ij}$ are terms in the element equations that multiply each of the unknown coefficients $a_i$ after substituting the assumed algebraic expressions into the governing equations. The $F_i$ are the forcing terms which represent the loading functions. The $a_i$ are the unknowns corresponding to the algebraic expression chosen for the solution. Each element will have a different set of $a_i$ but neighboring elements will have some $a_i$ in common. An FEA solution then consists of assembling the $K_{ij}$ and $F_i$ values in a large system of algebraic equations known as the system equations, and solving this system of equations for the $a_i$ coefficients. This matrix assemblage is usually sparse; most of the elements are zero except those near the diagonal band. Bandedness, which is a measure of how closely clustered the nonzero coefficients are to the diagonal, results in tremendous savings in computer time and cost, making even very complicated systems amenable to FEA analysis. For additional source material, the reader is directed to Reference 27, which provides a thorough explanation of the principles of FEA.

In plastic packaging of microelectronics devices, FEA is used to analyze both thermal and mechanical problems. Mechanical analysis is considerably more complicated so it will be discussed predominantly. Most investigators rely on relatively simple two dimensional analyses because the more realistic three dimensional models are more complex and costly. Lines of symmetry within the package are also exploited to further reduce the domain size and decrease the number of elements. This helps to contain costs, allowing a larger number of variations and design options to be evaluated. Figure 8-21 shows a typical cross section for evaluation, and the lines of symmetry that further reduce the solution domain. A typical mesh for this cross section with mesh refinement at the chip corner and the leadframe corner is provided in Figure 8-22. A two-dimensional model of an IC package can generate approximately 3000 system equations with 9,000,000 terms, whereas a three-dimensional solution can have 10,000 equations with 100,000,000 terms [28]. This underscores the need to carefully construct the mesh; using large simple elements in regions of little interest and low stress gradients, while providing sufficient refinement in areas prone to high stresses, gradients, and cracking. In plastic packages, the highest stresses are developed in the vicinity of the chip and paddle edges and the lead ends in the vicinity of the wire bonds, hence this is the region where the mesh should be concentrated to provide greater resolution. The degree of mesh refinement has to be moderated, however, by the higher costs associated with the greater number of elements and the declining improvement in stress resolution with smaller and smaller elements. This is especially true at geometric singularities such as sharp corners. Infinitely sharp corners generate infinite stresses based on linear elastic theory [31]. The stress level near a corner (within several elements) will, therefore, tend toward infinity as the mesh is refined. This is simply a limitation

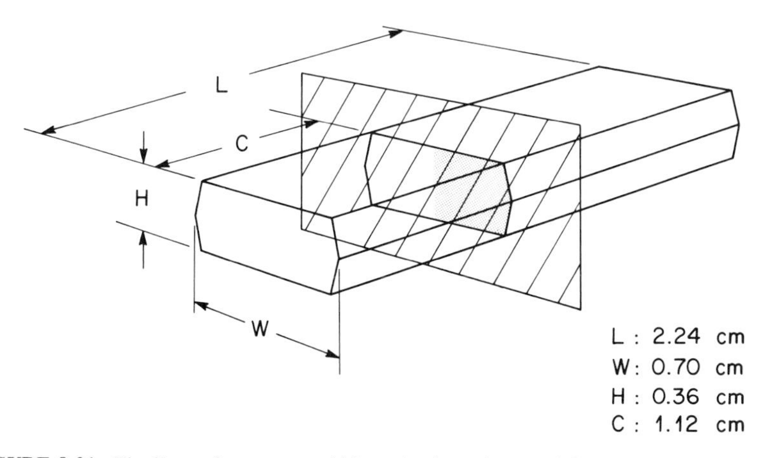


**FIGURE 8-21** The lines of symmetry within a plastic package and the cross section used often for finite element analysis. The shaded region indicates the area of the solution domain.

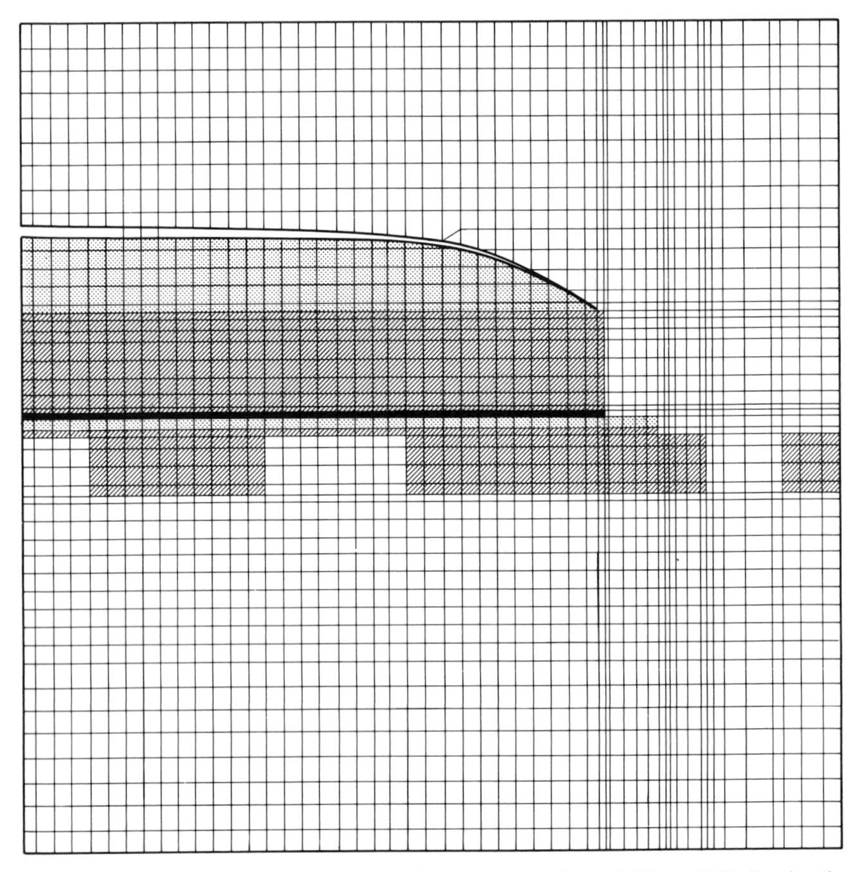

**FIGURE 8-22** Typical FEA mesh for the package geometry shown in Figure 8-21 showing the use of coarser elements in regions that do not experience steep stress gradients, and the mesh refinement in the vicinity of the chip edge that is most prone to package and passivation layer cracking. This mesh represents a minimal degree of mesh refinement. (Analysis by C. Virojanapa and S. C. Tighe, AT&T Bell Laboratories.)

of linear elastic analysis. In the actual part, infinite stresses are avoided because the geometry is not singular on a microscopic scale and localized yielding and cracking are likely to occur.

FEA is an important tool in assessing transient temperature gradients and the thermomechanical stresses that the thermal gradients cause in actual package geometries. Although this type of transient thermal analysis can be conducted with a custom finite difference solution, FEA is much more amenable to complicated geometries so that an FEA solution is usually more convenient. For conventional studies of temperature and stress profiles, one of the several commercial FEA packages can be used to further reduce the engineering effort and

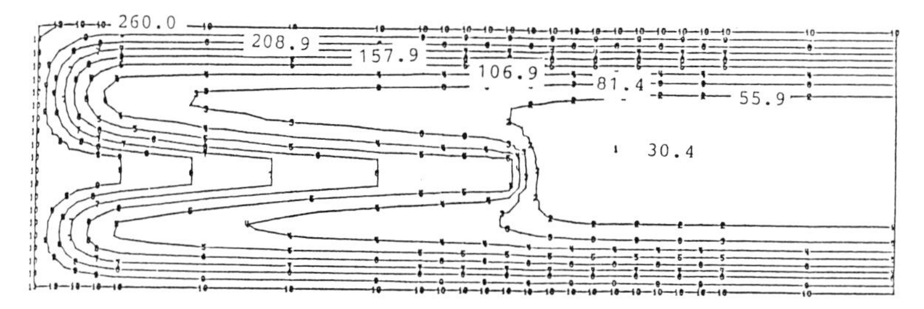


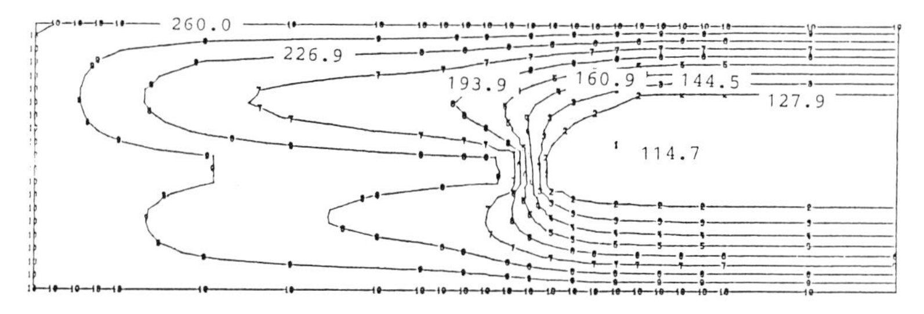


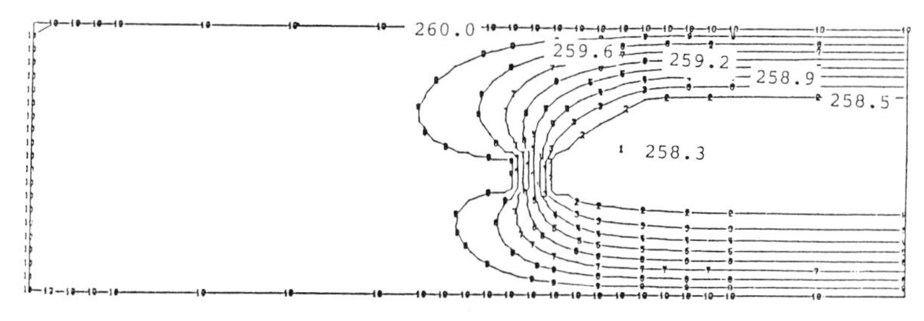


**FIGURE 8-23** Transient temperature profiles in a plastic package during solder dipping. The elapsed time after solder dipping began is shown above the temperature profile in each case. The package consists of a conventional epoxy molding compound, a copper leadframe, and a silicon die. (From Miyake, Suzuki, and Yamamoto, Reference 29. Copyright 1985 IEEE.)

reduce the time needed to obtain results. Figure 8-23 shows the results of an FEA solution of the temperature profile within a small plastic package under thermal loading conditions that simulate solder dipping [29]. These results show a very steep temperature gradient over the package volume in the first tenths of a second after exposure. The copper leadframe used in this case conducts the heat to the center of the package much more effectively and thereby sustains a much lower temperature gradient along the length of the lead. Most of the package has reached near the solder temperature after only four seconds of exposure. Results such as these are extremely useful in resolving packaging and assembly problems. For example, reductions in the solder dipping time intended to minimize the thermal induced stresses are ineffective if the exposure time is still more than the few seconds it takes the temperature throughout the package to approach the solder temperature.

The results of FEA for the thermochemical induced stresses within a DRAM package are shown in Figure 8-24. Referring back to the mesh for this analysis, it was assumed that there was an air space formed between the silicone rubber and the epoxy molding compound causing a complete loss of adhesion at this interface. The analysis indicates a maximum equivalent stress level at the chip corner with high stresses also found at the paddle support edge.

### 8.4.1 The Zero-Stress Condition

The conditions at which a zero-stress profile throughout the package is assumed have important implications for the stress levels and stress profiles predicted

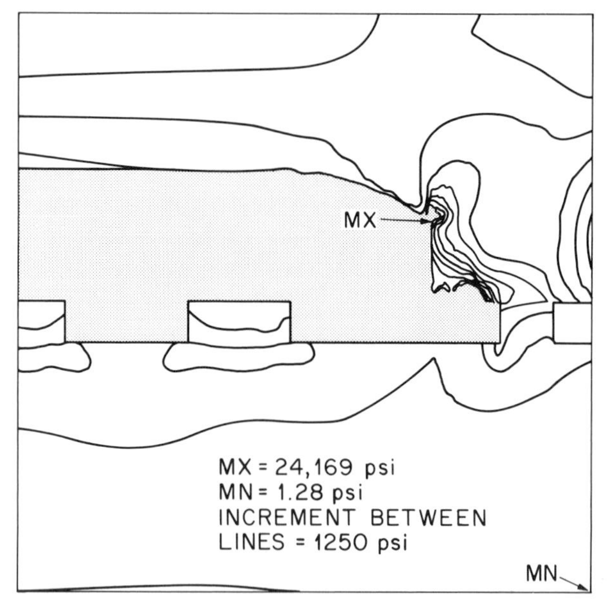


**FIGURE 8-24** Finite element analysis results for the package shown in Figure 8-21 with the mesh provided in Figure 8-22. In this result, the materials were assumed to be stress free at a uniform temperature of 170°C. It was then subjected to a thermal loading equivalent to cooling the package from 170° to −55°C. All interfaces were assumed to be rigidly connected except the silicone rubber-molding compound interface which was assumed to be completely unadhered, allowing the silicone rubber to shrink away from the molding compound forming an air space. (Analysis by C. Virojanapa and S. C. Tighe, AT&T Bell Laboratories).

under maximum loading. Most plastic packages are molded at 170–175°C, making this a logical choice for a zero-stress condition. There are occasionally mitigating circumstances that make a different condition more appropriate, however. If the epoxy molding compound has a glass transition temperature significantly below the molding temperature, then the glass transition temperature itself, known often in this context as the anchoring temperature, may be a more appropriate zero stress condition. The difference in stress level at full loading can be considerable because there can be significant stress relaxation in the rubbery state above $T_g$ that is unaccounted for in a linear elastic model that does not include stress relaxation effects. A published report [31] has found that a linear elastic FEA using the molding temperature as the zero stress state can significantly over predict the stress levels. This work indicates that $T_g$ is the more appropriate zero stress condition. Other considerations in choosing a zero stress condition concern the other materials and interfaces in the package and their zero stress conditions. For example, if the analysis is directed at stress levels responsible for die bond failures, then the conditions of the die attachment should be factored into the selection of a zero stress condition.

### 8.4.2 Mechanical Properties for Finite Element Analysis

The quality of the mechanical property data used in FEA has a profound effect on the accuracy of the results. In general, it is far more convenient to use constant room temperature values for the molding compound modulus and thermal expansion coefficient. The room temperature values are often provided by the material supplier, hence no additional experimental work is required to conduct an FEA when these constant values are employed. A recent study [31] compared FEA predictions obtained with constant room temperature values of modulus ($E$) and CTE ($\alpha$) with experimentally determined data of $E(T)$ and $\alpha(T)$, and then compared the results from both of these analyses with actual experimental measurements of the stress in plastic molding compound molded around low expansion inserts using a photoelastic technique. They concluded that although the shape of the stress profile was approximately the same, the stress levels found experimentally were in good agreement with stress levels predicted based on temperature-dependent material properties. Conversely, predicted stress levels for the constant $E$ and $\alpha$ values were about 50% too high as is shown in the plot provided in Figure 8-25.

### 8.4.3 Analysis of Interfaces

The analysis of the mechanical behavior of interfaces is one of the weakest aspects of FEA. Interfaces in actual packages are neither completely adhered nor completely free, but usually one of these two extreme conditions must be

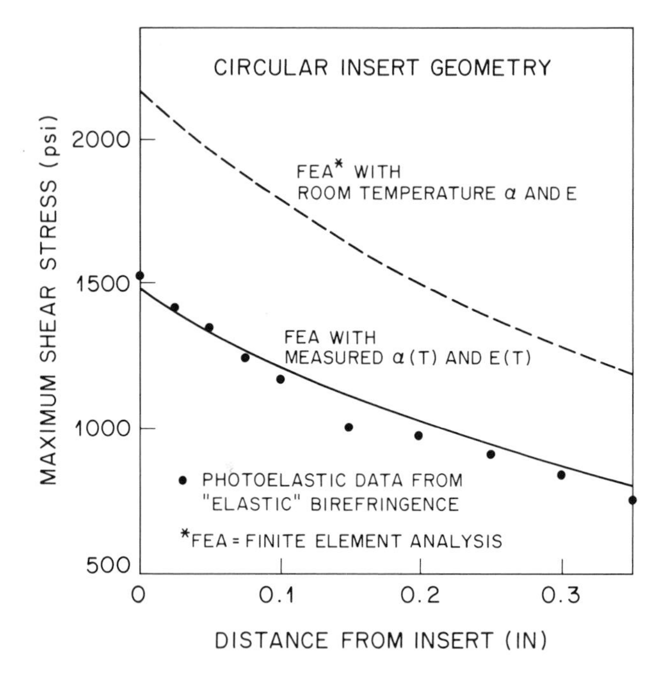


**FIGURE 8-25** Comparison of predicted stress levels for FEA using constant room temperature values of E and $\alpha$ with experimental data for $E(T)$ and $\alpha(T)$. Both of these predictions are then compared to measured values of the stress determined with a photoelastic technique on clear, unfilled molding plastic. (From Sullivan, Rosenberg, and Matsuoka, Ref. 31 Copyright 1988 IEEE.)

used in the mathematical treatment of interfacial behavior. This assumption has important ramifications for the predicted stress in a mechanical model, and negligible effect on temperature predictions. Most researchers assume a completely adhered interface since this is closer to the actual mechanics than complete lack of adhesion. Complete adhesion distributes the shrinkage stresses over the largest volumes of molding compound, silicon and leadframe and so provides the lowest predicted stress levels. Figure 8-26 shows the predicted shear stress levels around a rounded-corner insert determined through FEA with and without adhesion to the insert [31]. The experimental shear stress measurements made through a photoelastic technique are also provided on the same plot. These data indicate that the perfect adhesion assumption comes closer to the experimental values. In this particular case where the shrinkage forces are directed radially toward the adhered surface, the data are below the FEA results with perfect adhesion. For the case of adhesion to the leadframe or silicon device in an IC package, the shrinkage forces have a large component parallel to the adhered surface so experimental data are likely to fall between the predictions for perfect

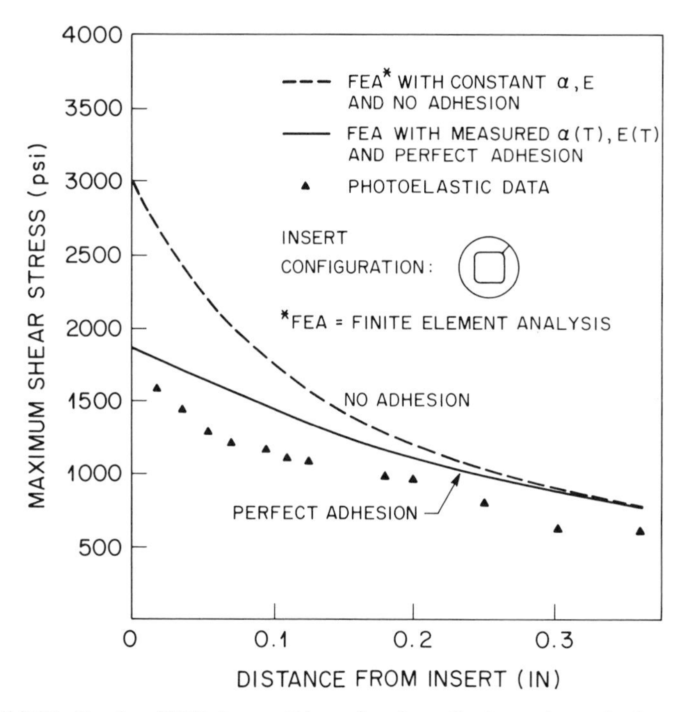


**FIGURE 8-26** Results of FEA for conditions of perfect adhesion and no adhesion of molding compound to a rounded corner insert that is shown on the Figure. The data presented are maximum shear stress plotted against the distance from the insert along the line shown in the insert configuration drawing. These results are compared with photoelastic data on the same specimen. (From Sullivan, Rosenberg, and Matsuoka, Ref. 31 Copyright 1988 IEEE).

and zero adhesion. The adhesion between interfaces in a molded plastic package breaks down with repeated temperature cycling. In general, this will cause the stress level to move from the values associated with complete adhesion toward the values of zero adhesion on repeated cyclic loading. There is no approach to FEA analysis of this progressive loss of adhesion that can be cited at the time of this writing.

### 8.4.4 Photoelastic Verification of Finite Element Analysis

One shortcoming of any theoretical approach to design and optimization work is that it can be difficult to verify the stress magnitudes and profiles that are predicted. With FEA of IC packages it is commonly accepted [28] that the stress profiles are fairly accurate in that cracks are usually found at the stress maximums, but the magnitudes of the stresses are much more uncertain. This makes it difficult to use FEA with confidence to predict package failures. Verification of stress levels obtained for FEA is an important aspect of the implementation of these methods. Birefringence [32] is a technique for measuring the stresses in materials that can be used to determine absolute stress levels and thereby calibrate finite element analysis. Several publications [31, 33] have made use of this technique to study orientation and stress in transfer molded plastics and to compare these to finite element models.

The birefringence method consists of passing light through the material. The light is diffracted by the slight differences in refractive index parallel and perpendicular to the orientation axis of the molecular segments. In a polarimeter, this diffraction is measured as optical retardation. The light intensity depends on the degree of optical retardation and the wavelength of light used. Therefore, in a monochromatic beam, a specimen with a stress gradient would produce a series of light and dark bands, known as a fringe pattern. A series of colored bands is produced when white light is passed through the sample. The stresses within the material can be related to the fringe pattern through the Stress Optic Law [32]:

$$\sigma_1 - \sigma_2 = \frac{N\lambda}{kt} \tag{8-30}$$

where $k$ is the stress optical coefficient, $\sigma_1$, $\sigma_2$ are the principal stresses, $N$ is the fringe order, $\lambda$ is the wavelength of the light, and $t$ is the thickness. Each fringe represents one wavelength of relative retardation of the light. If Hookean (elastic) behavior can be assumed, the principal strains can be related to the principal stresses determined in Equation (8-30):

$$\epsilon_1 - \epsilon_2 = \frac{\sigma_1 - \sigma_2}{E}(1 + \nu) \tag{8-31}$$

The refraction process requires that the stressed material be transparent to visible light and that it have a sufficiently large stress optical coefficient to provide a measurable level of optical retardation for the stress that can be imposed. This is a concern because birefringence cannot be applied to the molding compound itself which is usually highly filled, colored and opaque. There are means of circumventing this limitation, however. One is to allow the transfer molding compound to exert stress on a transparent material and infer the stress in the molding compound from the stress measured in this material. For example, the stress in an inner concentric cylinder made of a photoelastic material such as quartz is used in the standardized stress analysis test ASTM F-100 that was described in Section 4.1.4. A second approach is to use a clear, unfilled molding compound to calibrate the FEA using the mechanical properties of the unfilled material. This has the advantage that it can be used in the actual geometry of interest; for example, an IC package could be molded with clear, unfilled molding compound and the stress levels, which will be much higher with the unfilled material, read using birefringence [33]. The calibrated algorithm is then applied to the actual molding compound using its mechanical properties. These properties will be significantly different, but this does not appear to invalidate the verification [31].

The total stress in a specimen can be attributed to several sources. One interpretation [31] holds that there are two major contributors in plastic materials: stress birefringence and orientation birefringence. The difference between these two can be illustrated by considering a photoelastic material that has been molded at high temperature around a low CTE insert. When cooled to room temperature, a thermomechnical load will be applied that will generate stresses and attendant birefringence. If the insert is removed, the elastic stresses within the plastic dissipate since the plastic is now unconstrained and can assume its proper dimensions for the temperature. The birefringence, however, does not go to zero [31, 34]. The residual birefringence, which can be up to half of the total birefringence, remains after the insert is removed as evident in Figure 8-27, although this is found to decrease slowly with time at room temperature. This suggests that the birefringence decrease on removal of the insert is the elastic or Hookean stress whereas the residual stress is a viscoelastic phenomenon associated with molecular motions to accommodate the insert that then relax over much longer times in the glassy state.

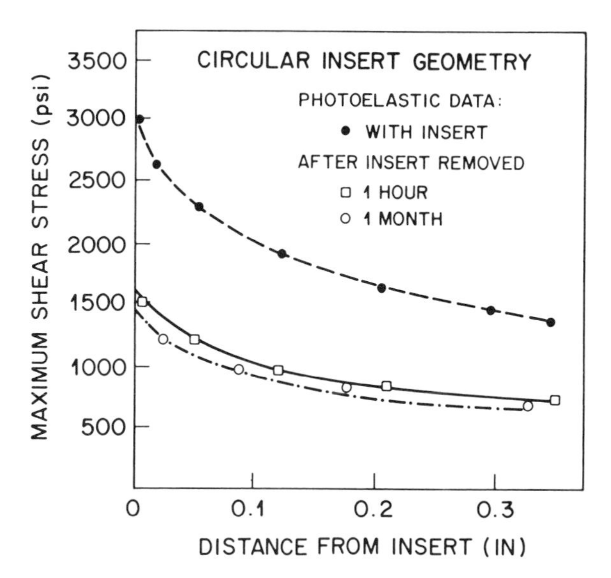


**FIGURE 8-27** Photoelastic data from birefringence measurements showing the maximum shear stress in the molding plastic surrounding a circular insert. Three lines are shown representing stress with the insert in place, one hour after the insert was removed, and one month after the insert was removed. (From Sullivan, Rosenberg, and Matsuoka, Ref. 31 Copyright 1988 IEEE.)

### 8.4.5 Strain Gauge Verification of Finite Element Analysis

Another approach to assess the mechanical stresses within a plastic package and to conduct verification studies of FEA is to use microelectronic strain gauges [35, 36] and piezoelectric devices [37, 38]. These can be special chips whose output signal or resistivity correlates to the stresses or strains imposed on the active area of the device. These devices are mounted on leadframes and encapsulated in the molding compound in the conventional process sequence. The strains or stresses at the chip surface can then be read during and after actual testing and temperature cycling. Both principal stresses and "in-plane" shear stresses can be determined, and newer devices can map the stress profile over the entire die surface [39]. The results must be interpreted carefully since the correlation to stress will change with temperature. Also, the correlation will be different for different specimens of the same model of strain gauge. Careful calibration of the sensor over the entire temperature and stress range that will be encountered is essential to the proper implementation of these devices. The devices themselves can be purchased from several of the major semiconductor companies.

## REFERENCES

1. K. L. Wong, "Unified Field (Failure) Theory: Demise of the Bathtub Curve," *Proceedings Annual Reliability and Maintainability Symposium*, 402 (1981).
2. *Microelectronics Packaging Handbook*, R. R. Tummala and E. J. Rymaszewski, Editors, Van Nostrand Reinhold, New York (1989).
3. J. T. Duane, "Learning Curve Approach to Reliability Monitoring," *IEEE Trans. Aerosp.*, **2,** 563 (1964).
4. W. J. Bertram, "Yield and Reliability," in S. M. Sze, Editor, *VLSI Technology*, Second Edition, McGraw Hill Book Company, New York (1988).
5. P. A. Tobias and D. C. Trinidade, *Applied Reliability*, Van Nostrand Reinhold, New York (1986), pp. 109–114.
6. L. Goldthwaite, "Failure Rate Study for the Log-normal Lifetime Model," *Proceedings of the 7th National Symposium on Reliability and Quality Control*, 208–213 (1961).
7. S. Groothuis, W. Schroen, and M. Murtuza, "Computer Aided Stress Modeling for Optimizing Plastic Package Reliability," IEEE/IRPS, CH2113-9/85/0000-0184 (1985).
8. S. Oizumo, N. Imamura, H. Tabata, and H. Suzuki, "Stress Analysis of Si-Chip and Plastic Encapsulant Interface," *Nitto Technical Reports*, 51 (Sept. 1987).
9. S. Oizumo, S. Ito, and H. Suzuki, "Analysis of Reflow Soldering by Finite Element Method," *Nitto Technical Reports*, 40 (Sept. 1987).
10. I. Fukuzawa, S. Ishiguro and S. Nanbu, "Moisture Resistance Degradation of Plastic LSIs By Reflow Soldering," IEEE/IRPS, CH2113-9/85/0000-0192 (1985).
11. T. Yoshida, "Surface Mounting Packages," *Semiconductor World*, **10,** 95–105 (1985).
12. T. Nishioka, S. Ito, M. Nagasawa, and M. Kohmoto, "Special Properties of Molding Compound for Surface Mounted Devices," *Electronic Component Technology Conference* (ECTC), Las Vegas, NV (May 21–23, 1990).

13. S. Okikawa, M. Sakimoto, M. Tanaka, T. Sato, T. Toya, and Y. Hara, ''Stress Analysis of Passivation Film Crack for Plastic Molded LSI Caused by Thermal Stress,'' *Proceedings Intl. Symp. on Test and Failure Analysis*, 275–280 (1983).
14. D. R. Edwards, K. G. Heinen, S. K. Groothuis, and J. E. Martinez, ''Shear Stress Evaluation of Plastic Packages,'' *IEEE Trans. on Components, Hybrids and Manufacturing Tech.*, **CHMT-12**(4), 618, IEEE Doc. #0148-6411/87/1200-0618 (1987).
15. T. Tabata, H. Suzuki, T. Hamada, and M. Yamaguchi, ''Internal Defect Observation of IC Package by Scanning Acoustic Tomography,'' *Nitto Technical Reports* (Sept. 1987).
16. R. E. Thomas, ''Stress Induced Deformation of Aluminum Metallization in Plastic Molded Semiconductor Devices,'' IEEE 0569-5503/85/0000-0037 (1985).
17. M. Isagawa, Y. Iwasaki, and T. Sutoh, ''Deformation of Al Metallization in Plastic Encapsulated Semiconductor Devices Caused by Thermal Shock,'' IEEE, CH1531-3/80/000-0171 (1980).
18. G. A. Lang et al., ''Thermal Fatigue in Silicon Power Devices,'' *IEEE Trans. Elec. Devices*, **ED-17,** 787–793, (1970).
19. J. E. Anderson et al., ''Prediction of the Temperature Field for an Electronic Device Operating Under Unsteady Electro-thermal Conditions,'' *5th Annual International Electronic Packaging Conference Proceedings*, 508–525 (1985).
20. J. W. Thornell, W. A. Fahley, and W. L. Alexander, ''Hybrid Microcircuit Design and Procurement Guide,'' available from National Technical Information Service, Springfield, VA, DOC AD 705974 (1972).
21. C. A. Harper, *Handbook of Thick Film Hybrid Microelectronics*, McGraw Hill, New York (1974).
22. A. D. Kraus and A. Bar-Cohen, *Thermal Analysis and Control of Electronic Equipment*, Hemisphere Publishing Corp., New York (1983).
23. *Reliability Prediction of Electronic Equipment*, U.S. Dept. of Defense, MIL-HDBK-217B, NTIS, Springfield, VA (1974).
24. R. B. Bird, W. E. Stewart, and E. N. Lightfoot, *Transport Phenomena*, John Wiley and Sons, Inc. New York (1960).
25. R. R. Tummala and E. J. Rymaszewski, Editors, *Microelectronics Packaging Handbook*, Van Nostrand Reinhold, New York (1989); another reference to the thermal conduction module is: A. J. Blodgett and D. R. Barbour, ''Thermal Conduction Module: A High Performance Multilayer Ceramic Package,'' *I.B.M. J. Research and Development*, **26**(1) (Jan. 1982)
26. G. N. Ellison, *Thermal Computations for Electronic Equipment*, Van Nostrand Reinhold, New York (1984).
27. D. S. Burnett, *Finite Element Analysis*, Addison-Wesley, New York (1987).
28. ''TI Finds a New Way to Predict Package Reliability,'' *Electronics* (March 31, 1988).
29. K. Miyake, H. Suzuki, and S. Yamamoto, ''Heat Transfer and Thermal Stress Analysis of Plastic Encapsulated ICs,'' *IEEE Trans. on Reliability*, **R-34**(5), 402, IEEE Doc. #0018-9529/85/1200-0402 (1985).
30. S. Groothuis, W. H. Schroen, and M. Murtuza, ''Computer Aided Stress Modeling for Optimizing Plastic Package Reliability,'' *Proc. 23rd Ann. Reliability Physics Symp.*, 182–191 (1985).
31. T. Sullivan, J. Rosenberg, and S. Matsuoka, ''Photoelastic and Numerical Investigation of Thermally Induced Shrinkage Stress in Plastics,'' *IEEE Trans. on Components, Hybrids and Manufacturing Technology*, **11**(4), 473, IEEE Doc. #0148-6411/88/1200-0473 (1988).
32. J. Javornicky, *Photoplasticity*, Elsevier Press, Amsterdam, The Netherlands (1974).
33. K. M. Lietchi, ''Residual Stresses in Plastically Encapsulated Microelectronic Devices,'' *Experimental Mechanics*, **25,** 226 (1963).
34. I. M. Daniel and A. J. Durelli, ''Shrinkage Stresses Around Rigid Inclusions,'' *Experimental Mechanics*, **24**(2), 240 (1962).

35. J. L. Spencer, W. H. Schroen, G. A. Bednarz, J. A. Bryan, and T. D. Metzgar, "New Quantitative Measure of IC Stress Introduced by Plastic Package," *Proc. Int. Reliability Physics Symp.*, 74–80 (1979).
36. H. Inayoshi, K. Nishi, S. Okikawa, and Y. Wakashima, "Moisture-Induced Aluminum Corrosion and Stress on the Chip in Plastic Encapsulated LSIs," *Proc. Int. Reliability Physics Symp.*, 113–117 (1979).
37. K. M. Schlesier, "Piezoresistivity Effects in Plastic Encapsulated Integrated Circuits," *RCA Review*, **43,** 590–607 (1982).
38. R. J. Usell and S. A. Smiley, "Experimental and Mathematical Determination of Mechanical Stress within Plastic IC Package and Their Effect on Devices During Environmental Tests," *Proc. Int. Reliability Physics Symp.*, 65–73 (1979).
39. S. A. Gee, W. F. van den Bogert, V. R. Akylas, and R. T. Shelton, "Strain Gauge Mapping of Die Surface Stresses," 39th Electronic Component Conference Proceedings, p. 343, ISSN 0569-5503 (1989).

# 9 Technology Trends in Plastic Packaging

Microelectronics technology is evolving so rapidly and the evolutionary changes are so significant that decision makers must be cognizant of the industry trends to properly anticipate both technical and business needs of the near future. Packaging falls in the middle of the process steps to produce an electronic system, and as such it has to respond to both upstream and downstream changes. These include wafer fabrication and device trends as well as interconnection technology and circuit board assembly trends. The rapid advance of device capabilities and the rapid changes in circuit board assembly are now placing greater emphasis on the performance limitations introduced by the plastic package. In device technology, significant increases in speed, power, and capacity occur at regular intervals. Packaging and interconnection have not been the technology limiting aspect of this evolution since essentially any high volume silicon device that could be made could be packaged without significant loss of performance. This assumption has been valid in that packaging and interconnection have not been impediments to system performance in all but the most sophisticated applications such as mainframe and supercomputers where there are extraordinary timing and power dissipation needs. The time in which packaging and interconnection can be taken for granted is now rapidly approaching an end. High production volume devices such as those used in personal computers and group work stations have interconnection density, power dissipation, and timing needs that often cannot be met by conventional plastic packages. Although the more expensive ceramic packages and pin grid packages can still meet most of these needs, the lower cost of plastic packaging is a significant benefit in these highly cost competitive products. This chapter addresses the important trends in microelectronics technology and assesses the implications these trends will have on plastic packaging.

## 9.1 INTEGRATED CIRCUIT TRENDS

The major integrated circuit trends that are expected to impact the packaging needs of the near future are summarized and briefly explained in the following sections.

### 9.1.1 Greater Number of I/O

The number of leads required for plastic packaged ICs has been increasing rapidly over the last decade as is shown in Figure 9-1. This number will continue to increase over the next five years reaching approximately 500 leads for single chip packages by the mid-1990s. Five hundred leads can be accommodated with 12.5 mil pitch on the outer leads, 4–5 mil pitch on the inner leads, and a package size of almost 2.5 square inches. Flow induced deformation on the leadframe is proportional to the fifth power of the lead length as described in Section 7.1.1 [Equation (7-2)], so a large package is much more susceptible to flow-induced yield loss. Accommodating the inner lead interconnection between the leadframe and the chip will also be difficult. The bond pad spacing on the device and the attendant wire bond spacing will have to decrease to prevent extraordinary die sizes which would exacerbate paddle shift problems (see Section 7.1.1 for a discussion of paddle shift). Figure 9-2 shows the die size as a function of lead count for several different bond pad spacings. This simple plot points out the importance of reducing bond pad spacing to minimize die size and flow-induced paddle movement. Larger dies and paddles, however, are advantageous when the leadframe manufacturing technology is stretched to its limit. A large paddle shortens the lead length and increases the spacing between leads at the lead tips. In some cases such as 196 PQFP die and wire leadframes, oversized paddles are required to provide a leadframe design that is manufacturable by either etching or punching.

Multichip modules, which offer very high interconnection densities on ceramic or silicon substrates, are likely to be used to cluster several high I/O chips, so that the number of leads that have to be accommodated by the package may stabilize or peak at the several hundred level. Eventually, very large multichip modules which are similar in size and function to present circuit boards

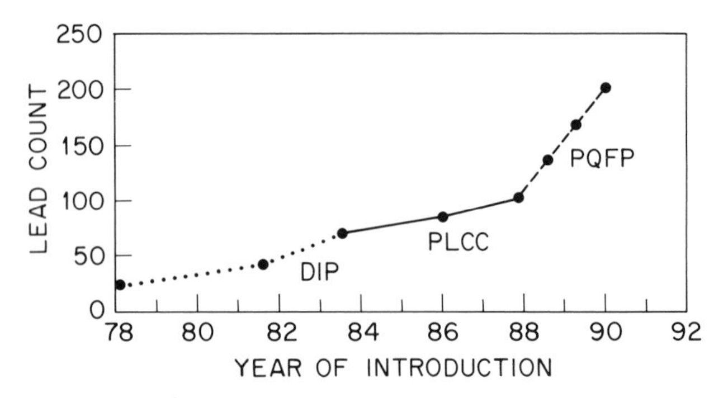


**FIGURE 9-1** Plot of the highest lead count in a molded plastic package versus the year of introduction. Results are for American semiconductor and assembly companies using NMOS and CMOS technology.

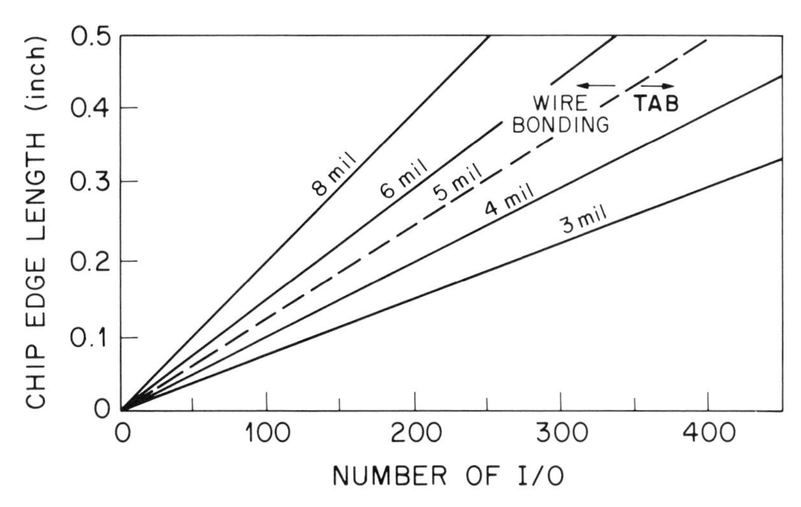


**FIGURE 9-2** Die size needed to accommodate a given lead count plotted for several bonding pad spacings between 8 and 3 mils.

will be developed. These modules with dozens of chips will have very high interconnection densities on the substrate but a much smaller number of leads will come off the module.

### 9.1.2 Power Dissipation

The power dissipation requirements will continue to increase significantly over the next few years [1] with some forecasts showing that the average power will reach 10 watts per chip by the end of the century as shown in Figure 9-3 [2]. The power actually goes up by the square of the feature size reduction ratio since power is based on number of features per unit area. A large movement toward gallium arsenide (GaAs) devices will accelerate the problem since they produce more heat than comparable CMOS devices. This is probably the most urgent of the trends affecting the future of low cost plastic packaging because some chips of the late 1980s already required active thermal management to be packaged in plastic. Current thermal management features of plastic packages, such as heat radiators or heat spreaders, are of limited value because the device is still largely surrounded by the plastic molding compound with its low thermal conductivity as discussed in Section 8.3.1.

Materials with significantly higher thermal conductivities will be needed to package CMOS devices of the near future. Thermal conductivities of the order of 60–100 cal/cm-sec-°C will be needed compared with present molding compounds of 15 cal/cm-sec-°C for fused silica filler, and 40 cal/cm-sec-°C for crystalline silica (quartz). Even with this substantial increase in thermal con-

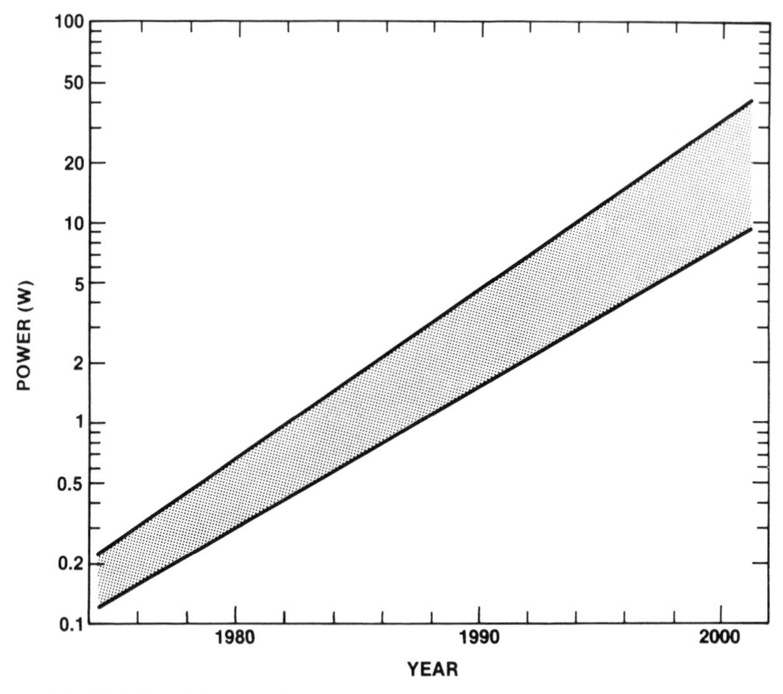


**FIGURE 9-3** History and forecast for power per CMOS chip versus year [2].

ductivity, heat spreaders and heat sinks will be required on the more powerful chips. Higher thermal conductivity of the molding compound can be achieved by using fillers with higher thermal conductivity than fused silica and using these fillers at higher loadings than have been achievable in the past. Figure 9-4 shows a bar chart of the thermal conductivity of several inorganic fillers that may be suitable for package molding compounds. Some of these, such as aluminum nitride, have been touted as important replacements for silica, silicon and alumina in electronic applications because of its superior thermal conductivity, but like most of the other filler materials it is not without problem when used in an epoxy molding compound. One of the primary problems in using nearly all of the high thermal conductivity fillers is that they are much more abrasive than silica to the point that they can damage the molding tool in less than a few hundred cycles. The problem is caused by the higher hardness of some of the materials and the filler particle shape. Also, many of these materials increase the viscosity of the molding compound far more than silica does at the same filler loading. The cause of this problem is not known with certainty but it is believed to be related to the shape of the particle and its wetability by the epoxy liquid. Most high-thermal-conductivity molding compounds have significantly reduced spiral flow lengths. Yet another problem is that some of the

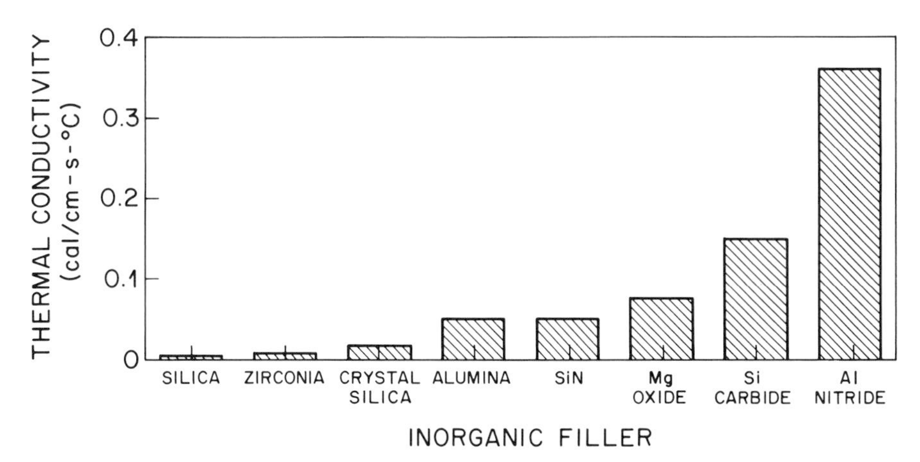


**FIGURE 9-4** Thermal conductivity of various inorganic filler materials that could be candidates for use in high thermal conductivity molding compounds.

fillers can have an interaction with the curing agent or catalyst in the molding compound and they retard the chemical reaction. Cost will also be a consideration in that these high thermal conductivity fillers are significantly more expensive than silica. It is apparent that significant technical hurdles remain prior to the development of high thermal conductivity molding compounds; yet the need is particularly urgent.

### 9.1.3 Smaller Device Features

Device features for CMOS technology have gone below one micron with further reductions expected. Most industry experts forecast device feature sizes of 0.35–0.3 micron by 1993 [3] with the possibility of 0.2 micron by the turn of the century. The development of X-ray resists would drive down feature sizes even further. The principal concerns of reduced feature size are the higher power, which was addressed above in Section 9.1.2, the faster clock rates and rise times that accompany reduced feature sizes at constant electric field strength as discussed in Section 9.1.5, and the deformation of the circuit pattern introduced by the shrinkage of the molding compound. With present materials, this differential thermal shrinkage between the molding compound and the silicon device can cause deformations of several microns, deformations that are roughly equal to the present conductor path widths. The problem is most often encountered near the periphery of the chip on aluminum conductor paths that run perpendicular to the shrink direction (see Section 8.2.5 for a discussion of aluminum line deformation). Without further reductions in the coefficient of thermal expansion of the molding compound, the deformation induced by the package will greatly exceed the conductor path widths and spacings as feature sizes and conductor widths continue to decrease. This would cause one conductor path to be pushed over an adjacent path or feature resulting in an electrical short, shown in Figure 9-5.

Passivation layer cracking is also an important concern (see Section 8.2.3). Unlike the ductile aluminum, the glass passivation layer will crack rather than deform under the shrinkage forces exerted by the molding compound. Passivation layers in double metal architectures are particularly vulnerable to this type of cracking because of the higher features with this geometry. Interestingly, further reductions in coefficient of the thermal expansion (CTE) of the molding compound that could eliminate conductor path deformation and passivation cracking may not be desirable, even if they were feasible. The CTE of some of the better molding compounds reached the CTE of copper leadframes which is 16 ppM/°C by the late 1980s. If they go below this value, then the molding compound will no longer shrink onto the leadframe, an important phenomenon that provides a good moisture seal along the leads. Movement away from copper is unlikely in the near future because its high thermal conductivity helps to resolve the power dissipation issue. In this way, many of the issues raised by the technology trends are not only interrelated, but also conflicting.

### 9.1.4 Larger Chip Size

The surface area of both memory and logic devices will continue to grow despite the significant reduction in device feature size. Large microprocessors were

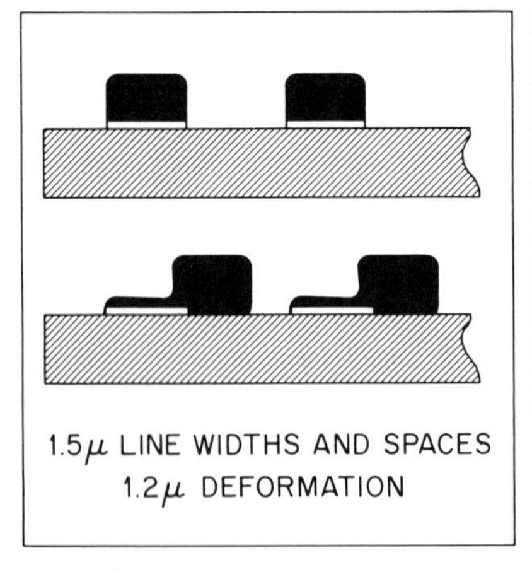


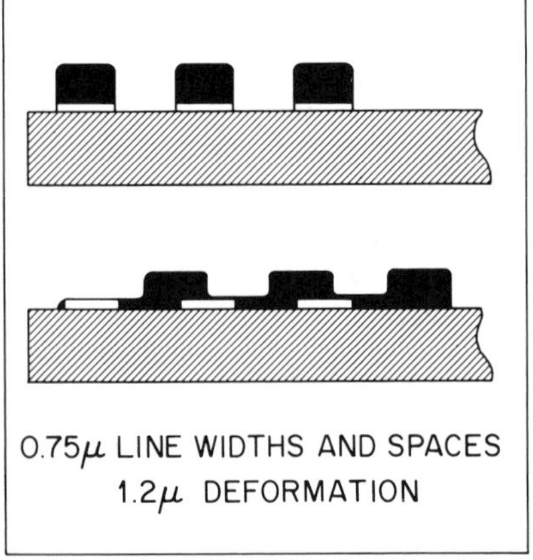


**FIGURE 9-5** Conductor path deformation due to disparity in coefficient of thermal expansion encountered when the device features are in direct contact with the molding compound: (a) conductor path deformation with 1.5 micron width; (b) conductor path deformation with 0.75 micron width. In both cases the deformation is 1.2 microns.

100 mm$^2$ in area in the late 1980s, with the rectangular DRAM devices approaching 100 mm$^2$ for the larger 4 Mbit chips. Chips that are much larger than 100 mm$^2$ should be expected if these trends continue as projected in Figure 9-6 [4]. These larger devices will generate higher thermal shrinkage forces because the dimensional differences on shrinkage between chip and molding compound will be greater. Larger chips require larger paddle supports making paddle shift problems more likely. Yet another implication of larger chips is the growing disparity in the expansion of the chip and the leadframe paddle support. On temperature excursions in oversized die designs, the dimensional differences between the die and leadframe can be large enough to warp the die and pad assembly, causing a buckling that can be apparent through the molded package. This problem is more severe with copper leadframes because of their larger coefficient of thermal expansion, and in thin body packages where warpage of the entire molded body can be observed. A resolution of this problem is not straightforward, although a compliant die attach material would be one way to absorb the dimensional disparities.

### 9.1.5 Clock Rates

The clock rates of microelectronic devices will continue to increase placing greater demands on the dielectric properties of the packaging materials to pre-

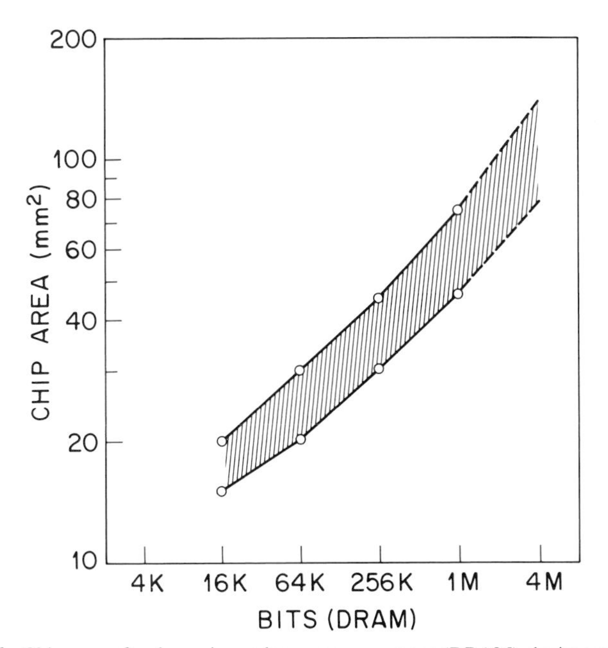


**FIGURE 9-6** Chip areas for dynamic random access memory (DRAM) devices plotted against memory capacity. (After Reference 4.)

serve signal speed and minimize signal dissipation [5]. A history and forecast of CMOS microprocessor clock rate is provided in Figure 9-7 [2]. The package degrades device performance if the clock rate is shorter than the propagation rate of signals through the package. In cases where the overall resistance of the conductor is low, signals propagate at the speed of light (electromagnetic radiation) in the medium which is proportional to $\epsilon'^{-1/2}$, but the geometry of the circuit has a strong influence on the transmission characteristics. Plastics actually have superior dielectric properties compared to ceramics and some other inorganic packaging materials as was shown in Table 2-5, but the package geometries used most often with plastic packaging, such as DIPs and PLCCs, are not the most conducive to propagating high frequency signals as is shown in the bar plot of propagation delays for the several different package options provided in Figure 9-8 [6]. This plot indicates that the PLCC design, and the similar PQFP, will be favored with regard to minimizing propagation delays, with TAB also favored in any design where propagation delay is an issue. This does not conflict with the greater utilization of these package types to meet higher I/O requirements since they offer far greater interconnection density

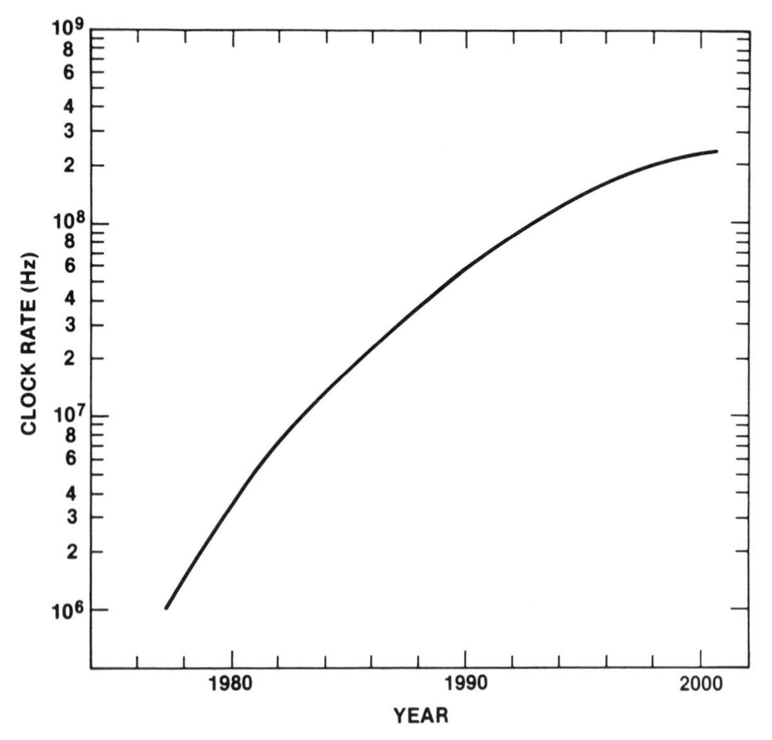


**FIGURE 9-7** History and forecast of the clock rates of CMOS microprocessors [2].

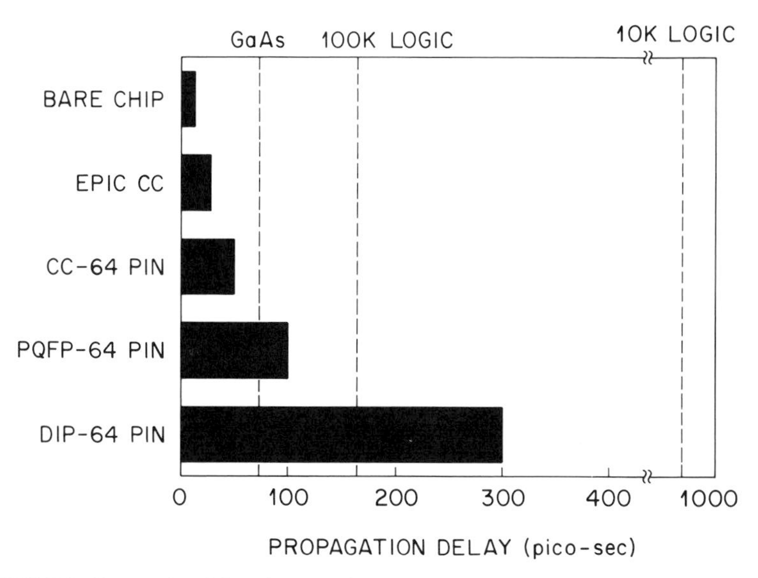


**FIGURE 9-8** Propagation delays for several different package options. The three vertical lines represent the clock speeds for several different device types. The more sophisticated devices cannot tolerate the longer propagation delays introduced by some packages. (Note: EPIC is an acronym for a premolded plastic leadless chip carrier with direct interconnection between the bond pads on the device and the fan out pattern in the cavity. CC is a ceramic chip carrier, PQFP and DIP are plastic packages.) (After Reference 6.)

than DIPs. In general, the gate delay scales as the inverse of the feature size reduction ratio at constant electric field strength [7] indicating that substantial increases in clock rate will accompany the substantial feature size reductions that will occur in the next decade. Although material selection can degrade device performance, the effects on system performance can be minimal if the transmission times of the signals are coordinated so that they all arrive at the same time. Nonetheless, plastics have an inherent advantage over ceramic packages with regard to dielectric properties, and this advantage may force greater acceptance of plastic in previous hermetic package preserves such as military and other performance-driven applications.

## 9.2 INTEGRATED CIRCUIT ASSEMBLY TRENDS

The IC assembly operations are influenced by the chip features as well as by the package requirements. The IC trends discussed above could have a significant influence on present assembly operations. These assembly trends are described below.

### 9.2.1 Hybrid Integrated Circuits

Hybrid integrated circuits (HICs) are not a recent development or an emerging trend in microelectronics. Rather, they are a fairly well established microelectronics technology that has occupied a niche in high end systems. One advantage of HICs is the high interconnection density that can be achieved by the direct chip attach of several active devices with fairly fine line conductors. Discrete devices such as film capacitors, chip capacitors and thick film resistors are also part of most HIC circuits. High performance is achieved since the discretes can often be precisely tuned to high tolerances by laser trimming operations. The new multichip module technology is simply an extension of HIC technology to higher interconnection density and more sophisticated silicon devices. The packaging of HICs has often been a shortcoming of the technology because most HICs were custom shapes and sizes that did not conform to standard package codes. The oversized ceramic or printed wiring board substrates were not considered amenable to post-molded plastic packaging due to the high incidence of package cracking in the thin edge of plastic that would surround the substrate. The trend in single chip packaging toward large devices in small outline packages, and the materials and process technology that was developed to package these large chips, opened up the possibility of extending molded plastic packaging to hybrid devices.

The technical hurdles to developing molded packages for HICs include the encapsulation of the substrate in molding compound in a way that allows all of the air in the cavity to be vented. The large substrate of a HIC will effectively divide the flow in the cavity into separate flow fronts above and below the substrate as is shown in Figure 9-9 with the possibility of one front enjoying a substantial lead over the other based on the individual flow resistances encountered on each side. Although this lead-lag problem is encountered in silicon IC packaging and was discussed in Section 6.4.4, the problem is more serious in HIC packaging because there is much less communication between the flow fronts as they are separated by a solid barrier instead of the semi-permeable leadframe. Analyzing the flow resistances on either side of the substrate using either flow simulation methods or experimental trials can often be used to balance the flows. Cavity halves of equal thicknesses do not always provide balanced flow fronts because of other incongruences in the flow resistance such as the location of the gate, temperature differences, and the flow restriction offered by the components which are most often placed only on one side of the substrate. Warping of the packaged HIC is also a major problem. Even slight thickness differences in the cavities above and below the substrate will cause warpage of the package and possible cracking of the substrate. The other major technical hurdle to achieve HIC packaging was the development of ultra low stress molding compounds. These materials also have exceptional adhesion that enables them to dissipate the thermomechanical shrinkage stresses over the en-

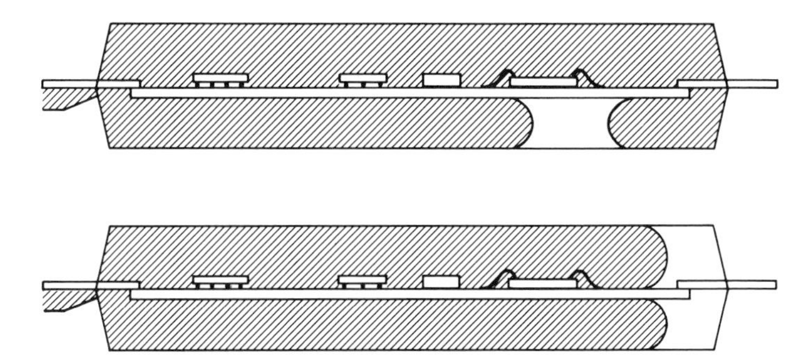

**FIGURE 9-9** Examples of the flow of molding compound into a HIC package cavity, showing how it divides into two separate flow fronts above and below the substrate. The first example shows excessive lead of the upper front resulting in it wrapping around the end of the substrate, sealing off the air vent at the end of the cavity, and forming an air void in the lower half. The second example shows more balanced filling above and below the substrate which causes both fronts to reach the vent simultaneously, pushing all of the air out of the cavity ahead of the flow fronts.

tire package volume by avoiding the stress concentrations that accompany delamination. They also have sufficient strength to withstand the high stresses created by the oversized substrate. There have been several reports of packaging hybrid circuits in conventional molded plastic packages using these low stress molding compounds [8, 9, 10]. This development is important in promoting more widespread use of hybrids since standardized HIC packages can be handled in all automated circuit board assembly equipment. It also provides a basis for extending molded plastic packaging to other large substrate devices such as the multichip modules that are discussed in the next section.

### 9.2.2 Multichip Modules

There is growing divergence between the feature sizes on the chip compared with the feature sizes on the circuit board as is shown in the plot of Figure 9-10 [2]. It will be increasingly more difficult to accommodate the very high interconnection densities that will accompany 500 I/O chips with the density that can be achieved by printed circuit boards, even advanced multilayer boards. Multichip modules (MCM) [11, 12], where the chip is interconnected by direct attachment to fine line, thin film conductors on the module, are an important means of bridging this growing gap in interconnection density. The MCM is essentially a mini-circuit board with chip on board interconnection between the devices and the board. They are not different in principle from hybrid integrated circuits (HICs) which combined active and discrete devices on a substrate using a mixture of both thin film and thick film technologies. The emerging MCM technology of the late 1980s differs from HICs in that it is largely multilayer

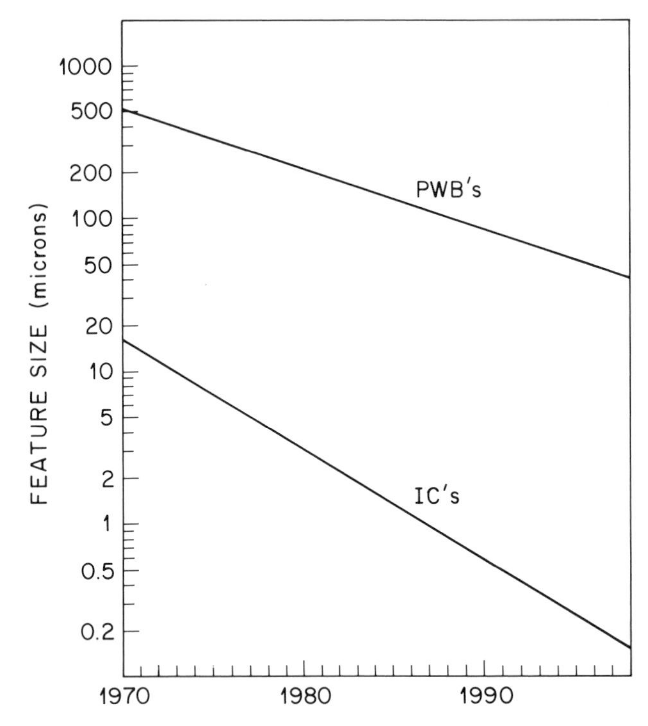


**FIGURE 9-10** Trends in interconnection density on the chip and on the circuit board plotted against the year that that density was introduced to production. The lines are a fit to data points collected from a large number of different systems [2].

with much higher interconnection densities. The interconnection pattern on the substrate is usually produced by the thin film deposition and etching techniques used for the chips themselves. Either aluminum or copper conductors can be used. Line spacings achievable are of the order of one thousandth of an inch compared to the five thousandths of an inch limit of most double sided and multilayer circuit boards. This increase in wiring density makes it possible to interconnect even the most dense structures with only two interconnect levels on the substrate compared to the 20 or more that can be required for a sophisticated multilayer board. The substrate material is often silicon because of its CTE match to the mounted chips, its excellent thermal conductivity, and the ability to use existing silicon wafer fabrication equipment to produce MCM substrates. Alumina can also be used as a substrate in fine line, thin film architecture if sufficient surface smoothness can be achieved. Polymeric films are used as the dielectric layers in most multilayer MCMs, and polyimide materials are often used in this capacity. The chips are interconnected to the conductor pattern on the substrate through wire bonds, solder bumps, TAB or a combination of these three technologies for the different chips on the module.

Early MCMs have been high end products which offered little motivation to use lower cost post-molded plastic packaging. As the technology matures and becomes more widespread, however, it will undoubtedly face cost competition from discrete component clusters and multilayer HICs. There will be interest and motivation to package MCMs in molded plastic to fend off these competitive pressures. There are significant technical issues in MCM packaging that may limit the extent to which plastic packages can be used. With several silicon devices on a single substrate, MCMs are likely to be large in surface area. To conserve area on the circuit board, the package outline is unlikely to be much larger than the substrate outline, least the interconnection density advantages of the MCM be squandered in an oversized package. A molded MCM package, a schematic of which is shown in Figure 9-11, is therefore likely to have only a thin edge of plastic around the substrate. The large size of the silicon or ceramic substrate will create high thermal shrinkage stresses in this thin layer making MCM packages more prone to cracking than other small outline single chip modules. Molding compounds with outstanding crack resistance, adhesion and CTE matching will be needed to package the MCM geometry. Another serious issue will be high power output as this will undoubtedly be a characteristic of these sophisticated systems. Molding compounds with excellent thermal conductivity will be needed, but materials with as much as four to five times the thermal conductivity of the conventional fused silica filler materials are still likely to fall short of being able to maintain the device surface temperature below the upper limit. This is probably the most serious challenge to MCM packaging: the high interconnection density of a number of high power chips will place a thermal loading on the plastic package that is without precedent in single chip configurations.

### 9.2.3 Wafer Scale Integration

Wafer scale integration (WSI) represents a significant jump forward in the degree of integration by incorporating devices and interconnection on the same piece of silicon. This silicon would be much larger than a conventional silicon

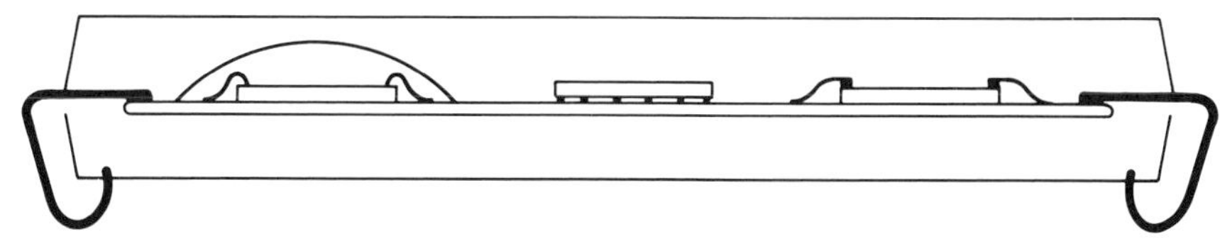

**FIGURE 9-11** Multichip module packaged in a post-molded plastic package showing the principle features of this design which include the multiple chips, a thin edge of molding compound surrounding the substrate, and the lead attachment to the substrate. The three chips on the module show three different attachment methods that can be used: wire bonding (left), solder bumps (center) and TAB (right). The wire bonded device is shown encapsulated in a glob top material, an option which may or may not be needed to prevent flow-induced wire damage.

device or even an MCM; most likely it would be wafer size. The interconnection density would be extremely high because the conductor paths between chips would be the same scale as the conductors on the chip. Several signal planes could be accommodated. It is uncertain whether the device functions would be segregated from interconnect functions as they are on conventional packaged devices on printed circuit boards, or whether device and interconnection would be melded seamlessly on the wafer. Other advantages of WSI would include high reliability since the system would be entirely aluminum metallization with no intermetallic junctions as are present in current technology. WSI structures could also be exceptionally fault tolerant because of inherent properties of redundancy and reconfigurability. There is also the opportunity for lower costs because of the elimination of several interconnection levels such as discrete packages and discrete printed circuit cards. Start-up and manufacturing costs could be low since the technology consists of silicon processing for which a wealth of expertise and equipment already exists. Unfortunately, the technical hurdles to WSI are as daunting as its advantages are attractive. Yields on the wafer would have to be very high even with some redundancy to provide a respectable yield of wafer size modules. Greater degrees of redundancy used to increase the yields would add signal layers and increase propagation delays, negating some of the inherent benefits and adding their own yield criteria. Although the promise of WSI is great, it appears that the difficulties and shortcomings are also very severe. It has not attracted the level of research and development activity that could make it a serious technology option in the near future. With regard to packaging, it is unlikely that any wafer scale device could be packaged in a molded plastic body. When and if WSI reaches a viable state, it will likely be packaged with more expensive custom modules that will incorporate advanced capabilities for power supply, heat dissipation and signal timing. A WSI module would be more closely related in form and function to a printed circuit pack than to a multichip module, hence its packaging would reflect this difference.

### 9.2.4 Tape Automated Bonding (TAB)

Tape automated bonding (TAB) is an interconnection technology where the interconnection or fan-out pattern is carried on a continuous metal tape that may or may not be supported by a dielectric film [13, 14, 15, 16, 17]. TAB was introduced in Section 1.6, but the technology and its implications are explored in greater depth here. There are several different types of TAB tape, and these are best illustrated through the use of the graphic provided in Figure 9-12 [13]. Supporting the pattern on film, which is more common for high I/O applications, allows very fine patterns to be accommodated. The absence of a wire bonding tool at the bond pad sites means that bond pad spacings on the chip

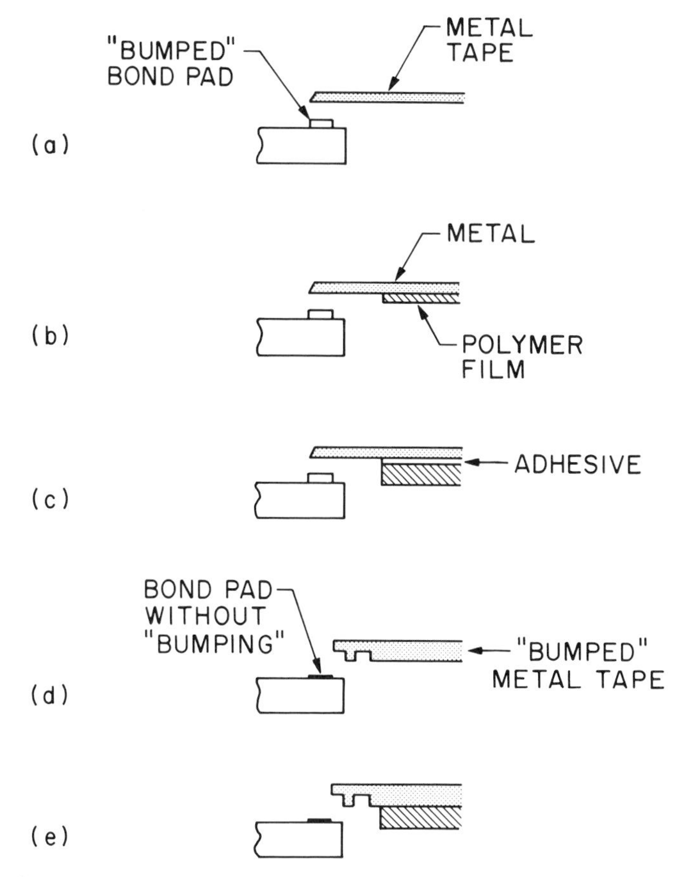


**FIGURE 9-12** Different types of TAB tapes. The single layer tape shown in (a) cannot be pretested and is suited for low cost applications. The tapes shown in (a), (b) and (c) require bumps on the wafer. Types (d) and (e) are bumped tapes. (After Reference 13.)

can be smaller. TAB can be used with 4 mil bond pad spacings, considerably less than the 6 to 8 mil spacings in use in the late 1980s. This allows reductions in chip size and increases the number of chips produced on a wafer in cases of bond pad-limited chips. The polymer film could be uniform or it could incorporate cut-outs to allow access to the leads from either side of the tape. Simultaneous attachment of all of the leads, known as gang bonding, provides for high production rates. Supporting the fanout pattern on a polymer dielectric, typically polyimide, allows testing and burn-in prior to final assembly. A list of the principal features of TAB is provided in Table 9-1. In TAB, the lead fingers are attached directly to the chip bond pads. Bumps on the lead end tips or on the chip bond pads formed through special metallurgy are needed to effect

**Table 9-1. Major Features of TAB**

| |
|---|
| Electrical properties are superior to wire bonding. |
| Thermal dissipation properties are superior to wire bonding. |
| Higher interconnection densities are possible. |
| Greater pull strength on bond pad connection. |
| Ability to test and burn-in device prior to assembly. |
| Higher productivity than wire bonding for gang attach TAB. |
| Requires extra bumps on lead ends or bond pads |

this attachment as is shown in Figure 9-12. The need for these bumps and the processes to manufacture them have been impediments to the more widespread use of TAB.

There are actually two ways of using TAB: as an assembly technique or as a packaging technique. As a packaging technique, TAB could be used as a direct chip attach option that would be much superior to chip on board (COB) because TAB would allow burn-in and testing, although fixturing would not be as convenient as with a packaged device. TAB would also compete with solder bump technology as an attachment method for multichip modules, particularly on substrates other than silicon where compliant leads would be an advantage in absorbing the mismatch in thermal expansion. TAB used as a packaging technology does not involve package molding, however, so it will not be considered further. As an assembly technique, TAB could be used instead of wire bonding for inner lead attachment. This application has good prospects as the bond pad spacing continues to decrease and approaches the limits of what can be achieved with wire bonded leadframes. The leadframe thickness decreases that have to accompany the lead pitch decreases may reduce leadframe rigidity below the useful range, thereby mandating a switch to TAB regardless of wire bonding capabilities. Both the electrical and thermal properties of TAB are superior to wire bonding as is shown in the comparison provided in Table 9-2 [14]. The electrical property advantages are derived from the larger cross section area of signal carrying conductor in TAB, and the rectangular cross section which provides lower lead inductances than round cross sections [14]. The ther-

**Table 9-2. Comparison of the Electrical and Thermal Properties of TAB and Wire Bonding (After Reference 14)**

| | *Wire Bonding* | | *TAB* |
|---|---|---|---|
| *Parameter* | *Aluminum* | *Gold* | *Copper* |
| Lead resistance, ohms | 0.142 | 0.122 | 0.017 |
| Lead to Lead Capacitance, pf, 6 mil spacing | 0.025 | 0.025 | 0.006 |
| Lead inductance, nh | 2.621 | 2.621 | 2.10 |
| Lead conduction °C/mW | 79.6 | 52.6 | 8.3 |
| Lead convection, free, °C/mW | 336.5 | 336.5 | 149.5 |

mal property advantages are also derived from the larger thermal conduction cross section and the larger surface area of the TAB lead compared to a wire bond. In addition, TAB, being a gang attachment method, would also offer important productivity advantages over the single wire at a time approach of wire bonding. These advantages all indicate that TAB should be the preferred interconnection technology for high I/O devices where electrical, thermal and productivity issues are far more important, at least for inner lead attachment which would still require a conventional outer package that could be molded plastic.

The significant advantages of TAB have not been deciding factors in promoting its further development and implementation by the late 1980s. A number of factors have slowed its acceptance. There is a large embedded base of wire bonding expertise and equipment, and wire bonding on leadframes can still accommodate the electrical, thermal and productivity needs of a very large fraction of production IC devices. Wire bonding has continued to improve and to be extended to new limits of interconnection density. It now appears that wire bonding may be used down to bond pad spacings of 5 mil centers. The time when a large segment of the market goes from the industry standard of 7 mils in the late 1980s to below 5 mils will be near the end of the century. Hence, there has not been sufficient motivation to invest in the equipment and expertise needed for the widespread implementation of TAB. For these reasons, the prospects for TAB in the 1990s will depend more on its advantages in attachment rate productivity and electrical performance, and the shortcomings of leadframe technology, rather than a capability advantage over wire bonding. Another major factor has been the difficulty in developing good low cost bumping technology, and the general difficulty in simultaneous attachment of large numbers of leads with the near perfect yields common in wire bonding. These do not represent any genuine limitations of the technology, but rather lack of sufficient motivation to invest in the development efforts that are sill needed. In view of the trends in device I/O requirements, thermal dissipation, and clock rates, it is apparent that there will be serious shortcomings with wire bonded leadframe technology that will provide the motivation to perfect and implement TAB.

Overall, very little is known about postmolded packaging of TAB using transfer molding. Flow-induced stress on the polyimide film and extra fragile lead fingers is expected to be a serious problem, however. Although wire sweep will not be relevant with TAB, a form of paddle shift which will manifest itself as flow-induced movement of the film support will clearly be an issue. For very high I/O designs with small bond pad spacings, the dense array of leads coming off the chip, shown in the photograph of Figure 9-13, will create a near impermeable surface that will retard the flow of the molding compound through the leads. The flow fronts below and above the tape will be more isolated and the distance between the flow front locations will increase, thereby increasing the magnitude of the pressure forces that can deform the assembly. The require-

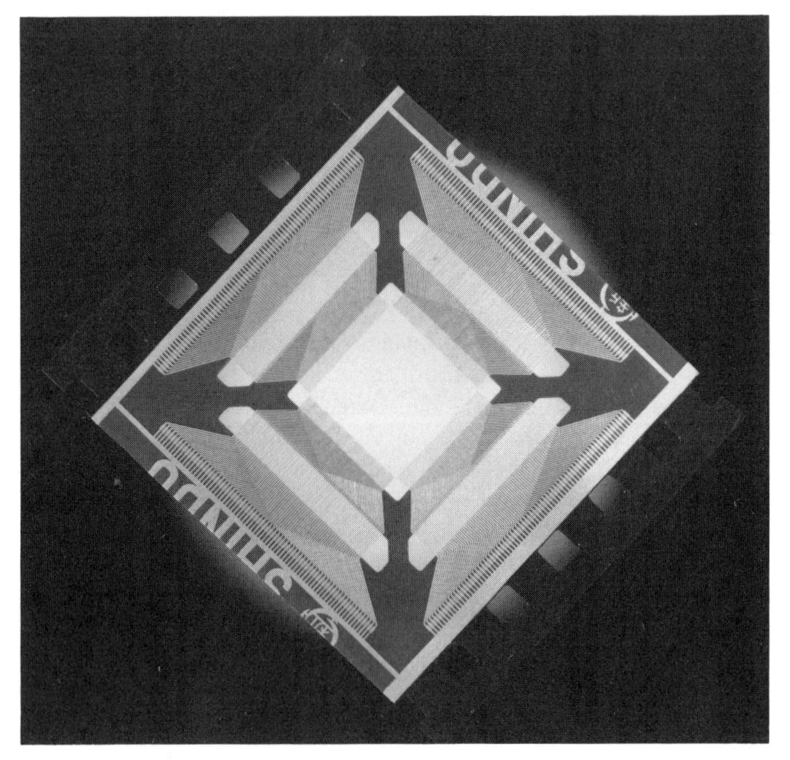

**FIGURE 9-13** Photograph of a TAB tape that provides 200 leads on 4 mil bond pad spacing. The dense array of leads coming off the chip provides very little permeability for the molding compound to move through the frame, thereby increasing the magnitude of the flow-induced forces.

ments that TAB will impose on the molding compounds and the molding processes are expected to be stringent. Low viscosity at the low shear rates that the molding compound flows over the tape will be needed. Adhesion to polyimide or other dielectric carrier will be as important as adhesion to the leadframe is in the present wire bonded leadframe technology.

### 9.2.5 Integrated Optics

The growing use of lightwave transmission of voice and data signals means that there will be much greater integration of photonic and electronic devices by the mid-1990s. Optical data link (ODL) devices which couple electrical and photonic signals will become widespread and subject to intense competitive pressure, both in terms of performance and cost. The multichip modules described previously are likely to be a preferred format for these ODLs since they can combine several electronic/photonic components within a single device. High levels of heat dissipation will be a major technical challenge to the design of these modules. Low cost plastic packaging will be an essential part of these devices from both performance and cost standpoints. Implementation of plastic packaging will require a design strategy to bring the fibers to the components inside the package while maintaining good moisture ingress seals. Avoiding expensive plug connectors would also be an advantage. Power dissipation strategies that go beyond those required for single chip modules will be needed. Cut-away views of two optical devices showing key aspects of the packaging are provided in Figure 9-14 [18].

The specific material requirements to accommodate fiber attachments are not completely known since they will be largely design dependent. There are some characteristics that are apparent, however. The molding materials and designs will require micron-level precision to accommodate fiber alignment. Flash-free

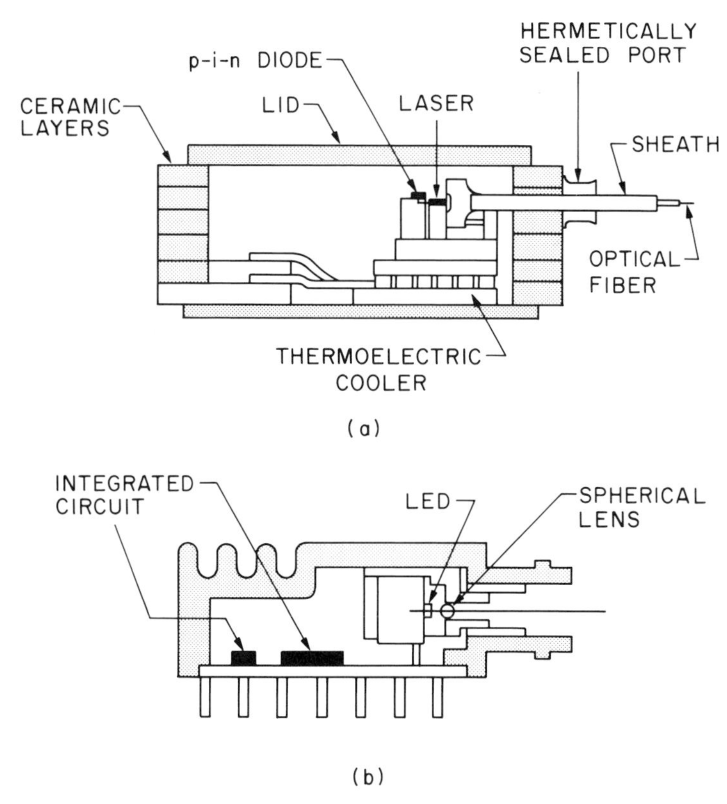


**FIGURE 9-14** Cutaway view of two optical devices: (a) a laser package, and (b) a light-emitting diode data link package [18].

molding will be another important advantage since the molding plastic will be required to seal around an optical fiber attachment feature. High thermal conductivity is another advantage since most optical devices are high power. In general, thermoset materials can achieve higher precision than thermoplastics, but new injection molding technology is rapidly closing the small gap that remains. Conversely, thermoplastic materials are more compliant than the highly filled thermosets and are, therefore, more amenable to bayonet-type fiber connector attachment.

### 9.2.6 Leadframe Technology

There are two primary trends in leadframe technology and both have important implications for the future of molded plastic packaging. The first is that copper alloys are displacing Alloy 42 (Iron-Nickel) as the major leadframe material for processor and logic chips due to the better thermal conductivity of copper. The drawbacks of copper are its poorer mechanical properties and its higher coefficient of thermal expansion, drawbacks that are serious enough to prohibit the use of copper in most applications where thermal dissipation is not an issue. This trend toward copper is not expected to change in the near future, barring some unforeseen resolution of the power dissipation problem. The second important trend is the difficulty in designing and manufacturing leadframes to accommodate the sharp increases in wire bond density needed for very high lead count PQFPs. In general, it is becoming more difficult to produce a tool that can punch 0.006 inch thick leadframes with less than 0.006 inch space between the lead tips. Very high I/O leadframes are increasingly available only as etched frames, but etching is not without its own limitations. Similar to punching, as the leads and the spacing between leads becomes smaller, the thickness of the lead must be reduced to preserve a lead cross section aspect ratio near one. The lead thickness cannot be greater than the lead width to prevent a mechanically unstable lead geometry. Therefore, lead thickness has had to be reduced to accommodate reductions in lead width. There is also the over-etching consideration. It is impossible to etch a lead that is significantly thicker than it is wide. The width of the lead will be reduced to nothing before the entire thickness is transgressed by the converging etch fronts moving from the top and bottom of the lead.

For the reasons described above, leadframe thickness has decreased substantially from 0.010 inch for packages such as 40 pin DIPs down to 0.006 inch for the fine pitch PQFPs (84–196 lead count). Leadframes for 164 and 196 PQFP can still be made, but etched frames are all that is available for these designs. In addition, the paddle is often oversized to shorten the lead length and thereby increase the spacing between the lead tips. There are several different types of lead tip layout [19] that offer different pad size, lead tip spacings and wire bond lengths. The pad-limited die, where the bonding pads on the chip are set to the minimum value allowed by the wire bonding tool, provides the smallest die pad, but the longest wire bond spans. Uniformity of the wire bond spans and the angle the wire makes with the lead tip are also important considerations for wire bond yield and molding yield [19]. The angle, often known as the angle of attack, is zero at the middle of the die pad and up to 45° near the corner of the die on some wire bond layouts.

There are serious repercussions on the molding technology of the thinner leadframes, larger die pads and longer wire bond spans. The flow-induced deformation depends on the inverse third power of the leadframe thickness as shown in Equation (7-2), paddle shift will increase proportional to the area of the paddle, and wire sweep will also increase linearly with wire bond length. Therefore, the flow-induced forces and the difficulty in molding the package will increase significantly. Even if these high lead count frames are moldable, the feasibility of producing die and wire leadframes with greater than 200 leads is not certain. It appears that leadframe manufacture will be the most important limitation to extending molded plastic packaging to very high lead counts. Inner lead bonding TAB is largely considered to be the best prospect for preserving molded plastic packaging at these high lead counts.

## 9.3 MOLDING PRODUCTIVITY TRENDS

Cost will continue to be an overriding consideration and a technology driver. Performance and reliability criteria will adapt to the cost constraints of the market. The 300–500 pin packages of the near future will require high productivity in assembly and molding operations in order to retain their cost advantage over premolded PQFP packages and pin (pad) grid arrays (PGAs). The expectations placed on the molding plastic to meet these productivity needs will include both cycle time and yield requirements. Present molding productivity for PLCC packages are approximately 800 packaged devices per hour, known as the estimated hourly output (EHO), and very near 100% yield. The molding compound and process characteristics to meet these standards are summarized in the following sections.

### 9.3.1 Low Viscosity at Low Shear Rates

Continued reduction in flow-induced stresses will be needed to achieve high yield on the fragile wire bonded leadframes used for high-lead-count packages or the TAB structures that may replace them. These stresses are proportional to the product of molding compound viscosity and velocity, hence reductions can be achieved by lowering one or both. The viscosity of the molding compound at the shear rate and temperature at which the material flows over the chip will have to continue to decrease to accommodate the reductions in leadframe thickness and increases in lead length that will occur in the next 5–10 years, but large reductions are unlikely given that the technology is approaching its limit. Processing improvements can reduce the need for some of this viscosity reduction.

### 9.3.2 High Production Rate

It is important that the production rate of molded packaging remain high to ward off competition from the more labor intensive PGAs and premolded packages. Large fragile packages will require smaller molds with fewer cavities to achieve high yields. Overall molding productivity should stay above 500 parts per hour. This will require molding compounds that can be molded at 99+ % yield with at least 24 cavities per cycle and a total cycle time of 2 minutes. In-mold cure times will have to be in the range of 60–90 seconds to achieve an overall cycle time of 2 minutes allowing enough time for mold cleaning and preparation. Materials that are amenable to automation will be much more likely to meet these productivity requirements.

### 9.3.3 Automation

Automation will play an increasingly important role in meeting the future requirements of molded plastic packaging. Not only will automation be needed to meet the productivity requirements, but it will be important in meeting quality requirements as well. Assemblies with 200 or more leads, wire bonds or TAB, will be exceptionally fragile to the point where any manual handling would present a serious opportunity for damage. Contamination of the devices by foreign debris in the molding area is also easier to control with automation rather than human operators. The third major advantage of automation is in lowering costs. As labor costs in the major IC manufacturing countries continue to rise, there will be increasing use of automation to remain competitive with less developed nations which can offer lower labor rates.

Several major types of automation are being implemented. One approach is partial automation of various labor intensive aspects of the process such as preform heating and handling. Automation of these procedures allows one operator to handle more than one transfer press; often as many as four or five presses can be worked by a single operator. Automated preform preheating and loading provide some of the greatest challenges to this type of automation, however. In most cases, it is impossible to heat the preforms to the full preheat temperature of 90°C because they become tacky and difficult to handle. Most systems settle for preheat temperatures in the vicinity of the glass transition temperature (50°C) so that the preform retains some firmness. A robotic arm places the heated preform into the transfer pot initiating the normal process sequence of transfer, curing and ejection of the molded leadframes. A diagram of a robotic preform heating and insertion system is shown in Figure 9-15.

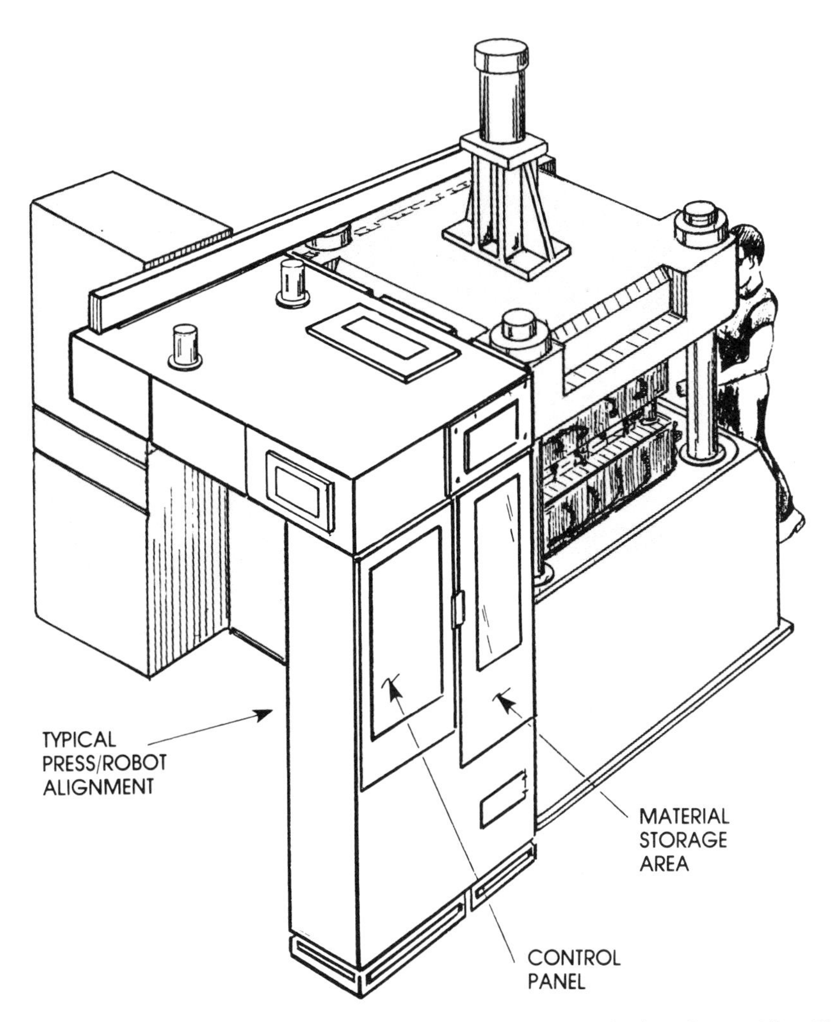


**FIGURE 9-15** Automated preform preheater and supply system for plastic package molding. The illustration shows the unit in light shading attached to the back of a conventional transfer molding press in dark shading. (Courtesy of Equipment for Semiconductors, Inc.)

There is also the total automation of a conventional single pot transfer molding press. In this technology, the process runs essentially without the aid of an operator, although in many cases an automated line will require 10–20% of an operator's time to simply move cassettes around and check into alarms and malfunctions. In this type of line, the assembled leadframes are loaded into cassettes and fed to a frame loader, similar to the many manual operations that have automatic frame loaders. The preforms are preheated and inserted in the transfer pot using the same automation principles described above. Transfer and packing are also fully automated requiring no operator involvement. A characteristic of these automated systems is that the performance is improved when a smaller molding tool with fewer cavities is used. The smaller molds with shorter flow lengths can work with faster cure molding compounds that recover some of the productivity lost in the fewer number of cavities. The gates are broken off from the molded body either by the action of the ejector pins or through a separate mechanism. Another robotic member then lifts the frames from the ejector pins and loads them into a second set of cassettes. Brushes emerge to clear off any debris and to scrub the cavities of any residual molding compound. A photograph of an automated molding system is shown in Figure 9-16. In most cases the system is completely enclosed to minimize the noise of the machinery and improve the cleanliness of the process. In a totally automated, in-line packaging system, the cassettes of leadframes would feed an automated trim and form press, and then the singulated packages would move on to a code marking station.

Another type of transfer molding automation is the multiplunger technology that has previously been described in Section 5.5.3. Multiplunger differs from the automated single pot system described above in that as many as 6 to 12 pots feed from 6 to 24 cavities: one to four cavities per transfer pot. Multiplunger also uses a partially heated or unheated preform to facilitate robotic handling.

**FIGURE 9-16** Photograph of an automated universal packaging system. (Courtesy of Towa Corporation.)

A major technical advantage is the short flow length which virtually assures a long cavity fill time, attendant low velocity, and a long packing phase that is very effective in minimizing voids and promoting high molding compound density. With the low number of cavities, multiplunger machines usually have production rates that fall below that of full size single pot molds, but the very fast curing molding compounds that can be used with this exceptionally short flow length compensate for some of the shortfall.

The molding compound trends to support automation will be toward materials that have shorter flow lengths and faster cure times. A growing segment of the molding compound market will be garnered by these fast cure materials, and they will begin to usurp a disproportionate amount of the suppliers' research and development activity. Manufacturers who do not pursue automated molding may find that they no longer have access to the best molding compound technology.

## 9.4 BOARD LEVEL ASSEMBLY TRENDS

Circuit board level assembly has an important influence on the selection of materials for molded plastic packaging. Therefore, the trends in these assembly operations have to be taken into consideration when future packaging strategies are considered.

### 9.4.1 Surface Mount

Surface mount technology allows greater interconnection density on the printed circuit board because the bonding pads and vias are not restricted to the 0.100 inch centers that are the minimum spaces that can be accommodated with drill-through holes. The leads on surface mount packages can therefore be 0.050 or 0.025 inch or less instead of the 0.100 inch that is standard for through-hole components. In addition, components can be placed on both sides of the board, effectively doubling the number of components that can be interconnected. There are numerous reports that predict the use of surface mount technology (SMT) will continue to grow rapidly into the 1990s. Industry consultants predict that about 70% of circuit boards will be designed to accept SMT by 1992 compared to only 43% in 1987 [20]. Despite the rapid growth of SMT, the fraction of boards that are completely surface mount was still small by the late 1980s. A large fraction of product are mixed technology boards which combine both through-hole and surface mount components. Although mixing technologies compromises many of the advantages of SMT, it is necessary because it is unlikely that all the components of a complex board will be available in surface mount packages. Surprisingly, the consumer products industry with its rapid product introductions and short product lifetimes has lead the way in the implementation of surface mount technology. In addition, many consumer electronic products, particulary personal and portable equipment, have strong market incentives to reduce size and weight. Products such as personal stereos, televisions, and video cassette players have been leaders in the innovation and implementation of important new interconnection and assembly techniques. As an example, a photograph of a high density surface mount board that is used in a personal compact disc player is provided in Figure 9-17.

In SMT, the entire package is subjected to the solder temperature, whereas in through-hole mounting only the leads themselves are subjected to the solder temperature. The trend in the type of reflow process used for SMT has also been away from vapor phase reflow and toward infra red reflow [20]. The principal ramification of surface mount attachment is the higher temperature excursion that the molded body experiences. Reflow solder temperatures are a minimum of 215°C with a maximum of 260°C for some processes and equipment. With through-hold mounting, the upper temperature limit was the mold and post-cure temperature, usually 175°C. Therefore, the thermomechanical

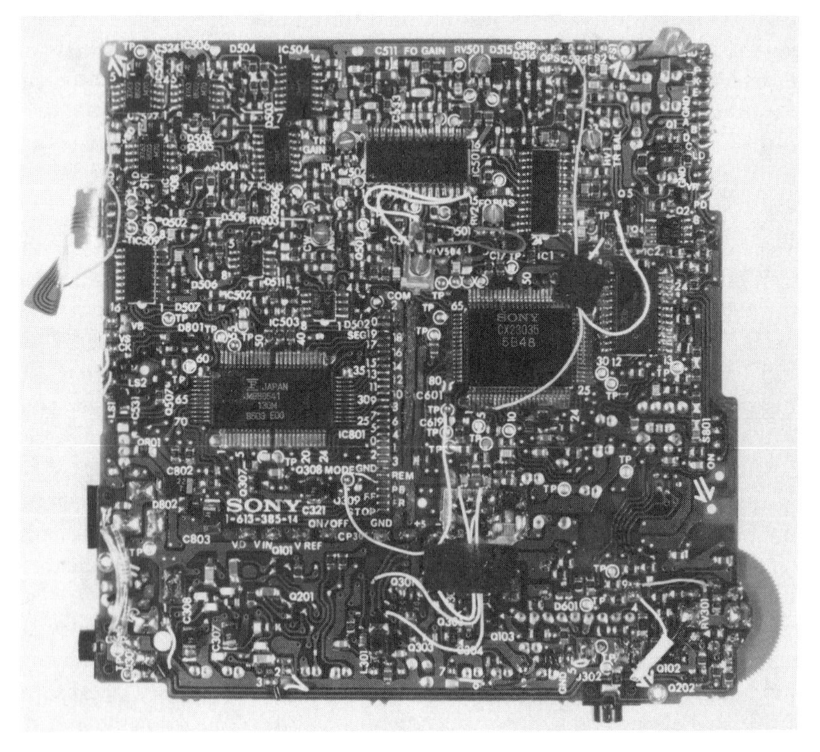


**FIGURE 9-17** A photograph of a high density surface mount board that is used in a portable compact disk player.

shrinkage stresses of SMT will be more severe since the overall temperature excursion is greater. This thermal shrinkage effect is minimal, however, because the reflow temperature is above the anchor temperature. (See Section 8.2.1 for an explanation of anchor temperature.) The shrinkage occurs in the rubbery region of material behavior, where the forces generated are small. The more serious problem is the volume increase of water absorbed in the molded body and clustered at the internal interfaces. Upon heating to the reflow temperature, this water vaporizes and expands, occasionally cracking the package through the mechanism described in Section 8.2.2.1. Improved molding compounds that either resist delamination at the paddle support interface, resist deforming at the reflow solder temperature, or resist cracking when deformed by the vapor expansion will be needed to eliminate this problem.

### 9.4.2 Circuit Board Materials

The use of conventional glass/epoxy printed wiring board will decline in the 1990s with most of the volume being taken up by glass/polyimide, thick film ceramic and co-fired ceramic, as shown in Table 9-3 [20]. In general, these non-epoxy materials have a lower coefficient of thermal expansion than epoxy/glass. The CTE of present epoxy molding compound is about three times greater than ceramic, hence a significant disparity in the coefficient of thermal expansion will be introduced when plastic molded packages are mounted on non-epoxy circuit boards. The larger size packages of the near future will be more severely affected by this disparity than today's packages. This trend in circuit board materials also underscores the need for molding compounds with very low CTE.

**9.4.2.1 Molded Printed Circuit Boards** Molded printed circuit boards are a relatively new technology [21, 22, 23] whereby the circuit board substrate is manufactured in an injection molding operation from high performance thermoplastic materials rather than from an epoxy-glass composite material in a compression molding process. There are only a small number of engineering thermoplastics that have sufficient thermal, mechanical and processability properties to be used for molded circuit boards. These include poly(sulfone), poly(ethersulfone), poly(etherimide), poly(arylsulfone), and poly(phenylene sulfide). The advantages of molded circuit substrates include:

1. Elimination of routing the board to size
2. Elimination of drilling
3. Ability to incorporate molded-in features such as connectors, standoffs, handles and bosses
4. Ability to incorporate molded-in alignment sites such as recesses for surface mount components

**Table 9-3. Volume Percent of Circuit Board Substrates Used with Chip Carriers**

| | *1987* | *1988* | *1992* |
|---|---|---|---|
| Glass Epoxy PWB | 68.35 | 56.53 | 40.35 |
| Kevlar Epoxy PWB | .17 | 0 | 1.70 |
| Graphite Epoxy PWB | 0 | 0 | 0 |
| Glass Polyimide PWB | 6.27 | 7.79 | 10.69 |
| Kevlar Polyimide PWB | .61 | .90 | 0 |
| Metal Core PWB | 2.45 | .23 | 2.58 |
| Multiwire PWB | .11 | .09 | .70 |
| Microwire PWB | .04 | .00 | .00 |
| Thin Film Ceramic | 2.05 | 3.24 | 4.29 |
| Thick Film Ceramic | 13.34 | 22.70 | 15.92 |
| Co-Fire Ceramic | .86 | 1.62 | 14.77 |
| Porcelain on Steel | 0 | 0 | 0 |
| COB | .94 | 1.49 | 1.94 |
| Other | 4.79 | 5.41 | 7.05 |
| TOTAL | 100.00 | 100.00 | 100.00 |

(From "Chip Carriers and Other Integrated Circuit Packages, 1988" Copyright by James D. Welterlen. Reprinted with Permission.)

5. Allows higher use temperature
6. Better dielectric properties than epoxy-glass boards
7. Lower *z*-axis expansion than epoxy-glass laminates which can improve reliability with through-hole components
8. Improved dimensional precision
9. The edge of the board can be contoured to facilitate insertion in edge connectors
10. Potential for full three-dimensional circuit pattern that could ultimately fuse circuity with other structural and aesthetic parts

Disadvantages include:

1. Higher startup and tooling costs
2. Longer startup time to allow for mold manufacture
3. More difficult to change the conductor path pattern since it is incorporated in the molding tool
4. Greater *x-y* expansion than epoxy-glass boards
5. More expensive for small production volumes
6. Mechanical properties of the board are usually inferior to epoxy-glass composite

The cost difference between a molded thermoplastic board and a conventional epoxy-glass board is not great. Lower costs are possible for molded boards, but large production volumes are needed to amortize the tooling cost. Circuit patterning technology where the conductor pattern is molded into the substrate can offer cost advantages compared to other types of masking and imaging. With regard to plastic packaging, the major implications of molded boards are the greater precision in registering surface mount components in recesses that could greatly improve solder attachment yields of fine pitch packages. This capability offers the opportunity of using even finer pitches on outer lead spacings, thereby further reducing the size of the package and reducing the problems of greater flow-induced stresses associated with large packages. A disadvantage, however, is the larger $x$-$y$ expansion which provides a greater mismatch in the CTEs of molded package and circuit board. This will cause greater stress on the package leads, particularly on large, high-pin-count packages.

### 9.4.3 Solvent Cleaning

Solvent cleaning operations are expected to remain a part of circuit board assembly, but there will be a growing movement to eliminate chlorofluorocarbons (CFCs) from industrial use because of the environmental damage they cause. New nonflammable cleaning solvents are under development, but it is impossible to predict which ones will emerge. Good solvent resistance will continue to be a need for any molded plastic package material.

### 9.4.4 Water-Soluble Flux

It appears that these fluxes will remain in production for the foreseeable future. These are acid-based fluxes which can be corrosive if they are able to enter the package. Most molding plastics are impervious to these fluxes over the exposure times and temperatures in question; it is the interfacial path along the lead which is most prone to ingress of the corrosive agents. Good sealing and adhesion along the leads is needed to prevent ionic impurity contamination during the fluxing operation. A common scenario is partial ingress of the acid during circuit board assembly, followed by gradual migration of the corrosives to the device with time and temperature cycling resulting in failures at a much later time.

### 9.4.5 Precision Alignment of Leads

The fine- and extra-fine-pitch devices of the near future will require precise location of the leads in order to achieve low assembly defect rates. There are several aspects of plastic package molding that affect the positioning of the leads. One of these is the disparities in shrinkage of the molding compound and the leadframe which can buckle the frame and compromise the trim and form operation. Close matching of the CTE of the molding compound and the leadframe is required to improve this precision. Preheating of the leadframes and dimensional compensation of the molding tool for the preheat temperature will be increasingly important to prevent misalignment of leads on the circuit board sites.

### 9.4.6 Conductive Polymers for Interconnection

Conductive polymer interconnect materials [24, 25] are organic materials that can be either isotropically or anisotropically conductive. The isotropic materials are conductive polymers that are not unlike die attach epoxies. Silver, nickel, or silver-coated glass are the common filler materials, whereas epoxies are the most common matrix materials. They are usually liquid or pastes that are used in place of solder to attach the leads of the component to the circuit board. They are applied either through a screen or are stamp printed. Anisotropically conductive materials are conductive in only one direction, typically the $z$ axis perpendicular to the surface of the board, insulating in the other directions. The anisotropy derives from achieving the proper filler alignment so that the inclusions bridge the smaller $z$-axis dimension providing conductivity, yet are below the percolation threshold in the $x$-$y$ plane providing isolation over distances that are of the order of the bond pad spacing. The anisotropic materials are often elastomers that incorporate aligned conductors in the $z$-axis surrounded by insulator to provide the $x$-$y$ isolation [24]. Anisotropic materials can be used for second level interconnection between the component and the circuit board, and they are also feasible for first level interconnection between the chip and the fan-out pattern. They are attractive for high density interconnections where conventional wire bonding may be inadequate.

This technology is too immature at the time of this writing to assess its prospects for replacing solder or wire bonding for first and second level interconnection. Widespread implementation of either solder bump technology or TAB could obviate the need for anisotropic conductive polymers for high interconnection density. The use of isotropic conductive materials for replacement of solder is probably more mature, but still not close to widespread commercialization. The implications for future packaging materials are significant, however, because elimination of solder would eliminate the high temperature requirements of the packaging materials, making many more plastics potential candidates. These lower temperature plastics often have lower viscosities than the high temperature materials, so flow-induced stresses may be reduced as well. It appears, however, that solder will remain the dominant interconnection medium until the turn of the century.

## 9.5 END-USE AND RELIABILITY TRENDS

The uses of integrated circuits will continue to increase in variety, and this variety will expose the devices to more severe ambient environments. One of the more important growing applications of IC devices is under the hood of an automobile. These chips control many of the engine functions such as air/gas mixture and ignition control, as well as many uses besides the drive train such as climate control, trip computers, and the audio system. The under-hood applications are clearly the most demanding in terms of package integrity and device reliability. Temperatures in this environment reach over 100°C. These prolonged high temperatures increase the thermal acceleration factor (see Section 8.3) as well as the physical aging of the epoxy [26] which can hasten embrittlement. In addition there is the opportunity of exposure to hydrocarbon liquids such as gasoline and motor oil. Other extreme environments that are growing in importance are aeronautics where the low temperatures, low pressures, and low humidities of high altitude flight alternate with ground conditions to place extra thermomechanical and moisture diffusion stresses on the molded body. Astronautical applications are even more demanding [27], but the volume of product is so low that it does not warrant much discussion here. In addition, devices used in satellites are in hermetic packages since these represent the ultimate in reliability-weighted criteria. Radiation hardening is an important aspect of reliability in military, space and some medical applications.

Reliability of nonhermetic packages will continue to be a key parameter in choosing among component suppliers. It is expected that requirements for the various reliability tests such as temperature cycling, thermal shock, and pressure cooker will continue to become more severe. These improvements in reliability performance will require the following molding plastic properties: (1) very low ionic impurity, less than 2 ppM of total ionic impurities, (2) excellent adhesion to the device and leadframe to prevent moisture ingress paths and delamination voids, (3) low modulus to reduce the thermal shrinkage stresses (flex modulus below 1000 $Kg/mm^2$), (4) low moisture uptake to prevent the moisture cracking problem during reflow solder attachment, (5) high temperature capabilities to withstand solder reflow attachment, and (6) matching of the coefficients of thermal expansion.

The trend of larger pieces of silicon surrounded by smaller amounts of molding compound will continue to drive the need for lower CTE of the molding compound until it approaches that of the silicon device ($2 \times 10^{-6}$ cm/cm-°C or 2 ppM/°C). Reduced feature sizes will place even greater constraints on the amount of thermal shrinkage mismatch that can be tolerated. The only caveat to this trend is the recent need to return to copper leadframes to facilitate heat removal. Shrinkage onto the leads is considered an essential part of preventing moisture ingress, but copper with a CTE of 16 ppM/°C can show more shrinkage than some ultra low stress molding compounds. Copper leadframes will not be able to forestall the need for more active thermal management, however, so its use may be temporary. Molding plastics with a CTE below 10 ppM/°C will be required to assure high reliability with the large silicon devices packaged in small outline packages. Thermal conductivity of the molding compound will have greater reliability implications as device power output continues to increase and strain the dissipative capacity of a molded plastic package.

## 9.6 MOLDING PLASTIC DEVELOPMENTS

The following is a summary of molding plastic developments, and the associated processes to support these developments, which have potential of meeting future molded packaging needs. Table 9-4 summarizes the advantages and disadvantages of each material.

### 9.6.1 Epoxy Molding Compounds

Epoxy molding compounds using transfer molding will remain an important part of molded packaging technology, but it is becoming increasingly uncertain whether they will be able to meet the yield, productivity and heat dissipation demands of future products. Yield of high-pin-count PQFP packages and heat dissipation are the most serious issues. Moisture cracking during reflow solder operations is also a problem, but likely to be more easily resolved. It is also uncertain how much more the CTE and viscosity of epoxy molding compounds can be decreased using present approaches of filler size, shape, size distribution, and loading level. Nonetheless, there is a large embedded base of equipment and expertise on epoxies, and these materials will remain the material of choice for the majority of product that is not at the leading edge.

### 9.6.2 Liquid Crystal Polymers (LCP)

Injection molded liquid crystal polymers [28] have unique properties that make them a contender for post-molded packaging. A key advantage is production rate. With a cycle time of only 20–30 seconds, LCPs can have a production rate that is four times greater than transfer molded epoxy. The injection molding process is also highly automated compared to transfer molding, and is amenable to total in-line process automation, further reducing labor costs. A unique property of LCP materials is that their viscosity is much lower than other polymers of similar molecular weight. Several major suppliers have developed [29] or are exploring new polymerization technology that would further reduce the viscosity to the low levels that are required for high pin count packages. Unfilled LCP can actually have a zero CTE in the flow plane, but most of this advantage is lost when the filler particles, needed for increased thermal conductivity, disrupt some of the zero shrinkage liquid crystal morphology. The cost of these materials is high, two to four times the cost of epoxy molding compound depending on grade and quantity, but it appears that some cost reduction could be realized because of the high filler loading and the anticipated high volumes that may develop.

**Table 9-4. Advantages and Disadvantages of Packaging materials**

| *Material* | *Advantages* | *Disadvantages* |
|---|---|---|
| Epoxy molding compound | Good adhesion. | Long cure time, low productivity. |
| | Amenable to chemical and physical modifiers for stress reduction. | Further viscosity reductions may be difficult to achieve. |
| | Low ionic impurities. | Further CTE reductions may be difficult to achieve. |
| | | Increased thermal conductivity needed. |
| Liquid crystal polymer | Short cycle time, high productivity, easily automated. | Adhesion poorer than epoxy molding compound. |
| | Low ionic impurities. | Viscosity increase on contact with colder mold and leadframe. |
| | Viscosity reductions are feasible. | |
| | Exceptional solvent resistance. | Increased thermal conductivity needed. |
| | Amenable to fiber attachment. | Poor weld line strength. |
| | | High molding temperature could cause solder reflow. |
| Poly(phenylene sulfide) | Very low melt viscosity. | Outgassing voids may be a fatal problem. |
| | Short cycle time, high productivity, easily automated. | Viscosity increase on contact with colder mold and leadframe. |
| | Good solvent resistance. | Ionic impurities are high in some grades. |
| | Amenable to fiber attachment | Reduction in CTE and increase in thermal conductivity required. |
| | | High molding temperature could cause solder reflow. |
| Liquid injection molding | Very low viscosity. | Long cycle time. |
| | Amenable to chemical and physical modifiers. | High impurity levels in present grades. |
| | | Early stage of development. |
| | Polymer chemistry and properties can be specially formulated. | More severe shelf life and pot life requirements. |
| | | Liquid systems can be more difficult to handle in factory. |
| | | Reduction in CTE and increase in thermal conductivity required. |

There are several major objectives in the development of a LCP molding plastic for packaging. Wire sweep and paddle shift are the principle impediments, despite the fact that the viscosity of packaging grade materials is relatively low. The incoming melt temperature is approximately 320°C, but in injection molding the molding tool and leadframe are at a lower temperature than the inlet melt temperature, so there is a significant viscosity increase on contacting the colder leadframe, which is at approximately 160°C. Process modifications such as hot runner systems or heated tips near the cavity gates could be used to boost the temperature and lower the viscosity just as it reaches the devices, but the abrasiveness of the glass-filled material is a concern when the flow passes repeatedly over an in-mold feature. Other issues in LCP molding compounds include low adhesion and the difficulty of using a large multicavity mold. Also, the high molding temperatures may not be compatible with TAB using solder attachment. The promise of this material is that the overall development of LCP for plastic packaging is at an early stage, and new advances are likely given the significant activity on LCP polymerization, rheology and morphology presently being conducted.

### 9.6.3 Poly(Phenylene Sulfide)

Poly(phenylene sulfide) (PPS) polymers have the longest history of any injection moldable plastic packaging material. These materials are in production for packaging of discrete devices, but no commercialization of post-molded IC packaging with PPS can be cited. PPS is attractive because it shares the same high productivity characteristics of the LCP materials, with the added benefit of a lower base resin viscosity. Grades of PPS developed specifically for IC packaging have shown viscosities of a few hundred poise at medium shear rates. The major problems with PPS materials are the high level of ionic impurities caused by the polymerization residuals, and the outgassing voids in low pressure molding. Using higher pressures to prevent or compress the voids can cause wire sweep and paddle shift making it difficult to locate an acceptable process window. New grades of PPS appear to eliminate the ionic impurity issue, but the outgassing voids will continue to be a problem with these low viscosity grades. High molding temperatures may inhibit use with solder attached TAB.

### 9.6.4 Liquid Injection Molding Materials

Liquid injection molding (LIM) and reaction injection molding (RIM) are reactive polymer process methods where very low viscosity monomers or oligomers are injected into the mold and polymerized therein to form the hard plastic [30, 31]. LIM usually refers to a one-part system, whereas RIM usually denotes a two-part system with dynamic mixing of the two components. LIM is preferable for IC packaging because the relatively small volumes of package molding would not be amenable to high-velocity dynamic mixing. The materials are usually either silicones, epoxies, or epoxy-silicone hybrids. The principal advantages of LIM are the very low viscosity of the base resin and the ability to custom formulate the material to meet specific requirements. High filler loadings to increase the thermal conductivity and lower the CTE can be accommodated while still achieving lower viscosities than epoxy molding compounds because the base resin is a low viscosity liquid at room temperature. High-thermal-conductivity fillers may be tolerated better in a LIM process because lower pressures and lower velocities could be used to minimize erosion. Principal disadvantages are the generally high level of ionic impurities in these liquid resins, long cure times, and the overall infancy of the technology of using LIM for electronics.

## REFERENCES


1. R. D. Hannemann, "Physical Design for VLSI Systems," *IEEE Trans.*, ICCD '86 (Oct. 6–9, 1986), ISBN 0-8186-0735-1.
2. D. W. McCall, "Materials for High Density Electronic Packages and Interconnection," Materials Research Society Meeting, San Diego, CA (April 1989).
3. R. J. Koop, "Device Architecture and Process Technology," *SEMI's Twelfth Annual Information Services Seminar Reprints*, Newport Beach, CA, (January 15, 1989).
4. K. M. Striny, "Assembly Techniques and Packaging of VLSI Devices," Chapter 13 in S. M. Sze, Editor, *VLSI Technology*, Second Edition, McGraw Hill Book Company, New York, NY (1988).
5. L. W. Schaper and D. I. Amey, "Improved Electrical Performance Required for Future MOS Packaging," *IEEE Trans. Components, Hybrids, and Manufacturing Technology*, **CHMT-6,** 283–289 (Sept. 1983).
6. N. Sinnadurai, "Advances in Microelectronics Packaging and Interconnection Technologies—Toward a New Generation of Hybrid Microelectronics," *Microelectronics J.*, **16(5),** (1985).
7. G. Baccarani, M. R. Wordeman, and R. H. Dennard, "Generalized Scaling Theory and Its Application to 1/4 Micrometer MOSFET Design," *IEEE Transactions on Electron Devices*, **ED-31,** 452 (1984).
8. C. Kurosawa and H. Suzuki, "Molding Compounds and Hybrid Technology," *ISHM*, **10(3),** 19 (1987); also in *Japanese Hybrids*, **3(1)**.
9. S. Paltikian, "Combining Technologies to Produce High Density, Surface Mountable Hybrid ICs," *Hybrid Circuit Technology*, 23 (October 1987).
10. J. O. Honeycutt and E. W. Mace, "Solution of Wire Creep Rupture Problem in a Thick-Film Power Hybrid," *IEEE Trans. on Components, Hybrids and Manufacturing Technology*, **CHMT-9(2),** 172 (1986).
11. C. J. Bartlett, J. M. Segelken, and N. A. Teneketges, "Multichip Packaging Design for VLSI-Based Systems," *IEEE Trans. on Components, Hybrids and Manufacturing Technology*, **CHMT-12(4),** (December 1987).
12. C. S. Ho, D. A. Chance, C. H. Bajorek, and R. A. Acosts, "Thin Film Module as a High Performance Semiconductor Package," *I.B.M. J. Research and Development*, **36(3),** 286 (1982).

13. R. L. Cain, "Beam Tape Carriers—A Design Guide," *Solid State Technology*, 53, (March 1978).
14. P. Burggraaf, "TAB for High I/O and High Speed," *Semiconductor International*, 72, (June 1988).
15. A. Bindra, "TAB Rescues IC Designers," *High Performance Systems*, 46 (March 1989).
16. D. B. Brown and M. G. Freedman, "Is There a Future for TAB," *Solid State Technology*, 173 (September 1985).
17. G. Dehaine and K. Kurzweil, "Tape Automated Bonding Moving into Production," *Solid State Technology*, 46 (October 1975).
18. D. S. Alles and K. J. Brady, "Packaging Technology for III-V Photonic Devices and Integrated Circuits," *AT&T Technical J.*, (January/February 1989).
19. W. E. Jahsman, "Lead Frame and Wire Bond Length Limitations to Bond Densification," *J. Electronic Packaging*, **111,** 289 (1989).
20. J. D. Welterlen, "Chip Carriers and Other Integrated Circuit Packages; 1988," Report from Welterlen, Inc., La Jolla, CA (Nov. 30, 1988).
21. L. T. Manzione, "Molded PC Boards: New Way to Fuse Function, Form," *Plastics World*, 53 (October 1983).
22. A. G. Osborne, "Molded Circuit Board Review," *Printed Circuit Fabrication*, (September 1987).
23. S. Chen, "Molded Circuits Assemblies at the Starting Line," *Electronic Products*, (July 15, 1988).
24. J. A. Fulton, R. C. Moore, J. C. Sekutowski, W. R. Lampert, and J. P. Mitchell, "Use of Anisotropically Conductive Polymers in Electronic Applications," Third Intl. SAMPE Electronic Material and Process Conference (June 20–22, 1989).
25. R. A. Dery, W. C. Jones, "Method of Forming an Electrical Interconnection Means," U.S. Patent No. 4,554,033 (Nov. 19, 1985).
26. S. Matsuoka, G. H. Fredrickson, and G. E. Johnson, "Lecture Notes in Physics, No. 277," in T. Dorfmuller and G. Williams, Editors, *Molecular Dynamics and Relaxation Phenomena in Glasses, Proceedings, Bielefeld 1985*, 188, Springer-Verlag, Heidelberg.
27. Materials Research Society Meeting, Symposium on Space Compatible Materials, April 25–26, 1989, San Diego, CA Final Program and Abstracts (1989).
28. P. D. Frayer, "High Temperature Liquid Crystalline Copolyester Composites. Part I: A Review of the New Data for the Self-Reinforced Matrix Phase," *Polymer Composites*, **8(6),** 379 (1987).
29. U.S. Patent No. 4,632,798, Eickman et. al., Assigned to Celanese Corporation (Dec. 30, 1986).
30. L. T. Manzione, "Reaction Injection Molding," in *Encyclopedia of Polymer Science and Engineering*, Vol. **14,** Second Edition, John Wiley and Sons, New York (1988).
31. C. W. Macosko, "RIM, Fundamentals of Reaction Injection Molding," Hanser Publishers (Oxford University Press in U.S.A.) (1989).

# Index

Donated by Hans-G. Elia